PREVENTING THERMAL CYCLING AND VIBRATION FAILURES IN ELECTRONIC EQUIPMENT

Dave S. Steinberg
Steinberg & Associates, and
University of California, Los Angeles

A WILEY-INTERSCIENCE PUBLICATION

JOHN WILEY & SONS, INC.

New York • Chichester • Weinheim • Brisbane • Singapore • Toronto

This book is printed on acid-free paper. ♾

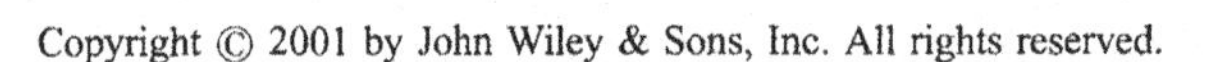

Published simultaneously in Canada.

Library of Congress Cataloging-in-Publication Data:

Steinberg, Dave S., date
Preventing thermal cycling and vibration failures in electronic equipment / by Dave S. Steinberg.
p. cm.
Includes index.
ISBN 0-471-35729-4 (cloth : alk. paper)
1. Electronic apparatus and appliances–Thermal properties. 2. Electronic apparatus and appliances–Vibration. 3. Electronic apparatus and appliances–Protection. 4. System failures (Engineering) I. Title.

TK7870.25 .S7324 2001
621.381′04–dc21

2001024704

To my wife Annette and to my two daughters
Cori and Stacie

CONTENTS

PREFACE

The information presented in this book was originally developed to help improve the reliability of automotive, aircraft, missile, communication, and entertainment electronic equipment used by commercial, industrial, and military electronics industries. These groups want to achieve a dramatic improvement in the readiness and reliability of their electronic equipment, while holding down the acquisition and support costs. One way to accomplish these goals is to implement an automotive and avionics integrity program, or AVIP. The requirements for such a program have been outlined in document MIL-A-87244, to ensure the electronics will achieve the desired reliable performance over the life of the equipment. This document outlines several critical areas and recommends detailed analysis and investigation to promote a better understanding of the various failure mechanisms resulting from high temperatures, thermal cycling, vibration, shock, and humidity. The areas of investigation include the electronic components, solder joints, circuit boards, interconnects, and assemblies.

An examination of many electronic failures has shown that most of them appear to be mechanical in nature. They typically involve cracks in the solder joints, lead wires, components, circuit boards, hermetic seals, cemented joints, connectors, and cables. These failures are often caused by various combinations of thermal, vibration, shock, humidity, salt, and dust environments combined, with poor manufacturing processes, poor design practices, and poor maintenance. These areas often contribute to radio frequency interference (RFI) and electrical magnetic interference (EMI) problems.

The methods shown in this book have been used successfully on a wide variety of large and small electronic systems, operating in many different combinations of severe environments for the past 30 years. Many opportunities were available to evaluate and test sophisticated electronic systems until they failed in various environments. In many of these programs, the reliability was extremely important since the lives of many people were involved. A large number of prototype models were fabricated, instrumented, and tested to destruction as well as later production models, to ensure the reliability of the design and the manufacturing methods. As the equipment costs began to increase rapidly due to the increased complexity, the electronic manufacturing groups began to look for ways to reduce these costs without reducing the reliability.

Today industry leaders have sharply reduced the costs of their electronic equipment, while they have increased the reliability at the same time. This has been accomplished through the use of powerful high-speed computers and new software programs. These new computers can perform tasks and solve complex problems in a matter of minutes that were previously impossible or would take too much time and money to investigate.

Computers have been so successful that some upper management people now believe that virtually all the answers to technical problems can be found in these computers. The belief by some is that there is no longer any need to build and test prototype models on new designs to obtain critical reliability data for operation in harsh environments. They think that a lot of money can be saved by eliminating the model building and testing of new products. The new Boeing 777 airplane is used as an outstanding example. This airplane was completely designed on computers, without the normal fabricating and testing of prototype models.

Many management people are not aware of the huge database Boeing has developed through its own experience with many similar aircraft, and the millions of dollars it spent to purchase the databases from other aircraft manufacturers. This information is available through its computer network to all of its design and test engineers, so the risks involved with a new, but similar, type of aircraft are sharply reduced.

When unusual new designs are being proposed, or when very severe environments are expected, or when unusual combinations of severe conditions are expected, then it is a poor policy to go into any large-scale production on a product that has not been tested in some way to verify the reliability of the product. This also applies to the use of new materials or processes or when significant weight reductions are demanded. Some type of test should be used to ensure the integrity of the product.

The purpose of this book is to show how to design, analyze, and evaluate electronic systems for lower costs and improved reliability in harsh environments, using hand calculations, without the aid of a computer. The methods shown have all been proven with extensive testing programs on prototype models and by qualification tests on production hardware. A good computer solution for a large complex system will always be better than a quick hand calculation. People who use computers know the expression GIGO, garbage-in garbage-out. It is very easy to make an error when you are tired, and several hundred data points must be entered into a computer program. One bad data entry can ruin a month or more of hard work when a computer model is being generated. Many computer users often look for a quick and simple way to determine if they have made a major error in their work. The quick methods shown in this book are often used for that purpose. Computer users call these quick methods a sanity check.

Field failure experience and testing experience has shown that every time a structure is subjected to a stress reversal, part of its life is used up. These stress reversals, or stress cycles, are typically caused by thermal cycling, vibration, and shock conditions found in everyday living. Every time the power is turned on in an electronic system such as a computer, radio, television set, cellular phone, automobile, or airplane, the temperature goes up. When the power is turned off the temperature goes down. This generates a thermal cycle stress. Every time a computer, radio, or television set is shipped from the manufacturer to the consumer, or every time an airplane or an automobile is used for transportation, vibration is experienced. This generates many vibration stress cycles. Every stress cycle will use up part of the fatigue life in the load-carrying structural elements of the system. When the total number of stress cycles reaches a critical level in the weakest element of the system, failures will occur.

This book concentrates on three important areas of investigation and analysis to ensure cost-effective and reliable electronic systems. These areas of investigation are as follows:

1. To understand how variations in the thermal coefficients of expansion (TCE) can effect the magnitude of the displacements, forces, and stresses that are developed in electronic assemblies during thermal cycling environments, and how these factors can affect the fatigue life of various structural elements
2. To understand how resonant conditions can effect dynamic displacements, forces, stresses, and fatigue life in electronic assemblies during different vibration and shock environments
3. To understand the concept of *damage accumulation*, and how it can be used to determine the approximate fatigue life of various electronic components and assemblies due to different combinations of fatigue accumulated in thermal cycling and vibration environments.

DAVE S. STEINBERG
Westlake Village, California
January 2001

SYMBOLS

A	Area (in.^{2}), amplification (dimensionless ratio)
a	length (in.)
ASIC	Application-specific integrated circuit
B	Length (in.)
BGA	Ball grid array
b	Fatigue exponent, width (in.)
C	Dynamic constant, length (in.)
CG	Center of gravity
c	Distance from neutral axis to outer fiber (in.), damping coefficient (lb·s/in.)
D	Plate stiffness factor (lb-in.), diameter (in.)
dB	Decibel
D_{XY}	Plate torsional stiffness factor (lb·in.)
d	Diameter (in.), length (in.)
DIP	Dual-in-line package
E	Modulus of elasticity (lb/in.^{2})
ESS	Environmental stress screen
F	Force (lb)
f	Frequency (Hz)
f_n	Natural frequency (Hz)
FEM	Finite element method
G	Shear modulus (lb/in.^{2}), acceleration in gravity units (dimensionless)
g	Acceleration of gravity, 386 in./s^2
H	Horizontal force (lb)
h	Height (in.), thickness (in.)
Hz	Frequency (cycles/s)
I	Area moment of inertia (in.^{4})
I_m	Mass moment of inertia (lb·in·s^2)
J	Torsional form factor (in.^{4}), shape of electrical lead wire
K	Linear spring rate (lb/in.), stiffness ratio (dimensionless), stress concentration factor
L	Length (in.)
LCCC	Leadless ceramic chip carrier
M	Bending moment (lb·in.)

m	mass (lb·s^2/in.)
N	Number of fatigue cycles to fail
N_0^+	Number of positive zero crossings (Hz)
n	Actual number of fatigue cycles accumulated
P	Force (lb)
PCB	Printed circuit board
PGA	Pin grid array
P, PSD	Power spectral density (G^2/Hz)
Q	Transmissibility (dimensionless ratio), coupled transmissibility
q	Uncoupled transmissibility (dimensionless ratio)
R	Ratio (Q/q), radius (in.), stress ratio (dimensionless), sweep rate (octave/min)
R	Miner's cumulative damage ratio (dimensionless)
R_n	Fatigue-cycle ratio (dimensionless)
R_C	Damping ratio (dimensionless)
R_Ω	Frequency ratio (dimensionless)
rms	Root mean square
r	Radius (in.), relative position factor
S	Stress (lb/in.^{2})
S_b	Bending stress (lb/in.^{2})
S_{CR}	Critical stress (lb/in.^{2})
S_e	Endurance limit stress (lb/in.^{2})
SMD	Surface-mounted device
TCE	Thermal coefficient of expansion (in./in./°C)
t	Temperature (°C), thickness (in.), time
V	Velocity (in/s), vertical force (lb)
W	Weight (lb)
X	Coordinate axis, displacement (in.)
Y	Coordinate axis, displacement (in.)
Z	Coordinate axis, displacement (in.)

Greek Symbols

α	Thermal coefficient of expansion (in./in./°C), angle (radians)
δ	Displacement (in.)
θ	Angle (radians)
μ	Poisson's ratio (dimensionless)
Δ	Difference
Ω	Angular velocity (radians/sec)
σ	Relation to rms stress

Subscripts

a	Aluminum
AV	Average
b	Bending
c	Component, ceramic

d	Dynamic
e	Endurance, epoxy
in	Input
max	Maximum
min	Minimum
n	Natural
0	Maximum
out	Output or response
ST	Shear tearout
st	Static
t	Tension
tu	Tensile ultimate
ty	Tensile yield
y	Yield

CHAPTER 1

Physics of Failure in Electronic Systems

1.1 FAILURES IN DIFFERENT TYPES OF ELECTRONIC ASSEMBLIES

Electronic systems utilize many different metals and plastics in the manufacturing and assembly processes. The physical properties of these materials must be known, as well as the operating environments, in order to produce a cost-effective and reliable product. The highest failure rates in electronic systems are generally associated with high component temperatures. When the power is turned on and a component begins to smoke, it can be concluded that a very high temperature is involved. Semiconductor devices are normally rated by their junction temperature capability, where typical values vary from about 150 to 200°C. A safety factor is normally applied where the maximum allowable junction temperature may be limited to a value that is 50°C below the maximum manufacturers rating.

In some applications, such as drilling for oil at depths of 20,000–30,000 ft, the coolant available for the electronics is the surrounding mud, which is at a temperature of about 200°C. The electronics used in oil drilling equipment must have a high rating, so the 200°C rated components are typically used. The power dissipations of these systems must be kept very low, to keep the junction temperatures below a value of about 204°C. The electronics are usually located at the end of the drill, just above the cutters. When the drill is cutting through rock, the electronics section can experience random vibration levels as high as 20 *G* rms. If there is an electronic failure, the entire 20,000–30,000 ft of drilling pipe has to be removed to replace the defective electronics section. Then the pipe has to be replaced down the hole. This can be a very time-consuming and a very expensive process, so the electronics must be reliable to save time and money.

Component junction temperatures for military and aerospace applications are usually limited to a value of about 100°C. In order to achieve an extra high reliability for the new Lockheed F-22 fighter aircraft, the junction temperatures for the semiconductor devices are being limited to about 65°C.

Hardware failure rate studies from the U.S. Air Force have shown that about 40% of all the failures in military aircraft electronic systems are found in the various electrical connectors located throughout the aircraft. Additional failures of 30% are found in the interconnects for the cables and harnesses. Another 20% are in the electronic component parts and 10% are due to other factors. Environment failure rate studies in military aircraft have shown that about 55% of all the electronic failures are related to thermal events such as high temperatures and thermal cycling. Another 20% are due to vibration, and 20% of the failures are due to humidity, with 5% due to dust [1].*

Automobile manufacturers design their underhood engine compartment electronics to survive continuous operation in an ambient temperature of 140°C. The electronics must be protected against water, which can splash up from the street during rainstorms, and humidity, which can cause water condensation on various electronic components.

1.2 AREAS OF ANALYSIS AND EVALUATION

Displacements, forces, and stresses must be obtained for the most critical mechanical structural elements in the electronic system. Experience has shown that the most failure-prone mechanical elements in electronic systems in thermal cycling and vibration environments are the electrical component lead wires and their solder joints [5]. These items must be examined for exposure to the various operating environments. The total fatigue damage accumulated in the various environments can then be added up using Miner's cumulative fatigue damage ratio, to obtain the approximate fatigue life for the most critical lead wires and solder joints. Fretting corrosion fatigue failures for the plug-in electrical printed circuit board (PCB) connectors are usually examined for the vibration environments only. Thermal cycling does not seem to have much effect on the reliability of the plug-in PCB electrical connectors. Random vibration causes most of the damage [1].

1.3 EFFECTS OF THERMAL CYCLING ENVIRONMENTS ON LEAD WIRES AND SOLDER JOINTS

Differences in the thermal coefficients of expansion (TCE) of the various materials used in electronic packages can result in very high stresses and strains in critical mechanical structural elements during thermal cycling events. Relative displacement differences of only 0.00050 in. between the PCB and components soldered to the PCB have been known to crack solder joints and lead wires. Most electrical lead wires will only fail when they are loaded in tension in thermal cycling events. Wires loaded in direct tension will fail when the ultimate tensile stress is exceeded. Lead wires loaded in bending during thermal cycling very seldom break. Even when the bending stresses in the wires exceed the ultimate strength and the wires experience plastic bending, testing experience shows the wires still do not break. There are very

* Numbers in brackets refer to references at the end of the book.

seldom enough thermal cycles accumulated in these events to cause a fatigue failure, unless there is a sharp and deep cut in the wire at a high-stress point. This can be demonstrated by bending a paper clip back and forth through a large displacement several times. The paper clip will experience permanent deformation since the bending is in the plastic range, but the paper clip does not break. This is because there are not enough stress cycles accumulated to produce a fatigue failure. It may take several dozen stress cycles to produce a fatigue failure with large displacements. It may take tens of thousands, and perhaps millions, of stress cycles to produce a fatigue failure with very small displacements. Most electronic systems will never develop that many thermal stress cycles in their lifetime [2, 3].

A computer that is turned on twice a day, every day for 15 years, will accumulate about 11,000 thermal fatigue cycles. A television set that is turned on 10 times a day, every day for 15 years, will accumulate about 55,000 thermal fatigue cycles. An automobile that is started 10 times a day, every day for 20 years, would accumulate 73,000 thermal stress cycles. A satellite in orbit around the earth experiences a thermal cycle about every 90 min. In 20 years it can accumulate about 117,000 thermal cycles. Electronic systems for aircraft are often designed to provide an operating life of about 10,000 h. In this period, a PCB with a natural frequency of 100 Hz will accumulate about 3.6 billion vibration fatigue cycles. The very large number of stress cycles associated with vibration means that stress risers in the form of small holes and sharp notches will be more sensitive in vibration than they will be in thermal cycling.

Transformers can carry high electrical current so their electrical lead wires are usually made of copper, with a large diameter to reduce the electrical resistance. When a transformer is surface mounted on a PCB, differences in the TCE between the body of the transformer and the PCB will force the wires to bend as shown in Fig. 1.1. Thermal cycling test data over a temperature range from −55 to +95°C shows that failures will occur in about 12 cycles. When most engineers not familiar with thermal cycling failures in electronic equipment are asked where the failures will occur, they say either in the lead wires or the solder joints. They are surprised when they find these are not the failure points in the system. They are more surprised when they find the failures are in the copper circuit traces on the PCB that are used as solder pads for making electrical connections to the PCB. The bending action of a very stiff wire can produce an overturning moment that will lift the solder pad off of

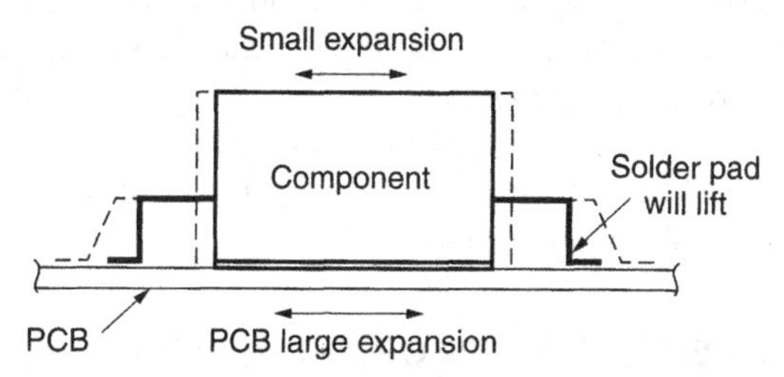

FIGURE 1.1 Large differences in the TCEs of the component and the PCB can force the solder pads on surface-mounted components to lift in thermal cycling environments.

the PCB. The solder pad is only held down to the PCB by an adhesive, which is not strong enough to resist the overturning moment action of a stiff lead wire. One way to reduce the expansion forces generated in thermal cycling events is to reduce the stiffness of the wire. The stiffness of the wire can be reduced by looping the wire to make it longer, or the wire can be coined to make it flat to reduce the moment of inertia, or the solder pad can be reinforced to solve the problem.

Copper-plated through-holes are often added to the solder pads for reinforcement to prevent them from lifting off of the PCB during thermal cycling events. Test data shows the fatigue life of the assembly with plated through-holes will now be increased to about 150 thermal cycles. There are only two places the failures can now occur, in the lead wire or the solder joint. Engineers not familiar with thermal cycling tests are surprised to find that bending failures almost never occur in the lead wires. When the calculated bending stresses in the wires are well above their ultimate tensile strength, they bend in the plastic bending range. They very seldom break as long as there are no sharp and deep cuts in the wire to act as a severe stress concentration.

1.4 EFFECTS OF VIBRATION ENVIRONMENTS ON LEAD WIRES AND SOLDER JOINTS

Vibration environments can often involve millions of stress cycles because natural frequencies in electronics can range from 50 Hz to well over 1000 Hz. Stress risers and stress concentrations can be very severe when millions of cycles are accumulated. Electrical lead wires loaded in tension or bending will fail very often in vibration events because a large number of stress cycles can be accumulated in a short period of time. Stress concentrations in the solder joints do not appear to have much effect on the fatigue life. The relatively soft solder tends to strain relieve the high local stresses associated with stress concentrations by plastically deforming in these areas [2].

Testing has shown that thermal cycling events typically produce many more solder joint failures than lead wire failures for surface-mounted and through-hole-mounted PCB components. Vibration events tend to produce many more electrical lead wire failures than solder joint failures in surface-mounted and through-hole-mounted PCBs. When a solder joint failure shows up during vibration, most of the time the crack initiation was due to thermal cycling. Since thermal cycling is slow, the crack propagation is also very slow. However, vibration can often develop several hundred stress cycles in a second, so the crack propagation is very rapid in vibration. The solder joint failure then appears to be due to vibration. An engineer with very little experience will immediately begin to investigate vibration-related stresses in the failed solder joint. An experienced engineer will first look for thermal-cycling-related stresses in the failed solder joint.

1.5 DIFFERENT PERSPECTIVES OF RELIABILITY

Most of the people associated with military reliability groups tend to use the mean time between failure (MTBF) method to evaluate the reliability of an electronic system. Tables and charts from MIL-HDBK-217 are then used to obtain the failure rates of the various different electronic components and parts for different vehicles and different environments. These values are used to calculate the mean time expected between failures. This method of evaluation means that a system is reliable when the number of failures during some specified operating period is at an acceptable level. Many computer programs have been written with software that permits a very fast calculation for the MTBF numbers based on the data obtained from the military handbook.

There appears to be something wrong with the probabilistic MTBF philosophy, especially for military programs, where the lives of many soldiers and civilians may depend on the reliability of their electronic equipment. Is a system really reliable when the number of failures during some specified operating period are at an acceptable level? For life-critical electronic systems, *there should be no failures during the operating life of the system.*

Consider a deterministic philosophy of reliability where the electronic system does not fail during the operating life or during the warranty period specified for the system. In other words, consider designing and manufacturing an electronic system with a failure-free operating period (FFOP) that extends through the operating life of that system. This is the goal of the avionics integrity program (AVIP) proposed by the Wright Patterson AFB in Dayton, Ohio, for the U.S. Air Force. The requirements have been specified in MIL-A-87244 [60]. The new Lockheed F-22 airplane incorporates the AVIP program in its electronic systems.

The AVIP used Miner's [61] cumulative damage ratio to add up all the damage from many different environments including thermal, thermal cycling, vibration, shock, and acoustic noise. A scatter factor (or safety factor) of two was used for the fatigue life evaluation. The electronic systems were evaluated for an FFOP of 20,000 h since the operating life of the airplane was expected to be 10,000 h. The fatigue life evaluation of the component lead wires and solder joints included the effects of manufacturing tolerances associated with the dimensional variations in the component lead wires and circuit boards, for the environments outlined above. Preliminary reports for the reliability of the F-22 electronic system have shown outstanding results.

The AVIP fatigue life evaluation requires special training in the art and science of analysis methods, using computers and hand calculations to obtain stresses and strains in the various load-carrying elements of the electronic system. The steady-state and transient properties of the materials used in the electronic equipment must be well documented. This requires more work than the MTBF method, so the AVIP method will be slightly more expensive. Many people believe the small increase in the cost is worth the big increase expected in the reliability.

1.6 CREEP AND STRESS RELAXATION IN SOLDER [3]

Solder has some strange properties that make it very difficult to accurately predict its fatigue life. For example, consider a cantilevered beam made of solder. If a constant external force is applied at the free end of the solder beam, it will bend and deform.

The beam, however, will continue to bend and deform as long as the external force is applied. This continued deformation with an externally applied force is called creep. It often occurs at a constant stress level. When the external force is removed, the solder beam displacement will not return to zero. If an external force is applied to the free end of the beam, the beam will displace and the stress level in the beam will rise. If the external force is removed, but the displacement is now held constant, the stress level within the beam will slowly decrease until it is zero. This is due to creep and stress relaxation. These creep and stress relaxation effects in solder often occur in electronic equipment when the power is turned on and off. When the power is turned on, the PCB and the electronic components heat up. Differences in the TCE of the various materials cause some materials to expand more than others. This expansion difference can produce high forces and stresses in the electrical lead wires and their solder joints on the PCB. When the power is turned off, the opposite happens as the system cools down. Now the materials shrink and the process is reversed, producing stress reversals, or stress cycles. When a large number of stress cycles have been accumulated, structural load-carrying elements can fail, causing electrical failures.

1.7 EFFECTS OF ALTERNATING STRESS CYCLES AND TEMPERATURE ON SOLDER JOINTS

Alternating stress cycles can occur in electronic equipment during exposure to thermal cycling and vibration conditions. Thermal cycling events are usually very slow compared to vibration events. The physical properties of solder are quite different for very slow stress cycles associated with thermal cycling and very rapid stress cycles associated with vibration. When a bar of solder is bent back and forth very rapidly, the alternating stress pattern will follow the displacement pattern very closely, as shown in Fig. 1.2. When a bar of solder is displaced and then held in the displaced position for a long time period, the stress pattern will initially follow the

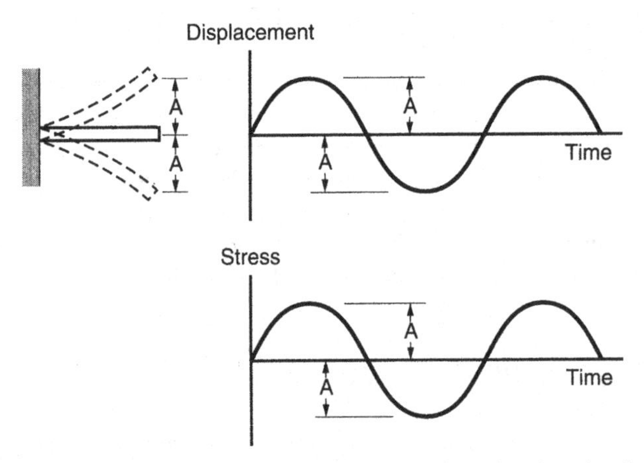

FIGURE 1.2 In linear systems the stresses will be proportional to the displacements.

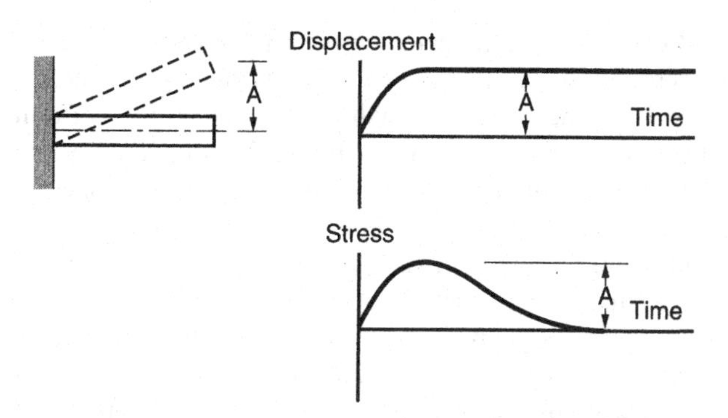

FIGURE 1.3 Solder can creep at elevated temperatures and relax stresses when displacements are held constant.

displacement pattern. If the bar is held in the displaced position, the solder will creep, causing the stress to slowly relax with time until it goes back to a zero stress condition, as shown in Fig. 1.3. The time it takes for the solder to creep and relax back to a near zero stress condition depends, to a great extent, on the temperature. The higher the temperature the more quickly the solder will creep and relax the stresses back to a near zero condition.

Consider the case where the solder bar is displaced back and forth, but held in the maximum positive and negative displaced positions for longer periods, as shown in Fig. 1.4. The solder can then creep and relax to a near zero stress condition every time the bar is held at the maximum displaced positions. If the bar is displaced through a displacement amplitude $+A$, the stress level will initially reach a value $+A$, directly proportional to the displacement amplitude. However, if the displacement

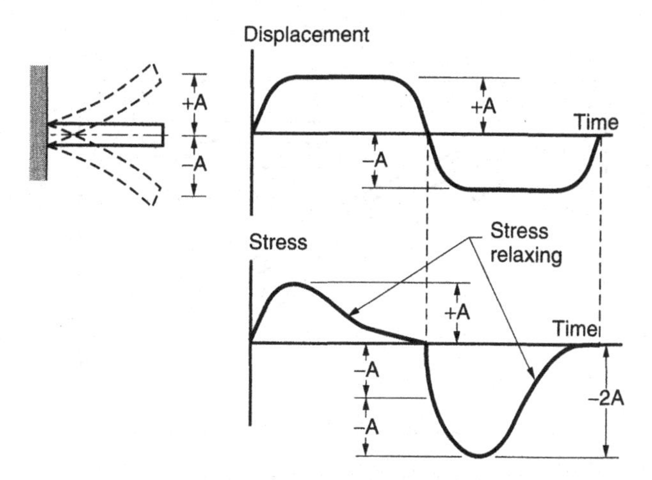

FIGURE 1.4 Solder creep and stress relaxing at elevated temperatures can cause an increase in the solder stresses during slow thermal cycling conditions where displacements are held constant for long periods of time.

amplitude is held constant for a longer time period, the stress will creep and relax back to a near zero condition. If the solder bar is rapidly displaced back to its original neutral position, the stress level will reach a value of $-A$. But if the solder bar is rapidly displaced to the $-A$ position, the solder will reach a stress value of $-2A$. This means that slow thermal cycle events will allow the solder to creep and relax stresses in a manner that results in higher solder joint stresses and strains then rapid thermal cycle events. Slow thermal cycle events, therefore, will result in more solder joint failures than rapid cycle events, over the same temperature range, because very slow thermal cycling events can *double* the solder joint strain and stress. So, the slower the solder is cycled, the weaker it gets because of the creep and stress relaxation properties of solder. The effects of temperature and the frequency of a mechanical cycling environment strongly influence the solder modulus of elasticity as shown in Fig. 1.5. Again, it shows that the slower the solder is cycled, the weaker it gets.

Solder creep at higher temperatures presents a problem in trying to establish a laboratory thermal cycling test program that produces the same types of solder joint failures that are experienced in the actual operating environment. Time is money. To reduce costs, accelerated laboratory life tests are usually run for short time periods that try to generate the same amount of damage that is generated for much longer periods in the actual operating environments. This is called accelerated life testing. The normal practice is to increase the temperature cycling range, along with a decreased temperature cycling time. A decreased temperature cycling time means a more rapid thermal cycle. The problem here is that the slower solder is cycled, the weaker it gets, so a more rapid thermal cycle will result in a much longer fatigue life. Therefore, the data gathered in the accelerated thermal cycling fatigue life tests must be examined very carefully to try to get better correlation with the fatigue failures produced in the actual operating conditions.

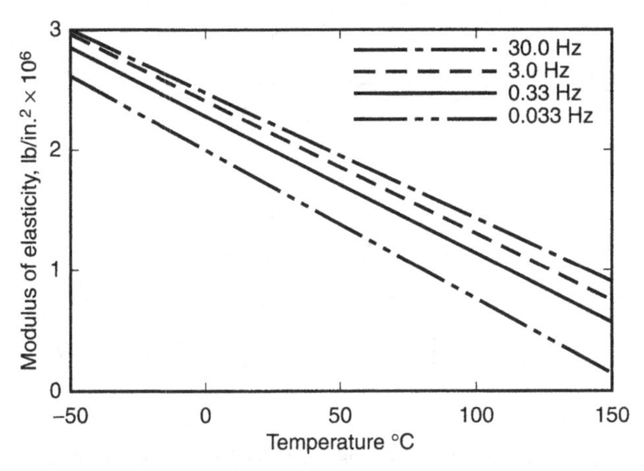

FIGURE 1.5 Creep properties of solder affect the modulus of elasticity, which depends on the temperature and the cycling frequency.

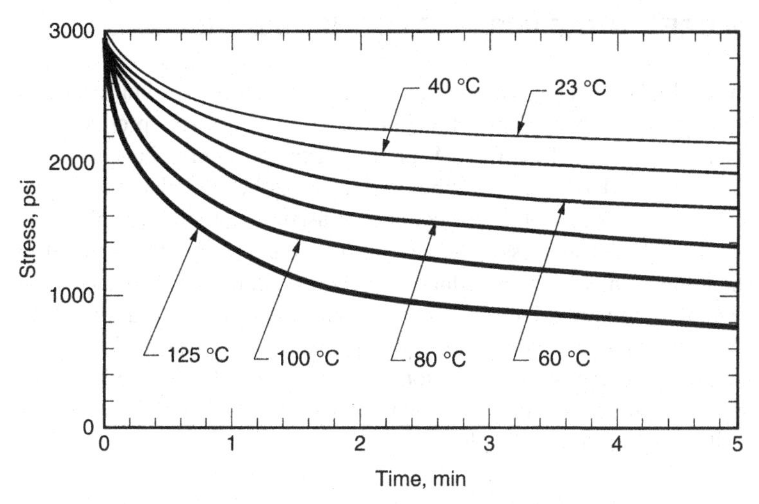

FIGURE 1.6 Rate of solder creep and stress relaxing depend on the temperature and stress level.

Solder tends to become more plastic with more creep as the temperature increases. At very low temperatures, below 0°C, solder is quite rigid, with very little creep and stress relaxation. The approximate time it takes for solder to creep and relax internal stresses at different temperatures is shown in Fig. 1.6. At temperatures around 125°C, eutectic solder (63% tin, 37% lead) can relax the stress levels from a value of 3000 psi to a value of about 1000 psi in a period of about 2 min.

Vibration, which has very rapid cycles, does not appear to have any creep effects on the solder. The electrical lead wires are usually made of copper, kovar, dumet, or nickel, so their mechanical properties are not effected by the normal high electronic temperature exposure. There appears to be no significant creep in the lead wires at the same elevated temperatures.

1.8 FAILURES IN THE PRINTED CIRCUIT BOARD STRUCTURE ITSELF

Very few thermal cycling or vibration-related failures occur in the PCBs themselves. Epoxy fiberglass and polyimide glass PCBs have TCEs that are very similar to the copper traces in the *x–y* plane, so the thermal cycling stress levels in the copper circuit traces are typically very low. Vibration can cause large dynamic displacements in the PCBs when their natural frequencies are excited, but this is very seldom a problem. Tests have shown that almost all of the electrical lead wires on the electronic component parts will fail long before there are any vibration fatigue failures in the circuit board structure itself [2].

1.9 FAILURES IN THE PCB PLATED THROUGH-HOLES

Copper in the PCB plated through-holes can fail in thermal cycling events, when there is not enough copper to carry the thermal expansion forces produced in the *z* axis, normal to the plane of the PCB. Vibration does not seem to have any effect on plated through-holes. Extra care must be used to strengthen the plated through-holes when materials such as copper–invar–copper are used to restrain thermal expansions in the *x–y* plane, when leadless ceramic chip carriers are surface mounted on the PCBs. The copper–invar–copper reduces the thermal expansions in the *x–y* plane, which substantially reduces the shear stresses developed in the solder joints on the leadless chip devices. But when the *x–y* expansions are reduced, Poisson's ratio comes into effect and the *z*-axis expansions are increased. This increases the stresses in the copper-plated through-holes. The copper can be made slightly thicker, which will make it stronger, to prevent plated through-hole failures.

1.10 EFFECTS OF SCATTER ON THE FATIGUE LIFE OF STRUCTURAL MATERIALS

Bending life tests are probably the most common type of fatigue tests that are run on many different types of materials. A large number of test samples are machined to close tolerances, from the same material, and subjected to different alternating stress levels. Each alternating stress level is then held closely until a failure occurs. The results of these tests are normally plotted on log–log curves, with the stress level *S* on the vertical *y* axis, and the number of cycles *N* to fail plotted on the horizontal *x* axis. These tests always result in a large amount of data scatter. A single straight sloped line is then drawn through the scattered data points to obtain the average results of the fatigue life, as shown in Fig. 1.7.

Even when the parts are machined and polished with close tolerances, and even when all of the test specimens are machined from the same forged billet, there will still be a large amount of scatter in the life test results. This is much more critical in

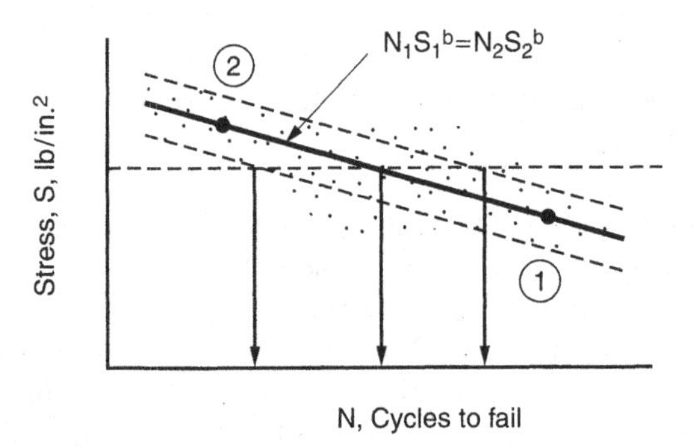

FIGURE 1.7 Typical log–log *S–N* fatigue life curve for structural materials where the best-fit straight line is drawn through the scattered test data points.

vibration conditions than in thermal cycling conditions because vibration almost always involves many more fatigue cycles than thermal cycling events. When structural elements are forced to bend and twist back and forth many million times, as they often do in vibration, four very important areas must be examined very carefully.

1. The very basic fatigue properties of the various materials used in the assembly
2. The effects of manufacturing tolerances on the physical part dimensions
3. The magnitude and nature of any induced alternating stress or stresses
4. The effects of stress concentrations due to small holes, notches, and small radii

Fatigue test data shows the fatigue life scatter for closely machined parts from the same basic material will often be in the range of about 10 to 1. This means that parts that are virtually identical will still have large variations in their fatigue life. Consider a condition where the average fatigue life of several hundred closely machined parts is about 3000 h in a defined vibration environment. An examination of a log–log plot for a fatigue curve with scatter data shows that there is still a possibility that one or more of the parts may last for about 10,000 h, and that one or more of the parts may last for only about 1000 h. This is the nature of fatigue.

When the effects of loose manufacturing tolerances are included in the fatigue life evaluation, the fatigue life scatter becomes even greater. Another factor that must be considered is the effects of stress concentrations, such as small holes and sharp notches. These geometric variations in a structural load-carrying element can raise the local stress levels very rapidly and shorten the fatigue life even more. These are the fatigue properties of almost all materials, which makes it very difficult to accurately predict the fatigue life of any complex structure. Therefore, it is a good practice to build prototype models that can be tested to establish the approximate fatigue life of a particular product, when the reliability of the product is very important. If no tests are run, then there will always be the danger of erratic bursts of high failure rates in the production units because of the large scatter associated with fatigue.

Engineers involved in the fatigue life analysis of electronic structural elements do not like to reveal the difficulties involved in fatigue life predictions to upper management personnel who are not trained in these areas. Personal experience with nontechnical upper management people is that they expect designers and engineers to be able to predict the fatigue life of the equipment to within ±20%. This is an almost impossible task with all of the variable combinations of materials involved. Some upper management people insist that all the information necessary to solve these problems are found in the expensive new computers that were just purchased. They have been brainwashed by the high-pressure computer sales staff that their new computers can perform these tasks. This is a half truth. The computer's ability to analyze these fatigue problems depends on the information entered into the computer. General fatigue data for many materials are now available in many computer software programs. Accurate fatigue life data, however, for a wide range of different geometric structural shapes in combination with different environments and materials are not available. This critical information can only be obtained from test data.

The method generally used to compensate for the inability to accurately predict the fatigue life is to use safety factors. These are sometimes called scatter factors. To ensure a good fatigue life for all of the mass-produced products, the structural load-carrying elements must be overdesigned to compensate for the wide scatter in the fatigue properties. The use of safety factors will always result in a slight increase in the size, weight, and costs of the product. This is usually a small penalty to pay for a significant increase in the fatigue life for the vast majority of the mass-produced products.

1.11 EFFECTS OF MANUFACTURING TOLERANCES ON THE FATIGUE LIFE

Tight manufacturing tolerances increase the costs of electronic equipment. Loose tolerances are desired to reduce costs. However, loose tolerances further reduce the accuracy of the fatigue life prediction for different environments. Therefore, when loose tolerances are used in the manufacturing processes, the electronic designs must be made more fault tolerant to keep the costs down and the reliability up. Typical variations in the physical sizes of the component body and electrical lead wires due to manufacturing tolerances are shown in Fig. 1.8. The most critical tolerance variation in the PCB itself is the thickness, which has a normal manufacturing tolerance of ± 0.0070 in. for multilayer boards. One effect of the PCB thickness

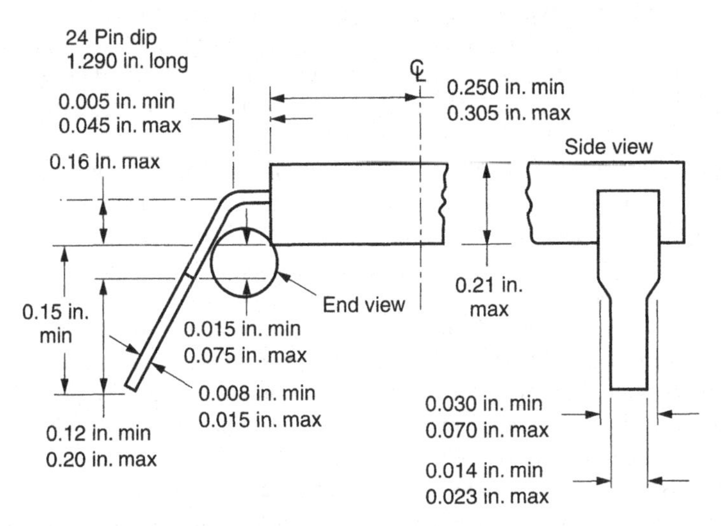

FIGURE 1.8 Effects of manufacturing tolerances on the dimensions of an electronic component.

variation is in the natural frequency of the PCB in vibration environments. The total thickness tolerance variation can change the PCB natural frequency by about 40%. The displacements and stresses are inversely related to the square of PCB natural frequency. The fatigue life of the lead wires and components mounted on the PCB may then be increased or decreased from the nominal fatigue life by a factor of about 8 times. This is a very large change in the PCB fatigue life, which may not be acceptable for a reliable program. Therefore, when loose tolerances are used in the manufacturing process, the electronic designs must be made more fault tolerant to keep the costs down and the reliability up. The geometric shapes of the structural members can often be modified to do this. Test data shows that the circuit board thickness variation due to manufacturing tolerances has very little effect on the fatigue life of the PCB in thermal cycling environments.

1.12 COMBINING DAMAGE DUE TO THERMAL CYCLING AND DAMAGE DUE TO VIBRATION

Anytime a structure experiences a stress reversal cycle, part of its life is used up. The two most common types of stress cycles that cause the most damage in electronic equipment are thermal cycling and vibration cycling. They can occur separately or they can be combined. Alternating stress cycles can be generated with a zero mean stress, as shown in Fig. 1.9. Alternating stress cycles can also be superimposed on a steady stress, as shown in Fig. 1.10. Here both stresses may be thermal, but most of the time the steady stress is thermal and the alternating stress is vibration. An alternating vibration stress cycle is often superimposed on an alternating thermal stress cycle, as shown in Fig. 1.11.

Predicting the fatigue life of electronic equipment exposed to any combination of an alternating stress superimposed on a steady stress is very difficult even with the

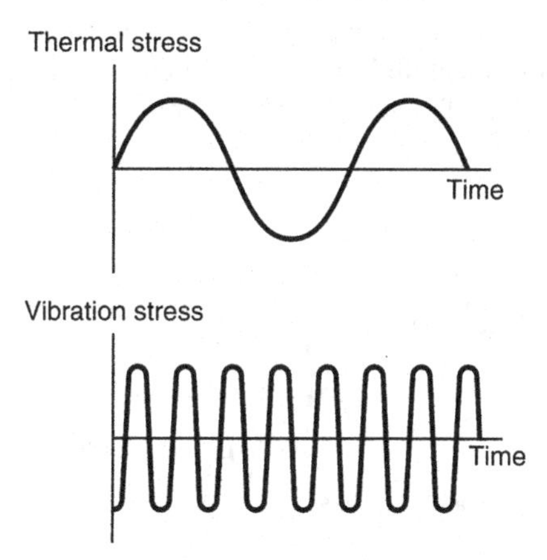

FIGURE 1.9 Thermal cycling stresses are usually slow and vibration cycling stresses are usually fast.

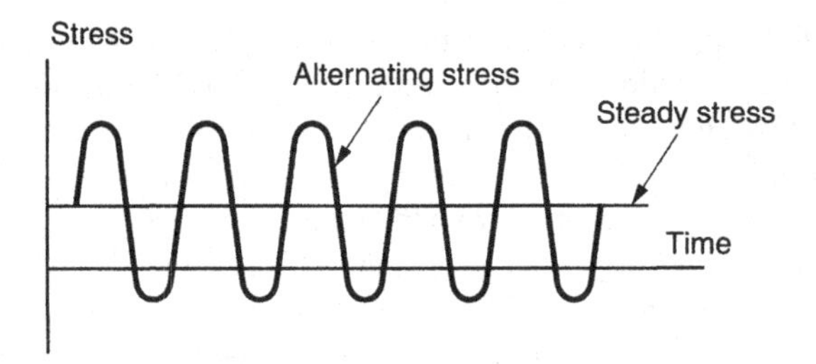

FIGURE 1.10 Alternating stresses are often superimposed on a steady stress.

use of computer programs. The fatigue life analysis from any computer evaluation, no matter how sophisticated, should not be trusted unless there are test data available from well-instrumented prototype test models that can be used to verify the accuracy of the computer evaluation.

Combined thermal cycling and vibration fatigue damage test data on PCBs with through-hole pin grid arrays (PGA) showed no damage when the vibration tests were performed at 95 and at 25°C. However, when the tests were performed at −55°C, many of the PGA wires fractured. The tests were always run with new assemblies to avoid any problems with possible accumulated fatigue damage. The following conclusions were drawn from the tests on the pin grid arrays:

1. Solder creep at elevated temperatures, and even at room temperature, allows the PGA lead wires to relax, which reduces the magnitude of the bending moments and bending stresses on the PGA wires. This increases the fatigue life of the wires.
2. Solder creep at low temperatures is sharply reduced, so high thermal stresses are locked in the PGA lead wires.
3. When the vibration is imposed at low temperatures, the PGA wires experience an alternating vibration stress superimposed on the sustained thermal stress, which increases the magnitude of the maximum stress acting on the PGA wires. This reduces the fatigue life of the wires.

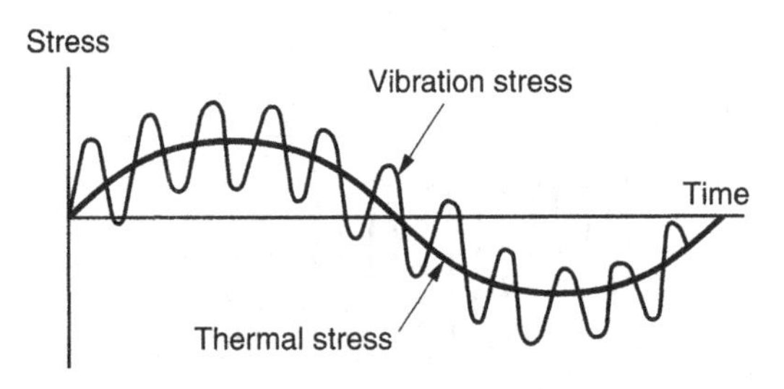

FIGURE 1.11 Fast alternating vibration stresses are often imposed on slow cycling thermal stresses.

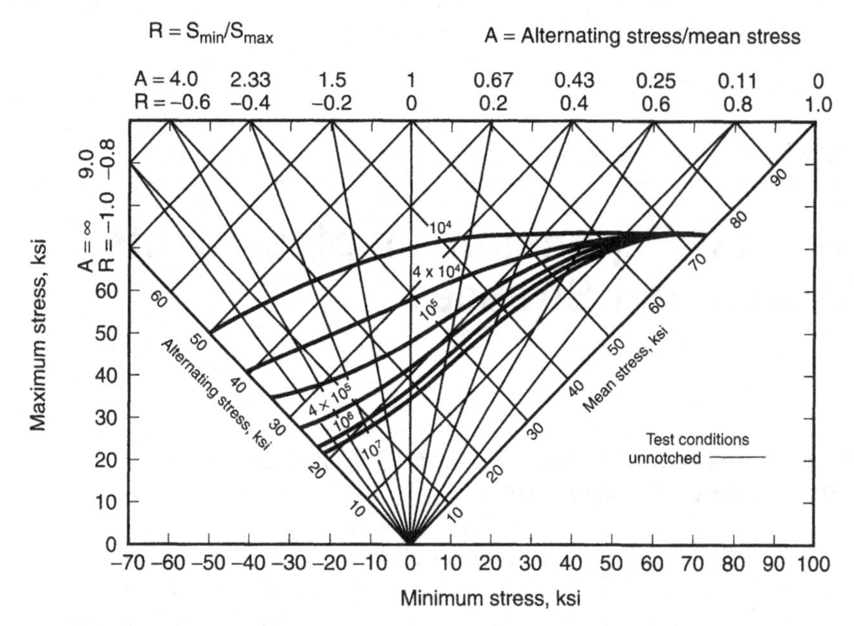

FIGURE 1.12 Fatigue life for 2024-T3 aluminum alloy for different combinations of alternating stresses superimposed on mean stresses for smooth polished specimens.

A good source for obtaining the approximate fatigue life of metal structures exposed to various levels of alternating stresses superimposed on steady stresses can be obtained from MIL-HDBK-5. A typical set of curves for 2024-T3 aluminum alloy is shown in Fig. 1.12 [4].

All of the thermal cycling and vibration fatigue analyses in this book are assumed to act separately. The thermal cycling fatigue damage is examined by itself and the vibration fatigue damage is examined by itself. The individual damages are then added together to obtain the total damage and approximate fatigue life.

CHAPTER 2

Thermal Expansion Displacements, Forces, and Stresses

2.1 MAKING STRUCTURAL ELEMENTS WORK SMARTER TO REDUCE FORCES AND STRESSES

Large displacements are often the cause of high forces and high stresses in thermal cycling and vibration environments. High stresses usually lead to rapid fatigue failures in electronic equipment, which are undesirable. Sometimes it is possible to alter the shapes of some critical structural elements to reduce these high forces and stresses. When a structural system is exposed to an environment that produces a constant displacement in the load-carrying members of that structure, it is often possible to reduce the forces and stresses by decreasing the spring rate of the load-carrying member. Consider the force P and the displacement Y in a simple linear spring system, as shown in Fig. 2.1. The spring rate K of this system is defined as

$$K = \frac{P}{Y} = \text{lb/in.} \quad \text{so} \quad P = KY = \text{lb} \tag{2.1}$$

This expression shows that when the displacement Y is constant, reducing the spring rate K will reduce the force P, which will reduce the stress and increase the fatigue life. Constant displacements can be produced in systems exposed to defined thermal cycling and defined vibration environments. Two basic types of structures are considered here that involve both the thermal cycling condition and the vibration or shock condition, as shown below [1, 5]:

1. Structures that experience axial thermal expansions or axial dynamic forces
2. Structures that are forced to bend by thermal expansions or by dynamic forces

Examine item 1 first for axial thermal expansions. Consider a beam or a wire that is cantilevered and exposed to a temperature cycling environment. Most materials will expand when the temperature is increased. When the temperature is decreased, most materials will shrink, or contract. The axial expansion and contraction magnitude X will depend on the material properties and the change in the temperature:

$$X = \alpha L \, \Delta t = \text{in.} \tag{2.2}$$

where α = in./in./°C, thermal coefficient of expansion (TCE) of the material
L = in., original length of the beam
Δt = °C, temperature change

For a given material with defined dimensions and a defined temperature change, the expansion and the contraction X as shown in Eq. 2.2 will be defined, or constant, so Eq. 2.1 can be applied.

When an axial force P is applied to the cantilevered beam, and the beam displaces through an amplitude X, the spring rate K will be defined by Eq. 2.1. The physical value for the spring rate K is defined as

$$X = \frac{PL}{AE} \quad \text{so} \quad K = \frac{AE}{L} = \text{lb/in.} \tag{2.3}$$

where A = in.^{2}, cross-section area of beam
E = lb/in.^{2}, modulus of elasticity of the material
L = in., original length of the beam

The spring rate K must be reduced in order to reduce the force P in the system, as shown by Eq. 2.1. An examination of the above relation shows that this can be done by increasing the length L, by reducing the area A, or by reducing the modulus of elasticity E. The probability of reducing the modulus of elasticity of the basic structural material is small because that would mean the material would have to be changed. This might be possible in a small number of cases. However, it is often possible to add a thin layer of a soft material, with a low E, in series with the basic structural material. For example, adding a thin interface layer of a soft rubber or silicone material, such as room temperature vulcanizing (RTV), with a low modulus of elasticity of about 1000 psi, can reduce the internally generated forces and stresses. This can solve many axial expansion thermal stress problems. Many

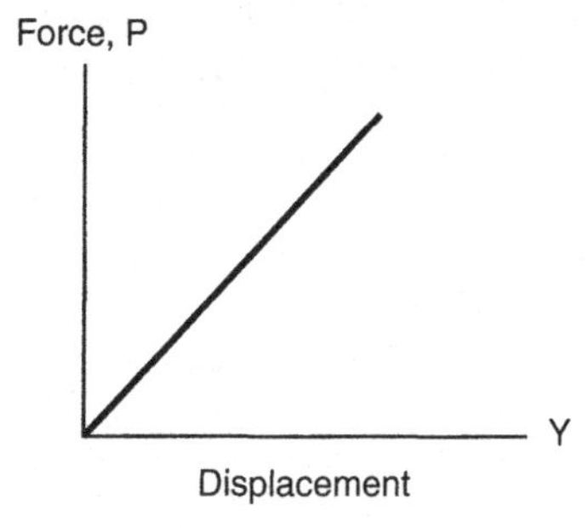

FIGURE 2.1 In a linear system the displacement is directly proportional to the force.

companies make thin silicone pads with a low modulus of elasticity that can be used to solve these expansion problems.

Sometimes the electrical lead wires can be made longer by using a small lateral offset in the wire, or by looping the wire, or by adding a small kink in the wire. These changes will force part of the wire to bend. The bending stiffness will be significantly lower than the axial stiffness, which can substantially reduce the spring rate and the forces in the wire.

Next consider item 2 above for beam or wire structures that are forced to bend due to thermal cycling or are forced to bend due to dynamic vibration induced forces. The general values for the spring rate K for a bending beam are defined as

$$Y = \frac{PL^3}{EI} \quad \text{so} \quad K = \frac{EI}{L^3} \tag{2.4}$$

where $E = \text{lb/in.}^2$, modulus of elasticity
$I = \text{in.}^4$, area moment of inertia of beam cross section
$L = \text{in.}$, length of beam

In a constant displacement system Eq. 2.1 shows that an easy way to reduce the force P in the system and prevent failures is to reduce the spring rate K. An examination of the above relation shows that this can be done by increasing the length L or by reducing the moment of inertia I. The probability of reducing the modulus of elasticity E is small because that would require a material change. This might be possible in a small number of cases. Again, it may be possible to add a thin layer of a soft rubber or silicone material with a low modulus of elasticity in series with the basic structural materials. This can reduce the spring rate and the forces in the wire.

Sample Problem: Axial Thermal Expansion in an Aluminum Bar

The aluminum bar shown in Fig. 2.2 will expand axially when the temperature increases. The bar, with a TCE of 23×10^{-6} in./in./°C (23 parts/million/°C) and a modulus of elasticity E of 10 million psi will experience a temperature change from −50 to +50°C. Find the following:

(a) The axial thermal expansion displacement X at the end of the bar
(b) The axial force P required to push the bar back to its original length
(c) The compressive stress S in the bar due to the above axial force

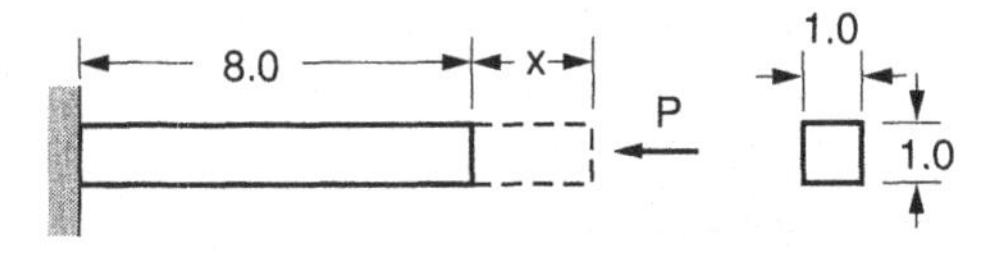

FIGURE 2.2 Cantilever beam expands when the temperature is increased, and an axial force is required to push the beam back to its original starting position.

Solution For part (a) use Eq. 2.2 to find expansion X with the following data:

$\alpha = 23 \times 10^{-6}$ in./in./°C, TCE of aluminum bar
$L = 8.0$ in., length of bar
$\Delta t = 50 - (-50) = 100$°C, temperature change

$$X = (23 \times 10^{-6})(8.0)(100) = 0.0184 \text{ in. expansion} \tag{2.5}$$

For the solution for part (b) use Eq. 2.3 to find the axial force P with the following data:

$A = (1.0)(1.0) = 1.0 \text{ in.}^2$ cross-section area of bar
$E = 10 \times 10^6 \text{ lb/in.}^2$ aluminum modulus of elasticity

$$P = \frac{AEX}{L} = \frac{(1.0)(10 \times 10^6)(0.0184)}{8.0} = 23{,}000 \text{ lb} \tag{2.6}$$

Solution for part (c) compressive stress S in bar:

$$S_c = \frac{P}{A} = \frac{23{,}000}{1.0} = 23{,}000 \text{ lb/in.}^2 \tag{2.7}$$

If the aluminum bar is restrained so it cannot expand, an axial force of 23,000 lb will be generated, which will result in an internal stress of 23,000 psi, as shown above. If the restrained bar is allowed to expand slightly, the internal forces and stresses can be reduced. A thin rubber pad with a low modulus can be placed at the end of the bar to allow the bar to expand slightly after the bar has been restrained at both ends. Now when the bar is restrained at both ends, it can still expand slightly into the soft rubber. This will reduce the internally generated force. The analysis for this condition is shown in the following sample problem.

Sample Problem: Reducing Axial Thermal Expansion Force in an Aluminum Bar

Add a 0.10-in.-thick rubber pad with a modulus of elasticity of 1000 psi to the end of the aluminum bar in the previous problem, then restrain the bar at both ends, as shown in Fig. 2.3. Find the axial force in the restrained bar for the same temperature cycle as the previous problem.

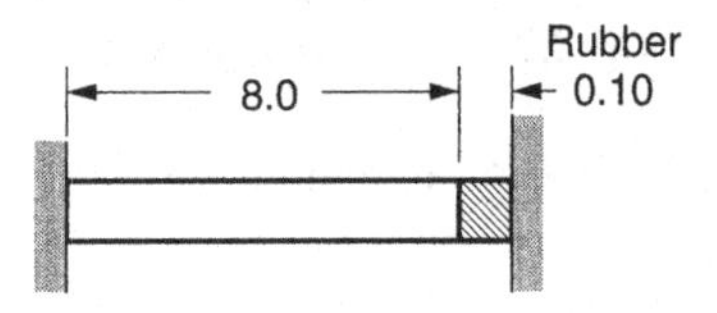

FIGURE 2.3 Thin strip of a soft material such as rubber can reduce internal thermal expansion forces in constrained structures.

Solution *Set Up an Equilibrium Equation for the System Displacements.* Since the system is restrained at both ends, any displacements developed must be the same for the aluminum bar and the rubber pad. A free-body evaluation will show that the aluminum bar and the rubber pad will both expand when the temperature is increased and there are no restraints. When the restraints are added, a compression force P will be generated that will be the same on the aluminum and the rubber. The balanced equation for the equilibrium condition is shown below. Subscripts a and r are for the aluminum and the rubber, respectively [5]:

$$\alpha_a L_a \, \Delta t - \frac{P_a L_a}{A_a E_a} = -\left[\alpha_r L_r \, \Delta t - \frac{P_r L_r}{A_r E_r}\right] \tag{2.8}$$

where $\alpha_a = 23 \times 10^{-6}$ in./in./°C, TCE of aluminum
$L_a = 8.0$ in., length of aluminum bar
$\Delta t = 100$°C, temperature change
$A_a = 1.0$ in.^{2}, bar cross-section area
$E_a = 10 \times 10^6$ lb./in.^{2}, aluminum modulus of elasticity
$\alpha_r = 110 \times 10^{-6}$ in./in./°C, TCE of rubber pad
$L_r = 0.10$ in., length of rubber pad
$A_r = 1.0$ in.^{2}, rubber pad cross-section area
$E_r = 1000$ lb/in.^{2}, rubber modulus of elasticity

$$(23 \times 10^{-6})(8.0)(100) - \left[\frac{(8.0)P_a}{(1.0)(10 \times 10^6)}\right]$$

$$= -\left\{(110 \times 10^{-6})(0.10)(100) - \left[\frac{(0.10)P_r}{(1.0)(1000)}\right]\right\}$$

$$0.0184 - 8.0 \times 10^{-7} P_a = -(0.00110 - 1.0 \times 10^{-4} P_r)$$

$$P_a = P_r = P$$

$$P = \frac{0.0195}{1.008 \times 10^{-4}} = 193.5 \text{ lb} \tag{2.9}$$

Comparing Eq. 2.9 with 2.6 shows that adding the thin soft rubber pad reduces the force on the constrained bar from 23,000 to 193.5 lb.

2.2 BENDING OF LEAD WIRES DUE TO *x–y* THERMAL EXPANSION DIFFERENCES IN PCBs

Materials such as epoxy fiberglass are a mixture of epoxy and fibers of glass. Epoxy has a TCE of about 52×10^{-6} in./in./°C and a modulus of elasticity of about 0.50×10^{6} psi at room temperature. At 100°C the TCE goes up to about 135×10^{-6} in./in./°C and the modulus goes down to about 0.150×10^{6} psi. Glass fibers have a TCE of about 5×10^{-6} in./in./°C and a modulus of about 10×10^{6} psi under all of the normal operating conditions for the electronics. The glass fibers are oriented horizontally in the x–y plane of the PCB. This mixture gives the PCB a TCE of about 15×10^{-6} in./in./°C in the x–y plane and an average flexural modulus of elasticity of about 2.0×10^{6} psi. Since the glass fibers are oriented in the x–y plane only, there is very little change in the TCE or the modulus of the epoxy in the PCB along the z axis, normal to the plane of the PCB. The TCE of the copper circuit traces on the PCB is about 17×10^{-6} in./in./°C, just slightly greater than the combined x–y TCE of the PCB, and copper is a fairly ductile metal. Therefore, thermal cycling has very little effect on the expansion or on the stresses in the PCB copper circuit traces.

2.3 EFFECTIVE LENGTH OF ELECTRICAL LEAD WIRES

Electronic components often have their electrical lead wires extending from a flat bottom surface of the component body. Many transformers are configured in this way. When this type of component is flush mounted on a PCB through-hole style, with no space between the bottom of the component and the top surface of the PCB, the length of the lead wire would appear to be zero. In a thermal cycling condition the expansion of the component and the PCB in direction along the vertical z axis will produce an axial load in the lead wires. An examination of Eq. 2.3 shows that the spring rate of the wire is an inverse relation to the length L of the wire. If the wire length is zero, the spring rate K would be infinite. This is impossible. Therefore, the wire must have some effective length even when there is a zero gap between the component and the PCB. A combination of tests and analysis has shown that the wire length does not stop at the surface of the component or at the surface of the PCB. The effective length of the wire extends slightly beyond these surfaces. The test data shows that axially loaded wires can extend about *two wire diameters into the component body and two wire diameters into the PCB*, as shown in Fig. 2.4. When the PCB thickness is less than two wire diameters, the PCB thickness should be used instead of the two wire diameters to obtain the effective length of the wire. The equivalent number of wire diameters should then be added to the free wire length to obtain the effective length of the wire.

This same approach was used for electrical lead wires that were forced to bend. The results showed that the effective length of a bending wire appears to extend about one wire diameter into the component and one diameter into the PCB. The

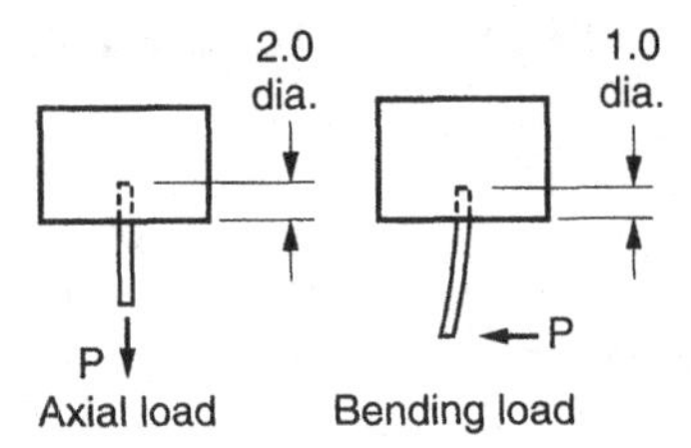

FIGURE 2.4 Effective length of an axial loaded wire will go into the component two wire diameters, and a wire loaded in bending will go into the component one wire diameter.

equivalent number of wire diameters should be added to the free wire length to obtain the effective length of the wire.

2.4 FORCES IN WIRES WITH FIXED ENDS FORCED TO BEND AND DISPLACE LATERALLY

Thermal expansion differences between the lead wires and the PCB often result in lateral displacements of the lead wires. When the lead wire extends from the bottom surface of the component and is through-hole mounted on the PCB, the wire appears to be fixed (or clamped) at the body of the component and fixed at the PCB solder joint. The wire is then laterally displaced so it is forced to bend through a distance equal to the expansion difference between the component and the PCB, as shown in Fig. 2.5. Expansions take place with respect to the center of mass. For a symmetrical component the center is at the middle of the component. In the case for thermal expansion, the effective length for evaluating the displacement will be one half of the measured component length.

Superposition can be used to find the lateral displacement and the lateral force in the bending wire. The bending wire has a point of inflection at the center of the wire. The section above the point of inflection and the section below the point of inflection will act as cantilever beams with half the length and half the displacement. Using the

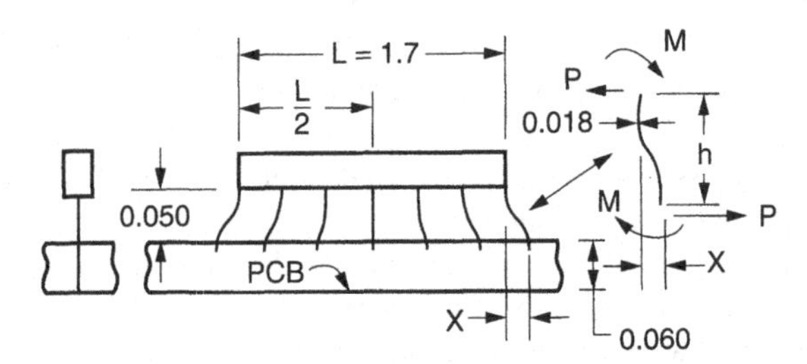

FIGURE 2.5 Differences in the TCEs of the component and the PCB can force long component wires to bend and displace in thermal cycling conditions.

equation for the deflection of two cantilevered beams end to end, the deflection for the fixed-fixed wire laterally displaced can be obtained as

$$\frac{X}{2} = \frac{P(L/2)^3}{3EI} = \frac{PL^3}{24EI} \quad \text{so} \quad X = \frac{PL^3}{12EI} \quad \text{wire displacement} \tag{2.10}$$

where $L =$ in., effective length of wire, which may include penetration into supports
$P =$ lb, lateral force developed by the lateral bending displacement of wire
$E =$ lb/in.^{2}, modulus of elasticity of wire
$I =$ in.^{4}, area moment of inertia for wire cross section

Sample Problem: Thermal Cycling Displacements, Forces, and Stresses in Lead Wires and Solder Joints for a Through-Hole-Mounted Component

A single-in-line package (SIP) electronic component 1.7 in. long is mounted through-hole-style to a 0.060 in.-thick epoxy fiberglass PCB, as shown in Fig. 2.5. Differences in the *x–y* TCE between the PCB and the SIP will force the lead wires to bend. The end lead wires will bend the most because they are the farthest distance from the neutral center point on the component. The axial stiffness of the PCB is so much greater than the bending stiffness of the lead wires that virtually all of the displacement due to thermal expansion differences will be reflected in the bending of the wires. Therefore, the axial stiffness of the PCB can be ignored here. The displacement difference will generate an overturning moment as the wire bends, which will induce bending stresses in the wire and shear tear-out stress in the solder joint. The system must operate a *rapid* thermal cycling environment where the temperature will cycle from −55 to +95°C. Find the following:

(a) Bending displacement of the lead wire
(b) Bending stress in the lead wire
(c) Shear tear-out stress in the solder joint

Solution

Part A: Lead Wire Bending Displacement Examine the end lead wires. They will bend the most since they are farthest away from the center of the component. Analyze the wires as beams fixed at both ends and laterally displaced through a distance X. The effective length of the wire in bending must include the penetration of the wire one wire diameter into the component and one wire diameter into the PCB. The displacement produced by the lead wire bending will be the same as the displacement difference between the PCB and the SIP component due to their different TCEs. The displacement difference between the PCB and the SIP can be obtained using Eq. 2.2. The bending displacement of the lead wire can be obtained from Eq. 2.10. Note that only half of the full temperature cycle of 150°C, or 75°C, is

TABLE 2.1 Correction Factor for Solder Creep (see Fig. 1.6)

Stress Relaxation (%)	Correction Factor	Equivalent Temperature Change (°C)
0	1.0 × 1.0 =	Baseline = 75.0 rapid cycle
25	1.25 × 75 =	93.75
50	1.50 × 75 =	112.5
75	1.75 × 75 =	131.25
100	2.0 × 75 =	150 slow cycle

used in the analysis since this represents the neutral point between the maximum and minimum temperatures for a rapid thermal cycle, where the solder does not have a chance to cold flow and creep. The 75°C temperature difference must *always* be used for all calculations involving thermal expansion difference and the bending force in the lead wire. The 75°C temperature difference can also be used to find the solder joint shear stress, but only for a rapid thermal cycle where no solder creep is expected. When some creep in the solder joint is expected in a thermal cycling condition, it will produce an increase in the solder stress. A correction factor must be used to account for the apparent increase in the solder stress. The correction factor shown in Table 2.1 is based on the amount of creep and stress relaxation expected in the solder joint, which is related to the solder temperature and time, as shown in Fig. 1.6. This figure shows how the solder creep reduces the stress levels. In slower thermal cycling environments there will be more solder creep. More solder creep means that the solder is always starting from a lower stress level in a thermal cycling condition, which results in higher solder stresses, as shown in Fig. 1.4.

The baseline for the solder joint stress is the rapid thermal cycle where the solder joint does not have time to creep and relieve itself. In this case the baseline solder stress would be analyzed for a temperature cycle of 75°C. For a very slow temperature cycle, where the solder has sufficient time to creep and strain relieve itself to a near zero stress condition, a correction factor of 2.0 must be used because the solder stress will be two times greater, as shown in Fig. 1.4. This is similar to using a temperature cycle of 150°C, and 150°C is two times greater than 75°C. The recommended correction factors are based on the amount of stress reduction expected in the solder due to its creep properties, as shown in Table 2.1.

Set the thermal expansion displacement equal to the bending displacement for the lead wire as shown below. Subscripts p, s, and w refer to the PCB, the SIP, and the wire, respectively:

$$(\alpha_p - \alpha_s)L_s\,\Delta t = \frac{P_w(h_w)^3}{12E_w I_w} \tag{2.11}$$

where $\alpha_p = 15 \times 10^{-6}$ in./in./°C, TCE of epoxy fiberglass PCB in x–y plane

$\alpha_s = 6 \times 10^{-6}$ in./in./°C, TCE of SIP

$L_s = \dfrac{1.7}{2} = 0.85$ in., half the length of the SIP component

$$\Delta t = \frac{95 - (-55)}{2} = 75^\circ\text{C}, \text{ for a rapid thermal cycle use half the full cycle } \Delta t$$

$h_w = 0.050 + 2(0.018) = 0.086$ in., wire length including two wire diameters

$E_w = 20 \times 10^6$ lb/in.^{2}, modulus of elasticity for kovar wire

$$I_w = \frac{\pi d^4}{64} = \frac{\pi(0.018)^4}{64} = 5.15 \times 10^{-9} \text{ in.}^4, \text{ wire area moment of inertia}$$

Substitute into Eq. 2.11:

$$(15.0 - 6.0) \times 10^{-6}(0.85)(75) = \frac{P_w(0.086)^3}{12(20 \times 10^6)(5.15 \times 10^{-9})}$$

$$5.737 \times 10^{-4} = 5.146 \times 10^{-4} P_w$$

$$P_w = 1.115 \text{ lb force in wire} \quad (2.12)$$

Part B: Bending Stress in the Lead Wire The bending moment M_w acting on the wire can be obtained from the sum of the forces and moments acting on the wire:

$$M_w = \frac{P_w h_w}{2} = \frac{(1.115)(0.086)}{2} = 0.0479 \text{ lb} \cdot \text{in.} \quad (2.13)$$

The bending stress S_w acting on the wire can be obtained from the standard bending stress equation as shown below. A stress concentration k value of 1.0 was used to show that no wire stress concentrations were considered here because thermal cycling conditions very seldom involve enough stress cycles to include the effects of fatigue accumulation:

$$S_w = \frac{kMc}{I_w} = \frac{(1.0)(0.0479)(0.018/2)}{5.15 \times 10^{-9}} = 83{,}709 \text{ lb/in.}^2 \quad (2.14)$$

The bending stress value is very close to the ultimate strength of the kovar wire and well above the wire yield strength of about 68,000 psi. However, no failures are expected in the lead wires. This is based on previous testing experience that shows that it takes several thousand thermal stress reversal cycles to produce fatigue failures in smooth unmarked wires at these stress levels. Thermal cycling conditions very seldom involve such a large number of thermal cycles. When there are sharp and deep cuts in the wire at high-stress points, fatigue failures can occur with only a few hundred stress cycles.

Part C: Shear Tear-out Stress in the Solder Joint Shear tear-out stresses will be produced in the PCB lead wire solder joints by the bending action and overturning moment in the lead wire, as shown in Fig. 2.6. The magnitude of the shear tear-out stress S_{st} is shown in Eq. 2.15. The solder joint height h is assumed to be equal to the thickness of the PCB. The area of the solder joint is taken at the diameter

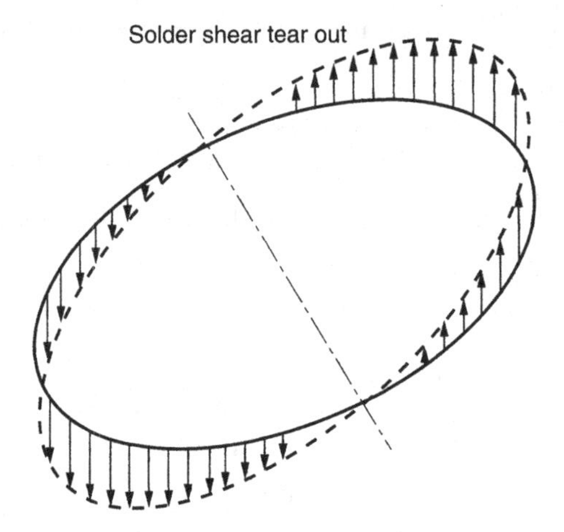

FIGURE 2.6 Differences in the TCEs of through-hole components and the PCBs can produce overturning moments in the wires that will produce shear tear-out solder joint stresses.

of the solder crack. Visual examinations of cracked solder joints show that they appear to occur in an area between the lead wire and the plated through-hole. Since this is difficult to locate accurately, an average diameter of the lead wire and the plated through-hole is used [6]:

$$S_{st} = \frac{M_w}{hA_s} \tag{2.15}$$

where $M_w = 0.0479$ lb-in., bending moment in wire; see Eq. 2.13

$h = 0.060$ in., solder joint height assumed to be same as thickness of PCB

$d_s = \frac{0.018 + 0.030}{2} = 0.024$ in., average solder joint diameter

$A_s = \frac{\pi(0.024)^2}{4} = 0.000452$ in.^{2}, average solder joint area

$$S_{st} = \frac{0.0479}{(0.060)(0.000452)} = 1{,}766 \text{ lb/in.}^2 \tag{2.16}$$

This solder joint shear tear-out stress is based on a rapid thermal cycle, where the solder does not have time to creep and to relax the internal stresses to a near zero stress condition. If the thermal cycling environment produces a long time exposure at the high temperature, the solder will have a chance to creep and relax its internal stress to a near zero stress condition. The slower solder is cycled, the weaker it gets. This would be typical for electronics under the hood of an automobile going to work or shopping several miles away, in heavy traffic. The high underhood temperatures will produce rapid thermal creep in the solder. When the automobile is parked, the solder cools down, but there is less creep when the solder temperature drops. When the engine is started again a few hours later for the return home trip, the process is again repeated. This results in a slow thermal cycle where the stress levels are increased and the fatigue life is reduced. The fatigue life reduction will depend on the thermal cycling period and the cooling down period. These areas are difficult to calculate because there is very little data available. Some approximate correction factors are shown in Table 2.1. A worst-case condition for a very slow thermal cycle will result in a correction factor of 2.0, which will double the solder joint stress level to a value of $3532\,\text{lb/in.}^2$. Doubling the solder joint stress can reduce the solder fatigue life by a factor of about 5.6.

Extensive experience with solder joint field failures and thermal cycling test failures has shown that the solder joint thermal cycling stress levels should not be allowed to exceed a value of about $400\,\text{lb/in.}^2$ in order to ensure a reliable long-term failure-free operating period (FFOP). The solder joint shear tear-out stresses here are well above the maximum desired value; thus some corrective action should be taken to reduce the solder shear stress levels. If the component height above the PCB can be doubled to 0.10 in., the solder joint shear tear-out stress can be reduced to a level close to $400\,\text{lb/in.}^2$ [6].

2.5 BETTER SAFE THAN SORRY

Because the fatigue life of any type of structure is very difficult to calculate, safety factors should be used to ensure the reliability of the product. Safety factors will increase the size, weight, and cost of the product slightly. This is a small penalty to pay for the increased reliability and customer satisfaction for a good-quality product. A satisfied customer is a repeat customer, which helps to ensure a profitable business.

It is often a good policy to assume a slow thermal cycle during any type of thermal cycling environment. Under these conditions the full temperature cycling range will be used for the thermal cycling environment to compute the solder joint shear stresses. For example, even with a rapid thermal cycling range from -40 to $+55°\text{C}$, where there is little creep and strain relief in the solder, the full temperature difference of 95°C can be used for the relative displacement and stress analysis. This is equivalent to using a safety factor of 2 for the analysis. When there is no creep and stress relaxation expected in the solder due to a rapid thermal cycle, the normal practice is to use half of the total temperature cycle or $95/2 = 47.5°\text{C}$ for the deflection and stress analysis.

CHAPTER 3

Vibration of Beams and Other Simple Structures

3.1 DYNAMIC FORCES DEVELOPED IN VIBRATION ENVIRONMENTS

Vibration usually means an oscillating motion where some structure moves back and forth. When the motion repeats itself for a period of time, it is called periodic motion. The simplest form of periodic motion is usually a sine wave or a single-frequency wave, as shown in Fig. 3.1. When a structure is excited by a sine wave or sine vibration that has a forcing frequency that is close to one of the natural frequencies of the structure, very high dynamic forces can be developed in the structure. High dynamic forces produce high dynamic stresses, which often lead to rapid fatigue failures. Low input acceleration levels can be magnified by a factor of 10 or 30 or 50 times or more. This magnification factor is called the transmissibility Q of the system. The Q is usually defined as the ratio of the output, or response, divided by the input. This can be in terms of the displacement ratio or the acceleration ratio; so the Q is dimensionless. In a simple spring and mass system, the transmissibility Q depends on the amount of damping in the system, as shown in Fig. 3.2. Notice that the transmissibility in this simple system will be infinite if the damping is zero. All real systems have some damping; so the transmissibility Q can never be infinite.

From the laws of motion, force (F) is defined as the mass (m) times acceleration (a), as shown in Eq. 3.1. It is convenient to define the acceleration in terms of the number of gravity units, or G's, as the ratio of the acceleration level (a) divided by the acceleration of gravity (g). Since these both have the same units, it means that the G term for gravity units has no dimensions; it is dimensionless.

$$F = ma = \text{lb} \tag{3.1}$$

where

$$m = \frac{\text{weight}}{\text{gravity}} = \frac{W}{g} = \frac{\text{lb}}{\text{in./s}^2} = \frac{\text{lb}}{\text{in.}}\,\text{s}^2$$

$$G = \frac{\text{acceleration}}{\text{gravity}} = \frac{a}{g} = \text{dimensionless}$$

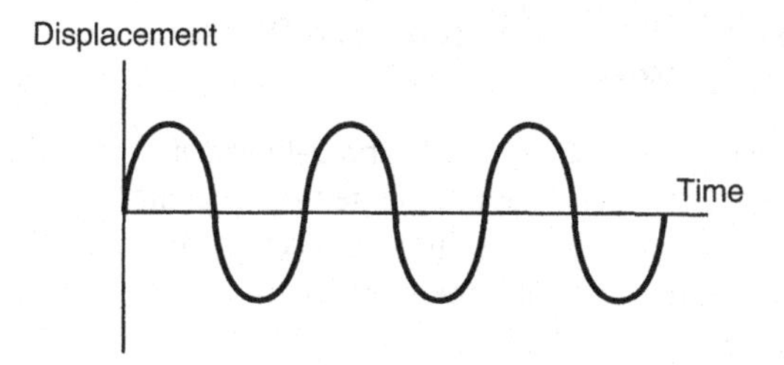

FIGURE 3.1 Single frequency sine vibration will produce harmonic motion.

$$\text{Dynamic force} \qquad F = WG = \text{lb} \tag{3.2}$$

This represents the dynamic force. If the input G level is used, it gives the input force. If the output or response G level is used, it gives the output or response force. Another way of writing this expression is to include the dimensionless transmissibility Q with the dimensionless input G_{in} level in gravity units to get the following response [2]:

$$\text{Output or response} \qquad F_{out} = WG_{in}Q = \text{lb} \tag{3.3}$$

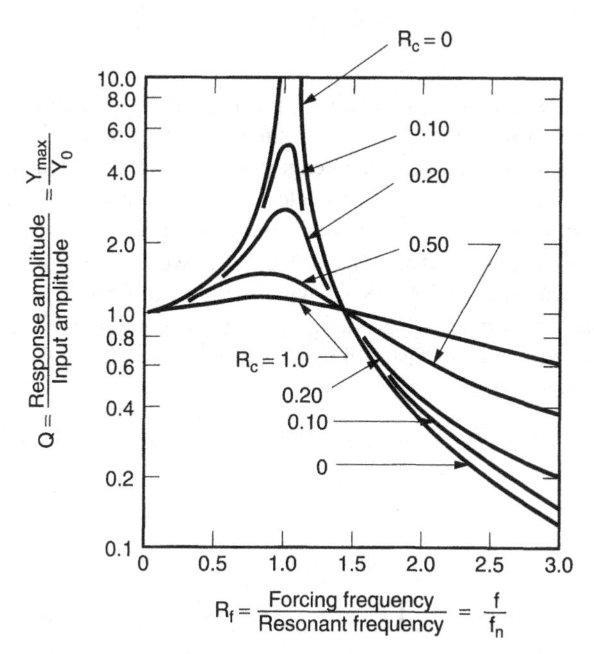

FIGURE 3.2 Response of a simple spring–mass system to sine vibration showing how the transmissibility Q changes with the damping.

Sample Problem: Vibration Forces and Stresses in Transformer Lead Wires

Determine the response force expected for a 3-lb transformer that must operate in a 5-G peak sine vibration environment mounted at the center of a PCB that has a Q of 20. Also find the dynamic tensile stress expected in the four 0.030-in. diameter electrical lead wires that hold the transformer to the PCB with soldered wires, as shown in Fig. 3.3.

Solution Substitute into Eq. 3.3 to obtain the dynamic force

$$F_{\text{out}} = (3)(5)(20) = 300 \text{ lb} \quad \text{on 4 wires} \tag{3.4}$$

Area of 4 wires:

$$A_w = \frac{\pi}{4}(0.030)^2(4 \text{ wires}) = 0.00283 \text{ in.}^2$$

Wire tensile stress

$$S_t = \frac{F_{\text{out}}}{A_w} = \frac{300}{0.00283} = 106{,}000 \text{ lb/in.}^2 \tag{3.5}$$

The above wires are loaded in direct tension. This condition can be very dangerous in a vibration environment when the wire dynamic tensile stress is more than about 50% of the ultimate tensile strength of the material. Fatigue accumulation associated with many thousand stress cycles can result in rapid fatigue failures. Transformer electrical wires are usually made of copper, which has an ultimate tensile strength of about 45,000 psi. The copper wires would be expected to fail very rapidly in this environment. Even kovar wires with an ultimate tensile strength of about 85,000 psi would not last very long under these conditions. The transformer must be bolted or cemented or fastened in some way to the PCB to reduce the forces in the wires to prevent failures in the wires.

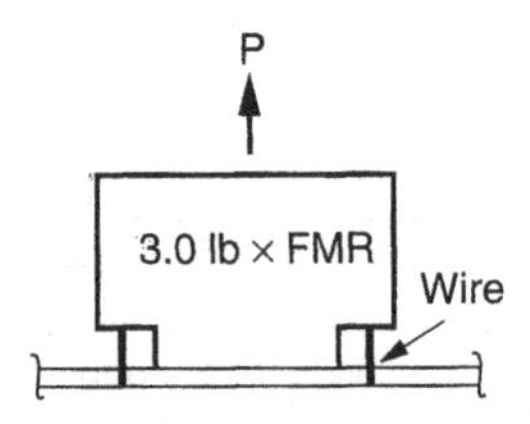

FIGURE 3.3 Vibration accelerations can produce high forces and stresses in component wires.

3.2 DETERMINING THE NATURAL FREQUENCY OF A SIMPLE STRUCTURE

The natural frequency is often called the resonant frequency. Sometimes people will classify them as the damped and undamped natural frequency. For all practical purposes with respect to electronic equipment, they are the same. The natural frequency can often be obtained by equating the kinetic energy of a vibrating structure to the strain energy of the vibrating structure without the damping. This will result in the most basic vibration equation that shows the natural frequency f_n for a simple single-degree-of-freedom spring–mass system can be related to the spring rate K, the mass m, the acceleration of gravity g, and the weight W, [1]:

$$f_n = \frac{1}{2\pi}\sqrt{\frac{K}{m}} = \frac{1}{2\pi}\sqrt{\frac{Kg}{W}} \tag{3.6}$$

This simple natural frequency equation can be written in a slightly different form by using the static displacement δ_{st} of the spring due to the action of the weight W, as

$$f_n = \frac{1}{2\pi}\sqrt{\frac{g}{\delta_{st}}} \tag{3.7}$$

The natural frequency equation shown above represents the fundamental or the lowest natural frequency of a simple vibrating structure that can be approximated as a spring and a mass that is restrained to move in only one direction. This is called a single-degree-of-freedom system because only one coordinate can be used to locate the position of the moving mass at any instant of time. A single-degree-of-freedom system can have only one natural frequency. When there are two masses in a system, then two coordinates are required to locate each mass at any instant of time; then it is called a two-degree-of-freedom system. When one mass on a spring is allowed to move in two different directions, it again requires two coordinates to locate the position of the mass at any instant of time; so it is also called a two-degree-of-freedom system. A two-degree-of-freedom system can only have two natural frequencies. The number of natural frequencies is always equal to the number of degrees of freedom.

Vibrating structures will always move in a direction that will yield the lowest strain energy. The direction of the vibrating force does not matter. When a cantilever beam is being vibrated along any one axis, the beam can respond in the same axis, or in any other axis or any combination of axes. The beam will always move in the direction that has the least resistance. The beam will often twist and bend at the same time. Finite element modeling (FEM) methods with a high-speed computer are very useful for determining how complex structures will move when they are being vibrated. Vibration tests with the proper equipment, including a strobe light, are very often used to observe the motion of vibrating structures. This helps to understand how different types of structures act in vibration and why they fail. Vibration theory says that all real structures will have an infinite number of natural frequencies. A cantilever beam, for example, can vibrate at its fundamental or lowest natural frequency, or its second harmonic frequency, or its third, or fourth, or fifth, up to infinity. Under some conditions, such as random vibration and shock, a large number of resonances can be excited at the same time.

Sample Problem: Natural Frequency of a Cantilever Beam with an End Mass

Find the natural frequency of a cantilever aluminum beam 8.0 in. long with a 6.0-lb transformer at the tip of the beam, as shown in Fig. 3.4. The beam is restrained to vibrate in the vertical direction only. When the beam weight is small compared to the end weight, the beam weight can be ignored with very little error.

Solution Find the static displacement of the cantilever beam; then use Eq. 3.7 to find the natural frequency. The static displacement at the end of the beam with a concentrated end load can be obtained from a structural handbook as

$$\delta_{st} = \frac{WL^3}{3EI} \tag{3.8}$$

where $W = 6.0\,\text{lb}$ end weight

$L = 8.0\,\text{in.}$, length of beam

$E = 10.5 \times 10^6\,\text{lb/in.}^2$, aluminum modulus of elasticity

$I = \frac{bh^3}{12} = \frac{(0.50)(2.0)^3}{12} = 0.333\,\text{in.}^4$, area moment of inertia

$$\delta_{st} = \frac{(6.0)(8.0)^3}{3(10.5 \times 10^6)(0.333)} = 2.93 \times 10^{-4}\,\text{in., static displacement} \tag{3.9}$$

Substitute into Eq. 3.7 for the natural frequency and note that the acceleration of gravity g is $386\,\text{in./s}^2$:

$$f_n = \frac{1}{2\pi}\sqrt{\frac{386}{2.93 \times 10^{-4}}} = 183\,\text{Hz} \tag{3.10}$$

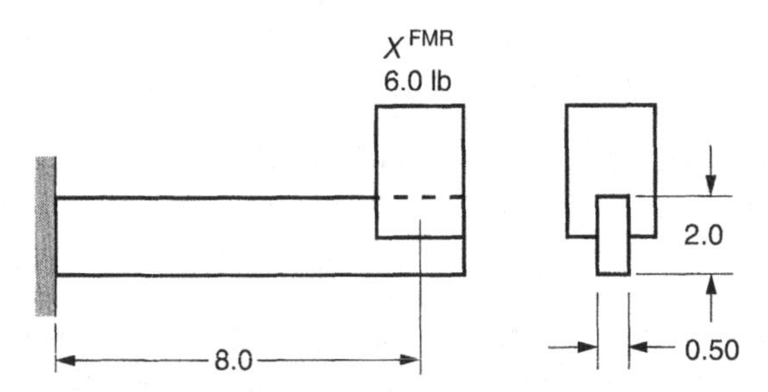

FIGURE 3.4 Transformer is mounted at the end of a cantilever beam.

The natural frequency of the cantilever beam can also be obtained from Eq. 3.6 using the spring rate K of the cantilever beam as

$$K = \frac{W}{\delta_{\text{st}}} = \frac{3EI}{L^3} \tag{3.11}$$

$$K = \frac{3(10.5 \times 10^6)(0.333)}{(8.0)^3} = 2.049 \times 10^4 \text{ lb/in., spring rate} \tag{3.12}$$

$$m = \frac{W}{g} = \frac{6.0}{386} = 0.0155 \text{ lb s}^2/\text{in., end mass}$$

Substitute into Eq. 3.6 to find the natural frequency:

$$f_n = \frac{1}{2n}\sqrt{\frac{2.049 \times 10^4}{0.0155}} = 183 \text{ Hz} \tag{3.13}$$

This shows the same results as Eq. 3.10.

3.3 RELATION OF DISPLACEMENT TO FREQUENCY AND ACCELERATION *G* LEVEL

Dynamic displacements that are developed during vibration conditions are often very important since they can be used to determine the dynamic stresses, which are needed for the calculation of the expected fatigue life of the critical electronic structural members. Dynamic displacements are very difficult to observe and to measure. Acceleration levels are easy to obtain with the use of small accelerometers, and frequencies are easy to obtain directly from the electrodynamic shakers performing the vibration. Therefore, if there is an easy way to obtain accurate dynamic displacements, it would reduce the amount of work required to calculate the expected fatigue life of critical electronic members. The following section shows how this can be done.

Consider a rotating vector that is used to describe simple harmonic motion of a mass suspended on a spring, as shown in Fig. 3.5. The vertical displacement Y of the mass can be obtained from the projection of the vector Y_0 on the vertical axis. The projection on the vertical axis is the displacement Y. It can be represented by Eq. 3.14, where Y_0 is the maximum single amplitude displacement from zero to peak:

$$Y = Y_0 \sin \Omega t \tag{3.14}$$

The velocity is the first derivative of the displacement:

$$V = \frac{dY}{dt} = \Omega Y_0 \cos \Omega t$$

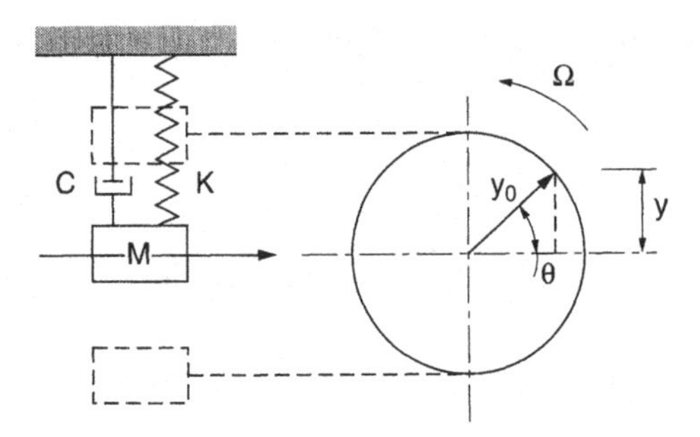

FIGURE 3.5 Vibrating spring–mass system can be represented by a rotating vector.

Acceleration is the second derivative of the displacement:

$$a = \frac{d^2Y}{dt^2} = -\Omega^2 Y_0 \sin \Omega t$$

The negative sign shows that acceleration acts in the opposite direction to displacement. The maximum acceleration will occur when $\sin \Omega t$ is one:

$$a_{\max} = \Omega^2 Y_0 \quad \text{where} \quad \Omega = 2\pi f \tag{3.15}$$

Dimensionless gravity units are desired using acceleration of gravity g at 386 in./s^2:

$$G = \frac{a_{\max}}{g} = \frac{4\pi^2 f^2 Y_0}{386} = \frac{f^2 Y_0}{9.8}$$

Rewriting the above equation results in the more convenient form shown below:

$$Y_0 = \frac{9.8\ G}{f^2} \qquad \text{single amplitude displacement} \tag{3.16}$$

When the *input* acceleration level is expressed in gravity units, or G's, the maximum single amplitude displacement Y_0 will also represent the *input* displacement. When the *output* or response is expressed in G gravity units, the displacement Y_0 will represent the *output* or *response* displacement. Notice that the frequency f in cycles per second, or Hertz, typically shown as Hz, is the frequency at any instant of time, unless the condition is a frequency dwell, where the frequency then remains constant.

The above equation is often considered to be the most important equation in the field of dynamics. It shows that the displacement, the acceleration level, and the frequency are locked together. They cannot be separated. Any two values automatically determine the third value. This equation is valid for sine vibration, random vibration, shock, and acoustic noise.

Sometimes it is convenient to express Eq. 3.16 in terms of the input acceleration level as G_{in}, but the maximum output or response displacement Y_0 is still desired. This can be done by including the transmissibility Q and by changing to the natural frequency of the structure as

$$Y_0 = \frac{9.8 G_{in} Q}{(f_n)^2} \quad \text{maximum response displacement} \tag{3.17}$$

Sample Problem: Finding Displacement from the Frequency and Acceleration

A sine vibration test was run using an input acceleration level of 2.0 G peak. A small accelerometer mounted at the center of a plug-in PCB showed a transmissibility Q of about 20 at the natural frequency of 200 Hz. What is the maximum expected dynamic displacement at the center of the PCB, and what is the expected relative displacement at the center of the PCB?

Solution The maximum single amplitude displacement expected at the center of the PCB can be obtained from Eq. 3.17 with the following information:

where $G_{in} = 2.0$ G's, input acceleration level in gravity units
$Q = 20$ transmissibility Q at the center of the PCB
$f_n = 200$ Hz, natural frequency of PCB

Substitute into Eq. 3.17 to find the maximum expected PCB response displacement:

$$Y_o = \frac{(9.8)(2.0)(20)}{(200)^2} = 0.0098 \text{ in.} \tag{3.18}$$

The maximum PCB response displacement shown above includes the input displacement as well as the response displacement. The bending stresses developed in the components on the PCB and the bending of the PCB itself will be related to the relative displacement of the PCB. The relative displacement is the output or response displacement minus the input displacement as

$$\text{Relative displacement} \quad Y_0 = \frac{9.8 G_{in}(Q-1)}{(f_n)^2} \tag{3.19}$$

$$\text{Relative displacement} \quad Y_0 = \frac{(9.8)(2.0)(20-1)}{(200)^2} = 0.00931 \text{ in.} \tag{3.20}$$

Comparing the results of Eqs. 3.18 and 3.20 shows there is very little difference between the absolute PCB displacement and the relative PCB displacement. This is typical for dynamic systems with high transmissibility Q's; so it is a common practice to use the absolute displacements instead of the relative displacement. In addition, there is so much scatter in trying to predict the fatigue life of electronic systems in vibration environments that it is common to use safety factors to be on the safe side of the prediction. Using the absolute dynamic displacement adds a small additional safety factor to the calculations.

3.4 DYNAMIC FORCES AND DISPLACEMENTS FOR BEAMS WITH CONCENTRATED LOADS

Beams with concentrated loads can often be approximated as simple single-degree-of-freedom systems. This approximation reduces the amount of work required to perform a dynamic analysis, but it also reduces the accuracy of the analysis. The reduction in the amount of work required is typically worthwhile, even with the reduction in the accuracy of the analysis. Any time a beam is evaluated as a single-degree-of-freedom system, the dynamic displacements and forces can be determined in two different ways. One way is to use the standard beam displacement equations with a concentrated load, and the other way is to use the displacement, frequency, and acceleration equation (3.16). This is shown in the following sample problem.

Sample Problem: Two Ways for Finding Dynamic Displacements and Dynamic Forces Acting on a Beam

A 3.0-lb inductor is fastened at the center of a 7.0-in.-long epoxy fiberglass beam that is bolted at both ends, as shown in Fig. 3.6. The structure must withstand a 4-G peak sine vibration input level. Use two different methods to find the dynamic force and two different methods for finding the dynamic displacement expected. Ignore the beam weight since it is small compared to the inductor weight.

Solution The natural frequency for a beam with a concentrated load can be obtained from the static displacement of a simply supported (hinged) uniform beam. Vibration test data on bolted beam structures used in electronic equipment shows that bolted joints act like simple supports or hinged joints. This is because there is never enough room for a large size bolt because all of the available area is usually occupied by the electronic components. Therefore, there is only enough room for small bolts, and small bolts are not capable of rigidly clamping the ends of a vibrating beam. The static displacement of a simply supported uniform beam with a

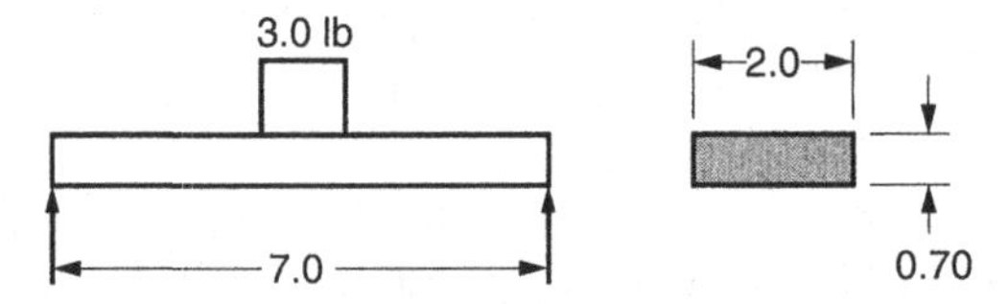

FIGURE 3.6 Concentrated mass is located at the center of a uniform beam supported at each end.

center concentrated load can be obtained from a structural handbook as

$$\delta_{st} = \frac{WL^3}{48EI} \tag{3.21}$$

where $W = 3.0$ lb, static load on beam
$L = 7.0$ in., length of beam
$E = 2 \times 10^6$ lb/in.^{2}, epoxy fiberglass modulus of elasticity
$I = \frac{(2.0)(0.70)^3}{12} = 0.0572$ in.^{4}, area moment of inertia

$$\delta_{st} = \frac{(3.0)(7.0)^3}{48(2 \times 10^6)(0.0572)} = 1.87 \times 10^{-4} \text{ in. static displacement} \tag{3.22}$$

Substitute into Eq. 3.7 to find the natural frequency

$$f_n = \frac{1}{2\pi}\sqrt{\frac{386}{1.87 \times 10^{-4}}} = 228.8 \text{ Hz} \tag{3.23}$$

A good approximation for finding the transmissibility Q for a beam is [1]

$$\text{Beam } Q = 2\sqrt{f_n} = 2\sqrt{228.8} = 30.2 \text{ dimensionless} \tag{3.24}$$

Solution Method 1 for finding the dynamic force using Eq. 3.3 is

$$F = WG_{in}Q = (3.0)(4.0)(30.2) = 362.4 \text{ lb} \tag{3.25}$$

Method 1 for finding the dynamic displacement using Eq. 3.22 is

$$\delta_{dy} = \delta_{st}G_{in}Q = (1.87 \times 10^{-4})(4.0)(30.2) = 0.0226 \text{ in.} \tag{3.26}$$

Method 2 for finding the dynamic force by substituting Eq. 3.26 into Eq. 3.21 and solving for W, which becomes the dynamic force F as shown:

$$W = F = \frac{(48)(2 \times 10^6)(0.0572)(0.0226)}{(7.0)^3} = 362 \text{ lb.} \tag{3.27}$$

Method 2 for finding the dynamic displacement using Eq. 3.17 is

$$Y_0 = \delta_{dy} = \frac{(9.8)(4.0)(30.2)}{(228.8)^2} = 0.0226 \text{ in.} \tag{3.28}$$

Comparing the dynamic force Eq. 3.25 with Eq. 3.27 shows good results. Comparing the dynamic displacement Eq. 3.26 with Eq. 3.28 shows good results.

3.5 NATURAL FREQUENCIES OF UNIFORM BEAM STRUCTURES

The natural frequency of a bending uniform beam can easily be determined by equating the strain energy to the kinetic energy of the vibrating beam. There are a number of ways in which the relationships can be set up. Some methods are approximate and some methods are quite accurate. Trigonometric functions are popular since they can be very accurate when the boundary conditions are accurately established. Polynomial functions can also be used but they are generally less accurate. The exact methods of solution require more work; so approximate methods are usually used that are satisfactory for engineering work. One very popular method of evaluation is the Rayleigh method, named after Lord Rayleigh who first proposed it. This method can be used to obtain the bending natural frequency for uniform beams with three different boundary conditions, restrained to move only in the vertical axis as

$$f_n = \frac{a}{2\pi}\sqrt{\frac{EIg}{WL^3}} \tag{3.29}$$

where $a = 3.52$ for a cantilever beam
$= \pi^2$ for a beam simply supported, or hinged, at both ends
$= 22.37$ for a beam clamped at both ends or a free beam with no supports
$E = \text{lb/in.}^2$, modulus of elasticity of beam material
$I = \text{in.}^4$, area moment of inertia of beam cross section
$G = 386\ \text{in./s}^2$, acceleration of gravity
$W = \text{lb}$, total weight of the beam
$L = \text{in.}$, length of the beam

Sample Problem: Natural Frequency of a Steel Beam and an Aluminum Beam

Determine the natural frequency of a uniform aluminum beam and a steel beam 6.0 in. long, simply supported at both ends, with a cross section that is 2.5 in. wide and 0.38 in. high, that is restrained to vibrate only in the vertical direction.

Solution Substitute into Eq. 3.29 to find the natural frequency of the aluminum beam first, with a modulus of elasticity of $10.5 \times 10^6\ \text{lb/in.}^2$ and a density of $0.10\ \text{lb/in.}^3$. Then find the natural frequency of the steel beam with a modulus of elasticity of $29.0 \times 10^6\ \text{lb/in.}^2$ and a density of $0.283\ \text{lb/in.}^3$.

$$f_n = C\sqrt{\frac{EIg}{\omega L^4}}\quad \text{where } C = \text{Modal constant}$$

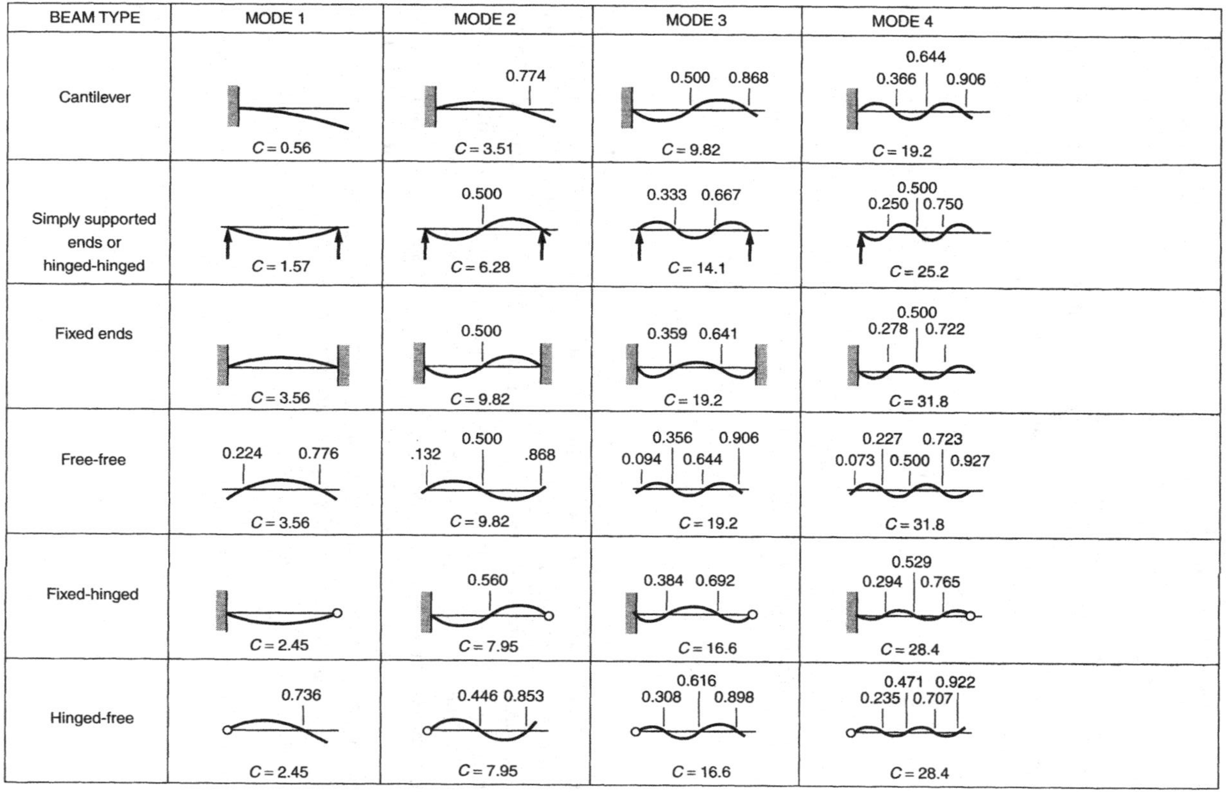

BEAM TYPE	MODE 1	MODE 2	MODE 3	MODE 4
Cantilever	$C = 0.56$	0.774 $C = 3.51$	0.500 0.868 $C = 9.82$	0.644 0.366 0.906 $C = 19.2$
Simply supported ends or hinged-hinged	$C = 1.57$	0.500 $C = 6.28$	0.333 0.667 $C = 14.1$	0.500 0.250 0.750 $C = 25.2$
Fixed ends	$C = 3.56$	0.500 $C = 9.82$	0.359 0.641 $C = 19.2$	0.500 0.278 0.722 $C = 31.8$
Free-free	0.224 0.776 $C = 3.56$	0.500 .132 .868 $C = 9.82$	0.356 0.906 0.094 0.644 $C = 19.2$	0.227 0.723 0.073 0.500 0.927 $C = 31.8$
Fixed-hinged	$C = 2.45$	0.560 $C = 7.95$	0.384 0.692 $C = 16.6$	0.529 0.294 0.765 $C = 28.4$
Hinged-free	0.736 $C = 2.45$	0.446 0.853 $C = 7.95$	0.616 0.308 0.898 $C = 16.6$	0.471 0.922 0.235 0.707 $C = 28.4$

FIGURE 3.7 Natural frequency equations and mode shapes for uniform beams with different

where $a = \pi^2$ for a simply supported beam

$E = 10.5 \times 10^6$ lb/in.^{2}, modulus of elasticity for the aluminum beam

$E = 29 \times 10^6$ lb/in.^{2}, modulus of elasticity for the steel beam

$I = \frac{bh^3}{12} = \frac{(2.5)(0.38)^3}{12} = 0.0114$ in.^{4}, area moment of inertia of cross section

$g = 386$ in./s^2, acceleration of gravity

$W = (6.0)(2.5)(0.38)(0.10 \text{ lb/in.}^3) = 0.57$ lb weight of aluminum beam

$W = (6.0)(2.5)(0.38)(0.283 \text{ lb/in.}^3) = 1.61$ lb weight of steel beam

$L = 6.0$ in., length

$$\text{Aluminum:} \qquad f_n = \frac{\pi}{2}\sqrt{\frac{(10.5 \times 10^6)(0.0114)(386)}{(0.57)(6.0)^3}} = 961.8 \text{ Hz} \tag{3.30}$$

Next find the natural frequency of the steel beam with the same dimensions, but with a different modulus of elasticity and a different density:

$$\text{Steel:} \qquad f_n = \frac{\pi}{2}\sqrt{\frac{(29 \times 10^6)(0.0114)(386)}{(1.61)(6.0)^3}} = 951.1 \text{ Hz} \tag{3.31}$$

The natural frequencies of the aluminum beam and the steel beam are very similar. This is because the ratio of the modulus of elasticity to the density is approximately the same for the aluminum and the steel. Therefore, when the physical dimensions are the same, the natural frequencies will also be about the same. These material property characteristics are similar for aluminum, steel, magnesium, and titanium. Beams of the same size with these materials, and supported in the same way, will all have approximately the same natural frequency. The same size beam made out of beryllium would have a much higher natural frequency because beryllium has a much higher modulus of elasticity and a lower density. The natural frequencies for uniform beams with different supports are shown in Fig. 3.7.

CHAPTER 4

Vibration of Printed Circuit Boards and Flat Plates

4.1 CHARACTERISTICS OF DIFFERENT PRINTED CIRCUIT BOARDS

Printed circuit boards (PCBs) can come in many different sizes and shapes with many different types of through-hole-mounted electronic components and many different types of surface-mounted electronic components. A wide variety of materials are used to fabricate the boards with different combinations such as epoxy glass, polyimide glass, epoxy kevlar, polyimide kevlar, epoxy quartz, polyimide quartz, and many different types of ceramics. Solder is still the most popular method used worldwide to attach the electronic components to the PCBs. Many different component attachment methods are used such as wave soldering, dip soldering, vapor phase, infrared, oven, and, of course, hand soldering. There is a big push in the United States to get rid of solder because of the lead content and its impact on the environment. Many different plastics and polymers are being made electrically conductive with a high fill of various powdered metals, such as silver. These are being tested and evaluated as a possible replacement for solder. However, there are still a few problems that have to be solved before they can gain wide acceptance.

Epoxy fiberglass appears to be the most popular material used by many different commercial, industrial, and military production organizations for the fabrication of PCBs. The shape of the PCB often depends on the available space. It may be round in the nose of an airplane and triangular in a wing tip. The rectangular shape is still the most common shape for the PCB since it can easily be adapted to the plug-in type of assembly with an interface electrical connector, which makes it convenient for maintenance and repair. Copper is the most common metal used for the etched circuit traces as well as for ground planes and voltage planes in multilayer PCBs. Copper has a high thermal conductivity; thus, it is often used to conduct the heat away from high heat dissipating electronic components. Thin sheets of aluminum are often laminated to PCBs in high heat dissipating areas because aluminum has a high thermal conductivity and it is much lighter than copper [1, 5].

The way in which the PCBs are supported within an enclosure can significantly effect the way the PCBs respond to vibration and shock environments. Most companies involved in the design and fabrication of electronic enclosures try to use a loose fit between the PCB and the enclosure to reduce the manufacturing costs. Costs are an important factor in the success of any organization. However, when plug-in types of PCBs are required to operate with a high degree of reliability in severe vibration or shock environments, then a loose fit between the PCB and the enclosure *must not be used*. Test data shows that a loose fit between the PCB and the enclosure can magnify the dynamic acceleration levels and cause premature failures in the electronic component lead wires, solder joints, and plug-in electrical connectors. To ensure the reliability of plug-in PCBs under these conditions, there must be a snug fit between the edges of the PCB and the enclosure. Many different types of PCB edge guides are available from different electronic equipment suppliers that will provide the required support at a reasonable price. These PCB edge guides are available in metals and plastics. The metal guides are typically used to allow the conduction of heat across the interfaces from the edges of the PCB to the enclosure. High interface pressures must be used between the PCB edges and the enclosure to provide a low thermal resistance for good heat transfer in space or high-altitude conditions. See Fig. 8.14 for the effects of a loose PCB edge guide.

4.2 EFFECTS OF VIBRATION ON CIRCUIT BOARD EDGE CONDITIONS

Circuit boards are typically fabricated using thin plastic materials, which make them act in a nonlinear fashion. This makes them sensitive to the acceleration levels produced in vibration environments. Vibration test data shows that the natural frequency of a plug-in type of PCB will often show an increase, when the input acceleration level is decreased. The tests also show that the natural frequency will often decrease when the acceleration level is increased. An accurate evaluation of the PCB natural frequency is desired since dynamic displacements, stresses, and the fatigue life are closely related to the natural frequency. An accurate natural frequency calculation requires an accurate evaluation of the edge conditions on the PCB. Only three normal edge conditions are usually shown for plug-in types of PCBs. These are free, simply supported or hinged, and fixed or clamped. The problem is that the PCBs are nonlinear; thus these edge conditions can change with a change in the input acceleration levels. A good way to evaluate changes in the PCB edge conditions during vibration is to use a strobe light to watch the board motions. This will show very quickly whether the edges are rotating or translating or a combination of both. Very often the edge conditions can be observed to change from simply supported with an input acceleration level of 2 G peak to free with an input acceleration level of the 10 G-peak. The natural frequency with the 2-G peak input level will then be higher than the natural frequency with the 10-G peak input level.

Sine vibration tests were run on a group of plug-in epoxy fiberglass PCBs vertically oriented, 7 in. high, 8 in. wide, and 0.062 in. thick, with a weight of about 1 lb. A 100-pin connector at the bottom of the 8-in. edge was plugged into the sockets of a mating connector mounted in a rigid vibration test fixture. The two vertical edges were restrained by a preloaded beryllium copper wavy spring in a channel type of edge guide. Three foam rubber strips were used to restrain the top edge of the PCB. Each strip was 1 in. wide and 0.125 in. thick, compressed about

TABLE 4.1 Effects of Input *G* Level on PCB Resonant Frequency and Transmissibility

Input *G* Level	Resonant Frequency (Hz)	Transmissibility Q
2	215	15.0
5	182	11.2
10	161	8.2

60% when installed. This was to simulate the action of a bolted top cover using similar strips. Strips of this type are convenient to use since they supply some support and damping, and they do not require expensive close tolerance assembly procedures. The disadvantage is they can wear out after many years of vibration; so the support and damping may be lost.

Tests using sine vibration peak input acceleration levels of 2, 5, 10, and 15 *G* were monitored with a strobe light to observe the motion of the PCBs during their resonant conditions. The 2-*G* input level showed all four edges of the PCB appeared to be simply supported, since all four edges appeared to be rotating with no translation. The 5-*G* test showed the sides of the PCB appeared to be supported, but the top edge at the rubber strips and the bottom edge at the connector showed some translation in addition to the rotation. The 10-*G* test still showed the sides appeared to be supported, but the top edge at the rubber strips and the bottom edge at the connector showed more translation than rotation. The 15-*G* test showed large amounts of translation at the sides with a little rotation, while the top and bottom edges appeared to be free because only translation could be observed. Resonant dwell tests with the 10-*G* peak input resulted in many broken connector pins. Table 4.1 shows a summary of the test results [1].

4.3 NATURAL FREQUENCIES OF PRINTED CIRCUIT BOARDS

Most PCBs can be approximated as flat plates with different shapes, different edge supports, and different loading conditions. The Rayleigh method used for one-dimensional beam structures can also be used for two-dimensional plate structures. General plate relations for kinetic energy can again be equated to the strain energy without damping, which will result in the plate natural frequency equation. A deflection curve is generated that satisfies the geometric boundary conditions, which are the deflection and the slope for a particular plate. Trigonometric functions or polynomial functions can then be used to obtain the required strain energy of bending and the kinetic energy of bending to obtain the natural frequency. A typical trigonometric deflection curve for the simply supported rectangular plate shown in Fig. 4.1, which will satisfy the required geometric boundary conditions, [7, 8]

$$Z = \sum_{m=1,3,5}^{\infty} \sum_{n=1,3,5}^{\infty} A_{mn} \sin \frac{m\pi x}{a} \sin \frac{n\pi y}{b} \tag{4.1}$$

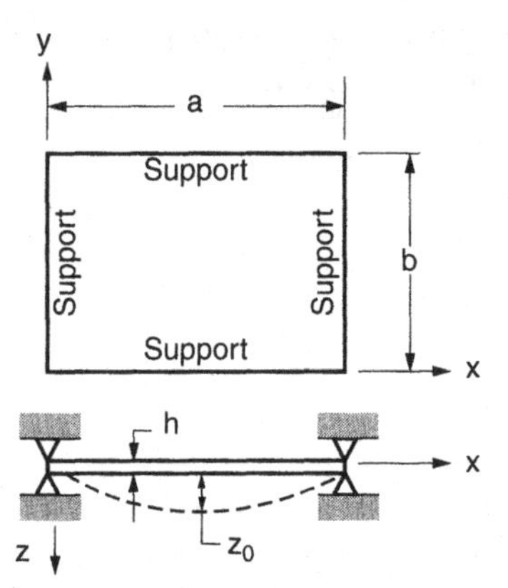

FIGURE 4.1 Uniform rectangular plate simply supported on four sides.

Considering only the fundamental resonant mode, where most of the vibration damage occurs, the above expression can be simplified by taking only the first term of the equation as

$$Z = Z_0 \sin\frac{\pi X}{a}\sin\frac{\pi Y}{b} \tag{4.2}$$

Setting the strain energy of the bending plate equal to the kinetic energy, when there is no energy dissipated, will result in the natural frequency if the uniform flat plate is simply supported on four sides. In this case the derived natural frequency equation is exact because the deflection curve used for the vibrating plate is exact for the boundary conditions examined. The resulting natural frequency equation is

$$f_n = \frac{\pi}{2}\sqrt{\frac{D}{\rho}}\left(\frac{1}{a^2}+\frac{1}{b^2}\right) \tag{4.3}$$

where $D = \frac{EI}{\text{inch}} = \frac{Ebh^3}{12b} = \frac{Eh^3}{12(1-\mu^2)}$ plate stiffness factor, lb · in.

E = modulus of elasticity, lb/in.^{2}

h = board thickness, in.

a = board length, in.

b = board width, in.

μ = Poisson's ratio, dimensionless

W = weight of assembled board, lb

g = acceleration of gravity, 386 in./s^2

$\rho = \frac{W}{gab} = \frac{\text{mass}}{\text{area}} = \frac{\text{lb}\cdot\text{s}^2}{\text{in.}^3}$

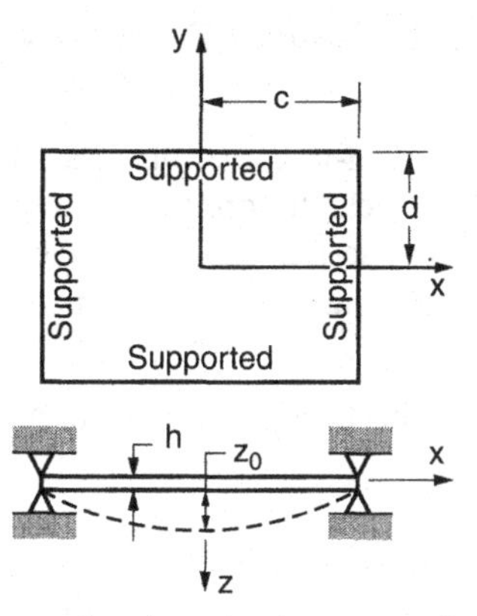

FIGURE 4.2 Uniform rectangular plate simply supported on four sides with the reference axes located at the center of the plate.

A typical polynomial deflection curve for the simply supported rectangular plate in Fig. 4.2 that will satisfy the required geometric boundary conditions is

$$Z = h \sum_{m=0}^{\infty} \sum_{n=0}^{\infty} \left(1 - \frac{X^2}{c^2}\right)\left(1 - \frac{Y^2}{d^2}\right)\left(\frac{X}{c}\right)^m \left(\frac{Y}{d}\right)^n A_{mn} \cos \Omega t \tag{4.4}$$

Considering only the fundamental resonant mode, where most of the vibration damage occurs, the above expression can be simplified by taking the first term of the equation as [8]

$$Z = Z_0 \left(1 - \frac{X^2}{c^2}\right)\left(1 - \frac{Y^2}{d^2}\right) \tag{4.5}$$

Setting the strain energy equal to the kinetic energy of the bending plate when there is no energy dissipated will result in the natural frequency equation for the uniform plate supported on four sides. The natural frequency obtained in this case will not be exact because the deflection expression was not exact; it was only an approximation. With the Rayleigh method, the resulting natural frequency using the polynomial will be slightly higher than the correct natural frequency:

$$f_n = \frac{5.31}{\pi} \left[\frac{D}{\rho}\left(\frac{1}{15c^4} + \frac{1}{9c^2 d^2} + \frac{1}{15d^4}\right)\right]^{1/2} \tag{4.6}$$

Sample Problem: Finding the Natural Frequency of a Plug-in Circuit Board

Find the natural frequency of a plug-in epoxy fiberglass PCB that measures 8.0 in. long by 6.0 in. wide by 0.062 in. thick, uniformly loaded with a total weight of 1.0 lb, simply supported on four sides. Use Eq. 4.3 to find the correct natural frequency; then use Eq. 4.6 to find the approximate natural frequency.

Solution The following data is needed to solve Eq. 4.3:

$a = 8.0$ in., board length
$b = 6.0$ in., board width
$h = 0.062$ in., board thickness
$E = 2.0 \times 10^6$ lb/in.^{2}, modulus of elasticity for epoxy fiberglass
$\mu = 0.12$ dimensionless, Poisson's ratio

$$D = \frac{(2.0 \times 10^6)(0.062)^3}{12(1 - 0.12^2)} = 40.3 \text{ lb} \cdot \text{in, board stiffness}$$

$W = 1.0$ lb, board weight
$g = 386$ in./s^2, acceleration of gravity

$$\rho = \frac{1.0}{(386)(8.0)(6.0)} = 5.40 \times 10^{-5} \text{ lb} \cdot \text{s}^2/\text{in.}^3\text{, board mass/area}$$

$$f_n = \frac{\pi}{2}\sqrt{\frac{40.3}{5.40 \times 10^{-5}}}\left(\frac{1}{8^2} + \frac{1}{6^2}\right) = 58.9 \text{ Hz} \tag{4.7}$$

Solution The following data is needed to solve Eq. 4.6:

$c = 4.0$ in., half of the board length
$d = 3.0$ in., half the board width

$$f_n = \frac{5.31}{\pi}\left[\frac{40.3}{5.40 \times 10^{-5}}\left(\frac{1}{15(4.0)^4} + \frac{1}{9(4.0)^2(3.0)^2} + \frac{1}{15(3.0)^4}\right)\right]^{1/2}$$
$$= 62.9 \text{ Hz} \tag{4.8}$$

The trigonometric base of Eq. 4.7 with the natural frequency of 58.9 Hz is the correct value. The polynomial base of Eq. 4.8 with the natural frequency of 62.9 Hz is approximate and slightly higher, which follows the Rayleigh theory.

The fundamental natural frequency equations for flat, rectangular, uniformly loaded plates with various free, supported, and fixed edge conditions are shown in Figs. 4.3–4.5.

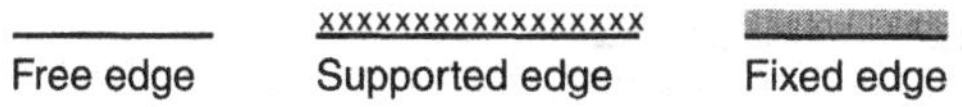

Free edge   Supported edge   Fixed edge

Equation	Plate
$f_n = \frac{\pi}{11}\left(\frac{D}{\rho}\right)^{1/2}\left(\frac{1}{a^2}+\frac{1}{b^2}\right)$	a, b
$f_n = \frac{\pi}{2}\left(\frac{D}{\rho}\right)^{1/2}\left(\frac{1}{4a^2}+\frac{1}{b^2}\right)$	a, b
$f_n = \frac{\pi}{2}\left(\frac{D}{\rho}\right)^{1/2}\left(\frac{1}{a^2}+\frac{1}{b^2}\right)$	a, b
$f_n = \frac{\pi}{5.42}\left[\frac{D}{\rho}\left(\frac{1}{a^4}+\frac{3.2}{a^2b^2}+\frac{1}{b^4}\right)\right]^{1/2}$	a, b
$f_n = \frac{\pi}{3}\left[\frac{D}{\rho}\left(\frac{0.75}{a^4}+\frac{2}{a^2b^2}+\frac{12}{b^4}\right)\right]^{1/2}$	a, b
$f_n = \frac{\pi}{1.5}\left[\frac{D}{\rho}\left(\frac{3}{a^4}+\frac{2}{a^2b^2}+\frac{3}{b^4}\right)\right]^{1/2}$	a, b
$f_n = \frac{\pi}{3.46}\left[\frac{D}{\rho}\left(\frac{16}{a^4}+\frac{8}{a^2b^2}+\frac{3}{b^4}\right)\right]^{1/2}$	a, b

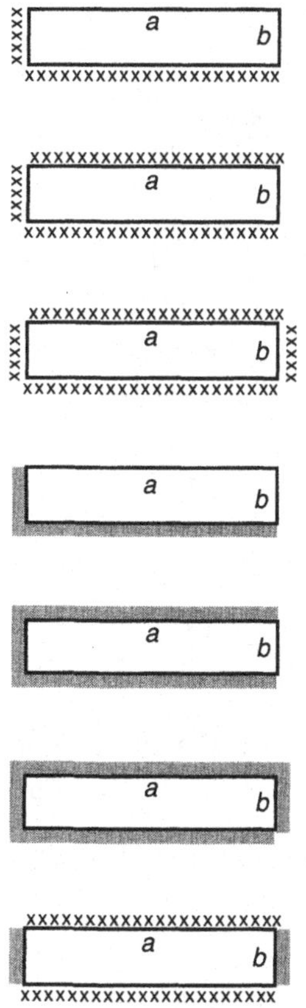

FIGURE 4.3 Natural frequency equations for uniform rectangular plates with various edge restraints.

4.4 CONDITIONS THAT INFLUENCE TRANSMISSIBILITY *Q*'s FOR VARIOUS STRUCTURES AND PLATES

Damping controls the transmissibility Q of simple systems, where high damping results in a low Q and low damping results in a high Q. In complex systems the phase relations between the forcing function and the responses of the various mass elements in the structure often have a greater influence on the transmissibility of the various masses than the damping in these elements. In areas where damping has a large influence on the transmissibility, high dynamic displacements usually mean

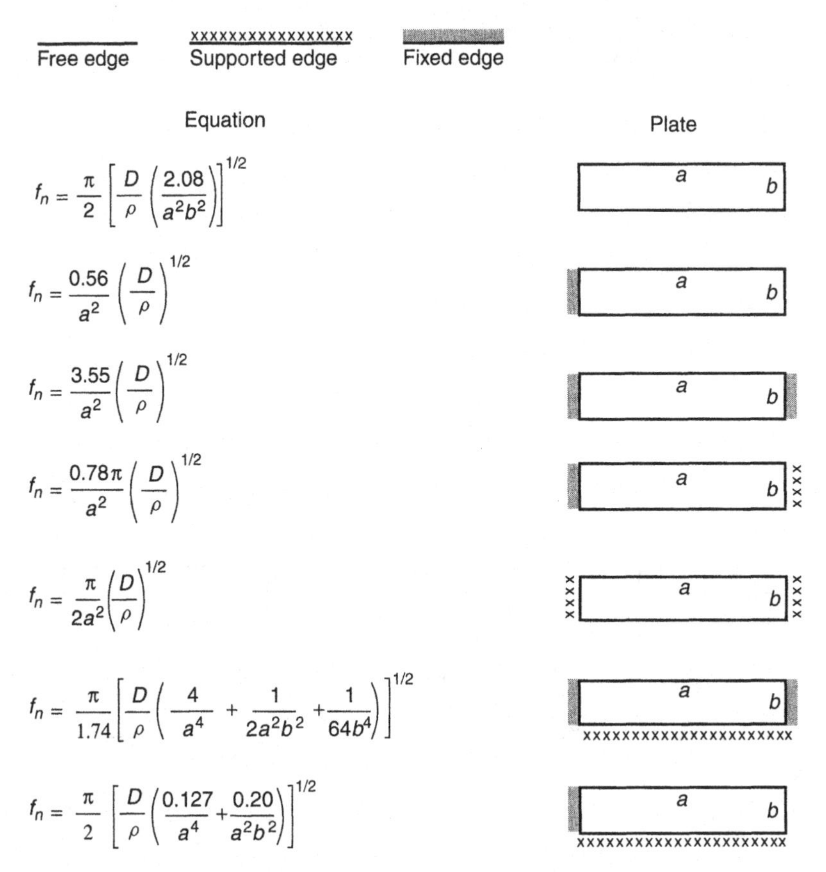

FIGURE 4.4 Natural frequency equations for uniform rectangular plates with various edge restraints.

high stress levels, and high stress levels mean high damping. Therefore, high input acceleration levels usually result in lower Q's because displacements, stresses, and damping are higher. Lower input acceleration levels usually result in higher Q's because displacements, stresses, and damping are lower. Since the transmissibility plays a very important part in the evaluation of the different stress levels, it must also play an important part in the determination of the fatigue life of the various structural elements in the system. Since the damping and the phase relations in many different types of structures are very difficult to estimate, tests should be run to obtain this data on critical programs.

Dynamic displacements in a structure are influenced by the natural frequency of the structure. A higher natural frequency with the same acceleration level results in lower dynamic displacements and lower stresses. Lower stresses result in lower damping, so the transmissibility will be higher. So higher natural frequencies usually result in higher Q's. The opposite is also true. Lower natural frequencies with the

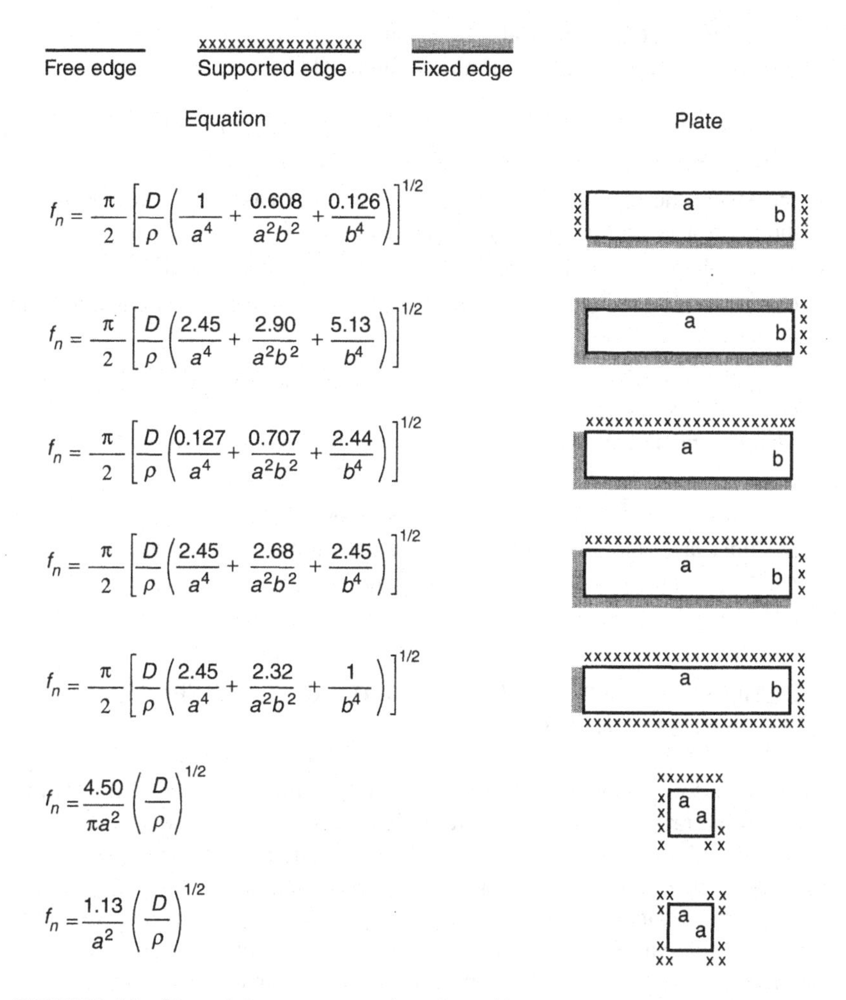

FIGURE 4.5 Natural frequency equations for uniform plates with various restraints.

same acceleration level result in higher dynamic displacements and higher stresses with higher damping. So lower natural frequencies usually result in lower Q's.

Low natural frequencies usually relate to frequencies below about 50–75 Hz. High natural frequencies usually relate to frequencies above about 300 Hz. PCBs used in large commercial mainframe types of electronic systems are often very large. They typically have low natural frequencies, often below 50 Hz. PCBs in the smaller personal and network types of computers are usually smaller. They typically have higher natural frequencies, often above 100 Hz. Military electronic programs for aircraft, missiles, ships, submarines, and tanks must have a more rugged design to withstand more severe environments. The PCBs for these electronic systems typically have natural frequencies between 150 and 250 Hz.

Some of the factors that generally contribute to the transmissibility Q of a circuit board operating in a vibration environment follow [1]:

1. Increasing the natural frequency will increase the Q, and decreasing the natural frequency will decrease the Q.
2. Increasing the input acceleration G level will decrease the Q, and decreasing the input acceleration G level will increase the Q.
3. More bolted joints will increase the damping and decrease the Q.
4. Circuit board edge guides that grip the edges firmly increase the damping and decrease the Q.
5. Plug-in types of electrical connectors with many pins increase the damping and decrease the Q.
6. Many large electronic components increase the damping, which decreases the Q.
7. Heat sink strips for better conduction increase the damping and reduce the Q.
8. Multiple layer circuit boards increase the damping and decrease the Q.
9. Conformal coatings increase the damping and decrease the Q.

4.5 ESTIMATING THE TRANSMISSIBILITY *Q* FOR DIFFERENT ELECTRONIC STRUCTURES

General rules for estimating the transmissibility values for different electronic structures in vibration environments are difficult to establish. This is because there are so many different types, arrangements, and combinations of electronic hardware being used in so many different commercial, industrial, and military applications. Extensive testing data collected over a period of several years on transmissibility values for a wide variety of electronic systems shows a lot of scatter. An average transmissibility equation was developed from the data that can be very useful in obtaining a good approximation of the expected transmissibility for some common structures used for packaging electronic equipment. This transmissibility equation is [9]

$$Q = A\left[\frac{f_n}{(G_{\text{in}})^{0.6}}\right]^{0.76} \tag{4.9}$$

where A = 1.0 for beam types of structure with end restraints
= 0.50 for plate and PCB structures with various perimeter restraints
= 0.25 for enclosures or boxes with lengths two or more times greater than the height and with various end restraints
f_n = Hz, natural frequency
G_{in} = input gravity units acceleration, dimensionless

Sample Problem: Effects of Natural Frequency and Input *G* Level on the *Q*

An engineer has an operational plug-in PCB that he does not want to damage. He requests a sine sweep test using a low input acceleration level of 0.25-G peak so the PCB will not be damaged. The resulting test data shows a transmissibility Q of 70 at a natural frequency of 285 Hz. The PCB must be able to operate in a 5.0-G peak sine vibration environment. The engineer multiplies $70 \times 5 = 350$-G peak, which he claims will destroy the PCB. Should he redesign the PCB to withstand 350 G's?

Solution Transmissibility Q with an input acceleration level of 0.25-G peak. Substitute into Eq. 4.9 to find the expected transmissibility Q for the 0.25-G input acceleration level with a natural frequency of 285 Hz:

$$Q = 0.50\left[\frac{285}{(0.25)^{0.6}}\right]^{0.76} = 69 \tag{4.10}$$

This is very close to the transmissibility Q of 70 obtained with the sine vibration tests run on the PCB; thus the data looks valid. However, remember that low input acceleration G levels have low dynamic deflections, low stresses, and low damping, which produce high transmissibility Q's. Higher acceleration input G levels have high dynamic deflections, high stresses, and high damping, which produce lower transmissibility Q's. Substitute the 5.0 input acceleration G level and find the expected PCB transmissibility Q.

Solution Transmissibility Q with an input acceleration level of 5.0-G peak:

$$Q = 0.50\left[\frac{285}{(5.0)^{0.6}}\right]^{0.76} = 17.6 \tag{4.11}$$

Multiply $5 \times 17.6 = 88.0$-G peak response. This is the acceleration G level the PCB must be able to withstand, rather than the 350 G level. A redesign may not be required. Vibration tests should still be run with the 5-G peak input to verify the integrity of the PCB if the application requires a high reliability.

4.6 NATURAL FREQUENCIES OF ODD-SHAPED CIRCUIT BOARDS

The natural frequencies for triangular, circular, rectangular, and hexagonal PCBs are shown in Figs. 4.6–4.9 with several different support conditions. The first three natural frequencies are shown for each support condition. The results were obtained using finite element methods (FEM) for the physical sizes shown in the figures. Natural frequencies for different size PCBs with similar shapes and supports can be

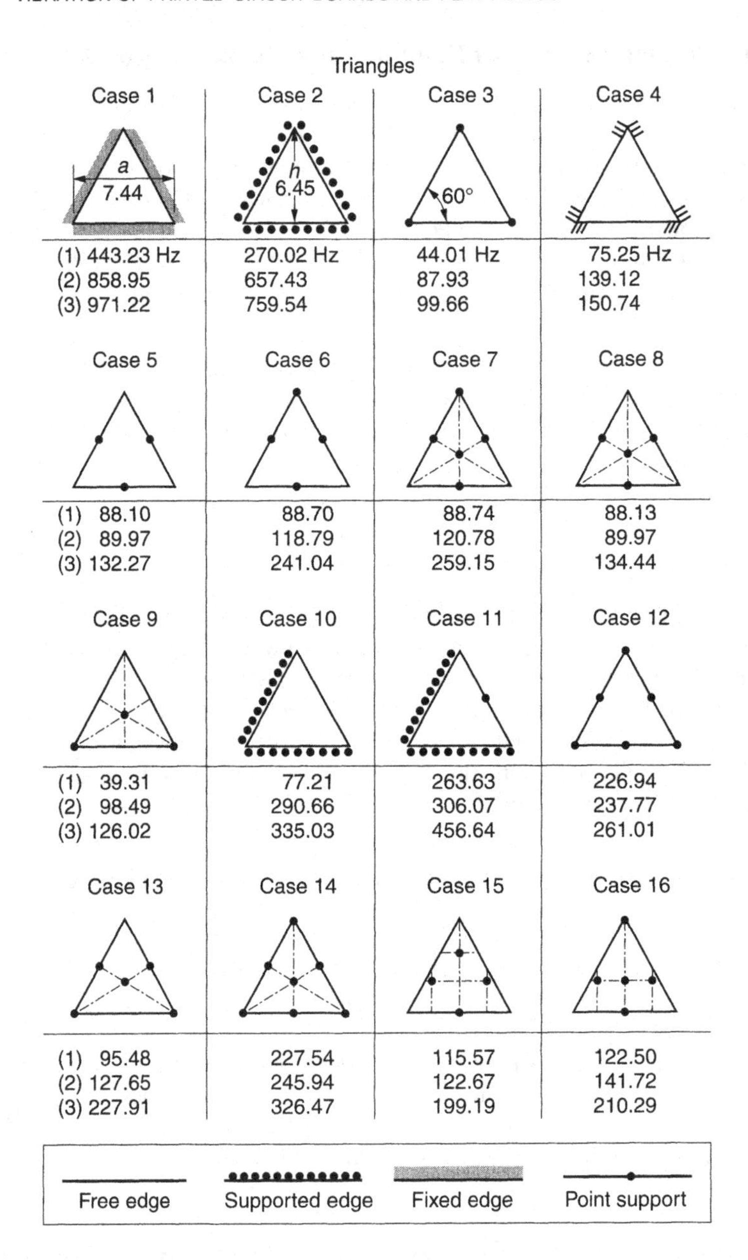

FIGURE 4.6 Natural frequencies for uniform triangular plates with various types of supports. (Reprinted with permission from *Machine Design Magazine*.)

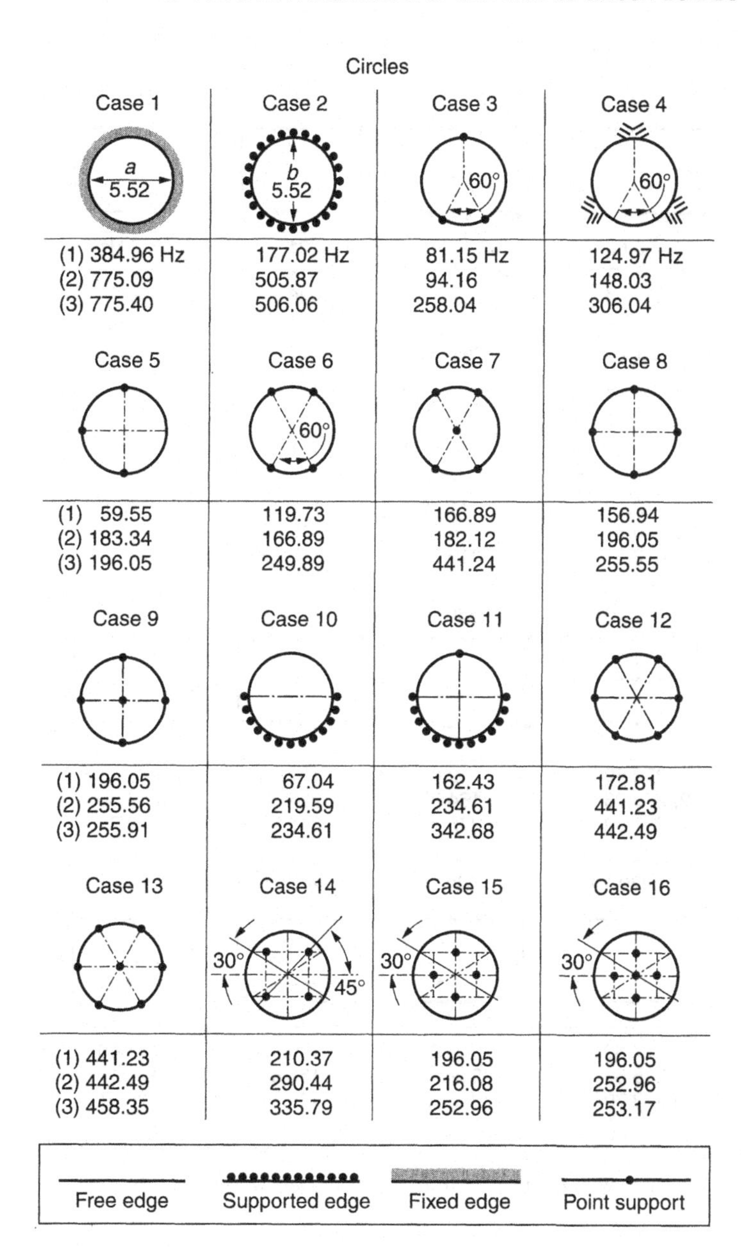

FIGURE 4.7 Natural frequencies for uniform circular plates with various types of supports. (Reprinted with permission from *Machine Design Magazine*.)

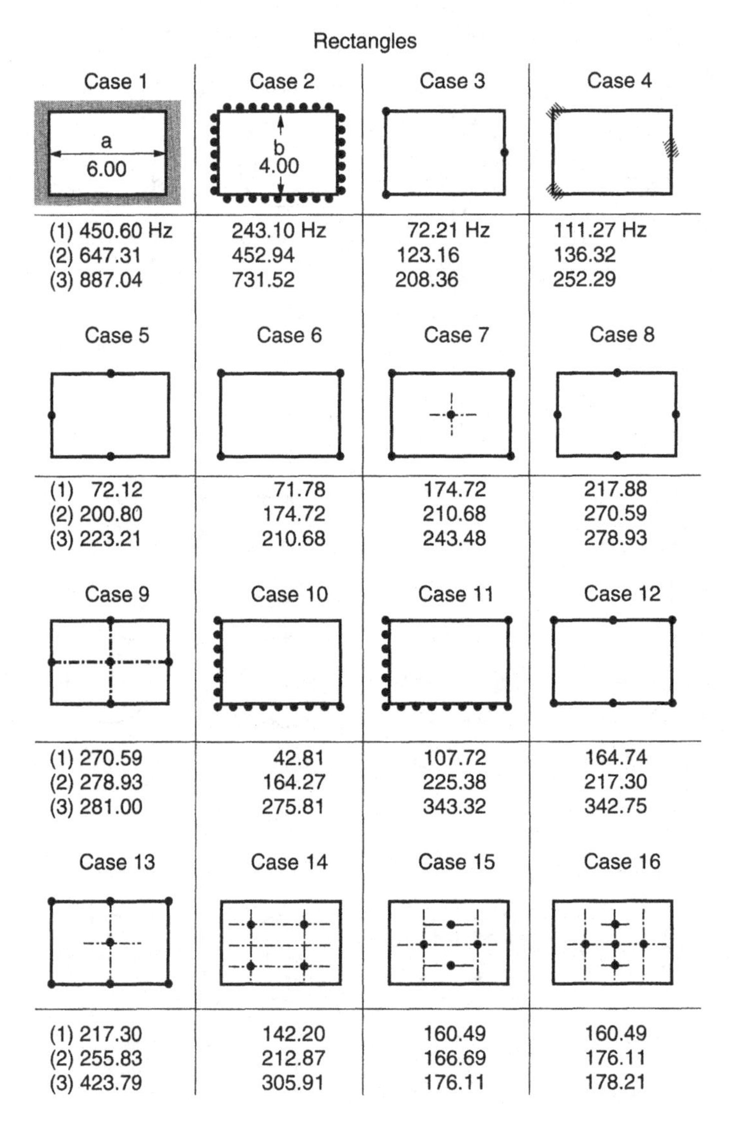

FIGURE 4.8 Natural frequencies for uniform rectangular plates with various types of supports. (Reprinted with permission from *Machine Design Magazine*.)

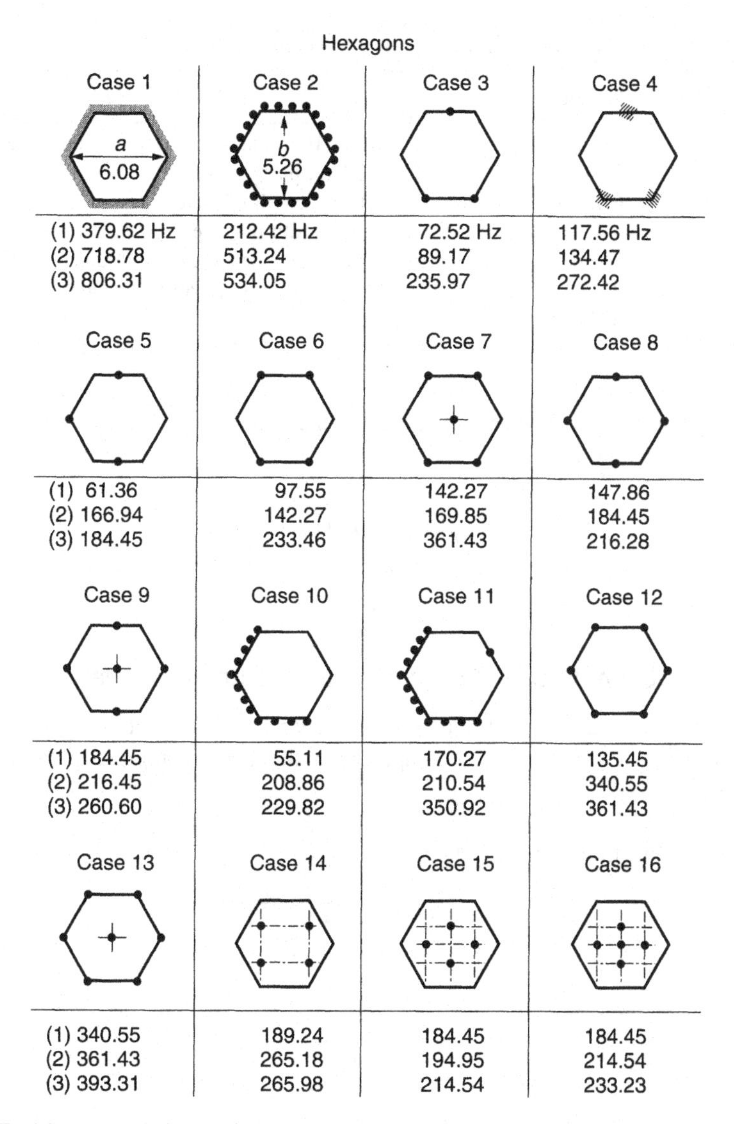

FIGURE 4.9 Natural frequencies for uniform hexagonal plates with various types of supports. (Reprinted with permission from *Machine Design Magazine*.)

TABLE 4.2 Old Specimen Values and New Specimen Values

Item	Old Value (1)	New Value (2)
$E =$	2×10^6	10.5×10^6 lb/in.^{2}, material modulus of elasticity
$h =$	0.10	0.150 in., plate thickness
$a =$	6.0	10.0 in., length of plate
$b =$	4.0	6.0 in., width of plate
$\gamma =$	0.208	0.180 lb/in.^{3}, density, total volume weight
$\mu =$	0.12	0.30 dimensionless, Poisson's ratio
$f_1 =$	217.88 Hz	$f_2 =?$

obtained using Eq. 4.12, where subscript 1 is the old value and subscript 2 is for the new specimen value [10]:

$$f_2 = f_1\left(\frac{a_1 b_1}{a_2 b_2}\right)\sqrt{\frac{E_2 h_2^2 \gamma_1 (1-\mu_1^2)}{E_1 h_1^2 \gamma_2 (1-\mu_2^2)}} \tag{4.12}$$

Sample Problem: Natural Frequency of a Rectangular Plate

Find the fundamental natural frequency of a new rectangular plate with the physical properties shown in Table 4.2, compared to the old rectangular plate physical properties.

Solution Substitute into Eq. 4.12:

$$f_2 = (217.88)\left(\frac{6.0 \times 4.0}{10.0 \times 6.0}\right)\sqrt{\frac{4.84 \times 10^4}{3.28 \times 10^3}} = 334.8 \text{ Hz} \tag{4.13}$$

CHAPTER 5

Estimating Fatigue Life in Thermal Cycling and Vibration Environments

5.1 FATIGUE FAILURES AND FATIGUE DAMAGE IN ELECTRONIC STRUCTURES

Stress reversals can cause failures in thermal cycling and vibration environments when the stress levels are high enough and there are many stress cycles. Experience has shown that the combination of high stresses and a large number of stress reversal cycles can destroy almost any type of structure. It appears that every stress cycle uses up part of the fatigue life of the structure. It does not matter if the stress cycle is generated by thermal, vibration, or shock. Every stress cycle generates some strain. This is known as Miner's cumulative damage criteria. It uses a ratio of the actual number of stress cycles n divided by the number of cycles required to produce a failure N, or n/N. When the sum of all the ratios equals 1.0, the part is expected to fail. If the strain is small, the stress will be small and the amount of damage generated will be small. However, if many stress cycles are accumulated over an extended period, failures may occur. Materials can fracture when they experience repeated stresses that are considerably less than their ultimate static strength. The failure appears to be generated by microscopic cracks that slowly grow with every stress cycle until they grow into a visible crack. This leads to a sudden and complete rupture without any warning [11].

The growth of fatigue cracks in any material depends on the stress distribution, the properties of the material, and possible corrosive properties of the environment. Cracks usually grow more slowly in their early stages and then more rapidly in the advanced stages. The appearance of a crack does not always mean that a failure will occur. Cracks may simply stop growing due to structural changes, or they may grow so slowly that a failure does not occur during the life of the equipment. Since there is no way of knowing if an observed crack will continue to grow, it is best to be safe and to consider any crack as a possible failure, so some corrective action can be taken [12–14].

Small cracks will have a very small radius at the tip, which acts like a stress riser or a stress concentration that drives the crack deeper with every stress cycle. Cracks can often be stopped by increasing the radius at the tip of the crack. This may be accomplished by drilling a small hole at the tip of the crack.

Fatigue damage accumulation can be used to estimate the approximate fatigue life of electronic systems operating in thermal cycling and vibration environments. The fatigue properties of the various structural materials used in the electronic systems must be known to calculate the approximate fatigue life of any load-carrying element. If the fatigue properties of any critical materials are not known, then research must be done using sources such as the material manufacturer, textbooks, technical magazines, or libraries to obtain the required information. If the information cannot be obtained, then fatigue life tests must be run on several samples to obtain the required data. Fatigue life testing is not an easy task since it is often a combination of science and art. Tests must be carefully planned, and the instrumentation selected must not influence the test results. The tests must be run at high enough alternating stress levels to ensure that some of the test samples will fail. The test samples must be instrumented carefully to obtain accurate stress levels at critical points. Instrumentation must also be able to accurately record the number of stress cycles generated. Different constant alternating stress levels should be run with enough test specimens to obtain a life history that can be plotted on a log–log curve with the stress S plotted on the vertical axis and the number of cycles N to failure plotted on the horizontal axis. A single straight line can then be drawn through the scattered test data to obtain the average fatigue curve, called the S–N fatigue curve for the material. A typical plot of a S–N fatigue curve is shown in Fig. 5.1.

5.2 RELATING THE FATIGUE PROPERTIES TO THE SLOPE OF THE FATIGUE CURVE

Every point on the S–N fatigue curve represents a point of failure. Any two well-separated points on the log–log fatigue curve shown as points 1 and 2 in Fig. 5.1 can be selected to write the equation of the fatigue curve as [15]

$$N_1 S_1^b = N_2 S_2^b \tag{5.1}$$

where N = number of fatigue cycles required to produce a failure at points 1 and 2
S = stress level associated with the failures at points 1 and 2
b = slope of the fatigue line when it is plotted on a log–log curve

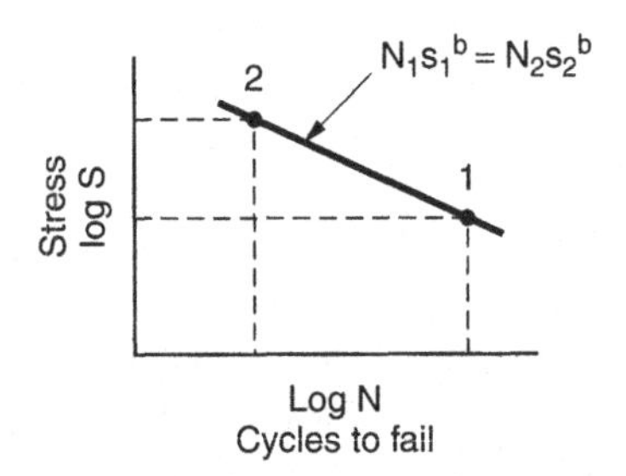

FIGURE 5.1 Typical log–log S–N fatigue life curve with a down sloping straight line.

Many of the structural materials in electronic systems, such as electrical lead wires and solder joints, will be forced to operate in their nonlinear ranges in the thermal cycling and vibration environments. This condition makes it very difficult to calculate their approximate fatigue life. Therefore, it is often convenient to assume these structural elements are linear so their approximate fatigue life can be calculated when they are required to operate in different environments. In linear systems the stress S will be proportional to the displacement Z and also proportional to the acceleration G level in dimensionless gravity units. Also in a linear system exposed to vibration, the time T of the vibration exposure will be proportional to N the number of stress cycles developed during the vibration exposure. This allows Eq. 5.1 to be written in several more convenient ways depending on the information desired:

$$N_1 Z_1^b = N_2 Z_2^b \tag{5.2}$$

$$N_1 G_1^b = N_2 G_2^b \tag{5.3}$$

$$T_1 G_1^b = T_2 G_2^b \tag{5.4}$$

The slope of the vibration fatigue line b can be obtained from the general fatigue properties of most metals. This shows that the fatigue properties are related to the ultimate stress and not to the yield stress. There appears to be no fatigue damage noticed in nonferrous alloys such as aluminum for the first 1000 stress cycles. Smooth polished specimen test data on nonferrous metals shows the endurance limit stress is typically one third of the ultimate tensile stress at about 100 million stress reversals. Since real systems are being examined, a stress concentration k must be used to account for holes, notches, sharp corners, and cross-section changes in the electronic structures. Using a typical stress concentration $k = 2$, will result in a vibration fatigue endurance limit that is one sixth of the ultimate tensile stress for nonferrous metals. This information is shown in Fig. 5.2, and it can be used to find the fatigue exponent slope b for nonferrous metals with a slight modification of Eq. 5.1 as

$$\frac{N_1}{N_2} = \left(\frac{S_2}{S_1}\right)^b \tag{5.5}$$

where $N_1 = 10^8$ cycles to fail
$N_2 = 10^3$ cycles to fail
$S_1 = S_e =$ endurance limit $= \frac{1}{6} S_{tu}$ or $\frac{1}{6}$ ultimate tensile stress when $k = 2$
$S_2 = S_{tu} =$ ultimate tensile stress

Substitute into Eq. 5.5, take the log of both sides, and solve for the b exponent slope:

$$\frac{10^8}{10^3} = \left(\frac{S_{tu}}{\frac{1}{6} S_{tu}}\right)^b \quad \text{so} \quad 10^5 = (6)^b \quad \text{so} \quad b = \frac{\log_{10} 10^5}{\log_{10} 6} = \frac{5}{0.778} = 6.4 \tag{5.6}$$

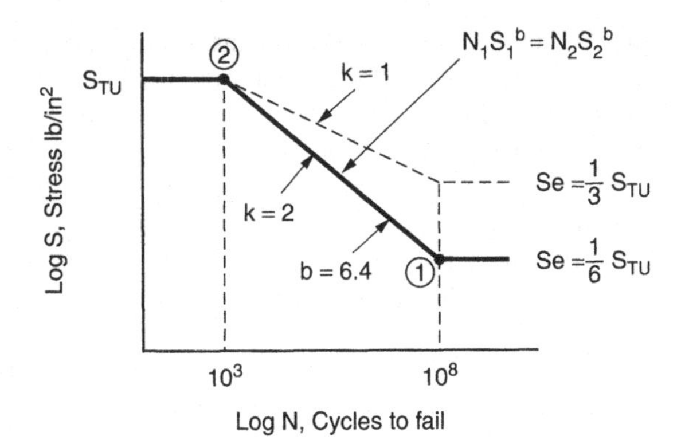

FIGURE 5.2 Typical log–log *S–N* fatigue life curve for nonferrous alloys where a stress concentration *k* value of 2.0 results in a *b* exponent slope of 6.4.

The *b* exponent slope of 6.4 includes a stress concentration $k = 2$ for metals that have a well-defined yield point, such as aluminum, magnesium, copper kovar, dumet, and metals that may be used for electrical lead wires. Sometimes this stress concentration is used directly in the bending stress equation. This is acceptable, as long as the stress concentration factor is *used only once*. The *b* exponent for solder is different because of the unusual elasto plastic and creep properties of solder, especially at high temperatures. The solder fatigue exponent *b* is shown later to be 2.5, with the use of Fig. 5.3.

Sample Problem: Change in Vibration Fatigue Life of Inductor on a PCB

A 0.6-lb inductor supported with four 0.030-in. diameter copper wires is mounted at the center of a plug-in printed circuit board (PCB) with a natural frequency f_n of 175 Hz. Find the approximate fatigue life of the wires based on the dynamic force generated by a 5-*G* peak sine vibration resonant dwell that loads the wires in tension. The wires are screwed, not soldered, to the PCB, so solder is not a problem here.

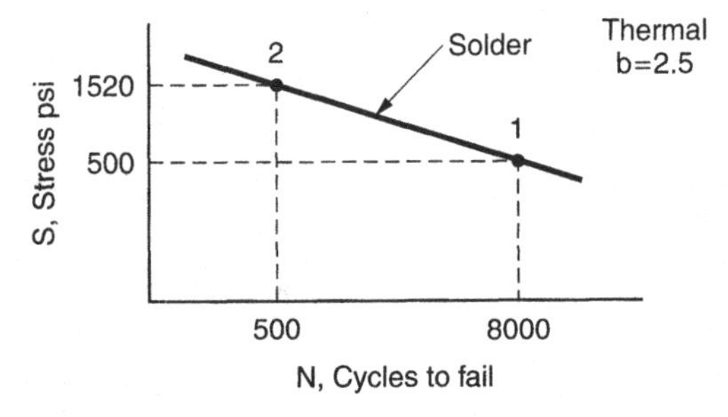

FIGURE 5.3 Typical log–log *S–N* fatigue life curve for 63/37 Sn/Pb solder during a rapid thermal cycle so there is no creep that results in a *b* exponent slope of 2.5.

Also find the estimated fatigue life in the wires when the acceleration input level is increased to 6-G peak. Use three different methods to solve for the approximate wire fatigue life.

Solution 1 First calculate the fatigue life based on the wire dynamic tensile stress for a 5-G peak input and the fatigue properties of the copper wire with a stress concentration of 2 as shown in the S–N curve in Fig. 5.4. The dynamic force F acting on the four wires can be obtained as

$$F = WG_{\text{in}}Q \qquad \text{(Ref. Eq. 3.3)}$$

where $W = 0.6$-lb weight of inductor
$G_{\text{in}} = 5.0$ peak input acceleration

$$Q = 0.5\left[\frac{175}{(5.0)^{0.6}}\right]^{0.76} = 12.2 \text{ transmissibility (Ref. Eq. 4.9)} \tag{5.7}$$

$$F = (0.6)(5.0)(12.2) = 36.6 \text{ lb, dynamic force} \tag{5.8}$$

Next find the dynamic tensile stress S_t on the four copper wires:

$$S_t = \frac{F}{A_w} \tag{5.9}$$

where $F = 36.6$ lb

$$A_w = \frac{4\pi}{4}(0.030)^2 = 0.00283 \text{ in.}^2 \text{ area of 4 wires}$$

$$S_t = \frac{36.6}{0.00283} = 12{,}933 \text{ lb/in.}^2 \text{ tensile stress on 4 wires} \tag{5.10}$$

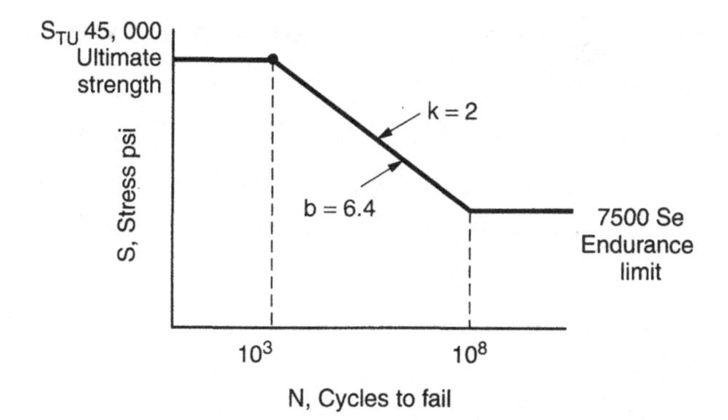

FIGURE 5.4 Typical log–log S–N fatigue life curve for 99.6% drawn copper wire with a geometric stress concentration k value of 2.0, which results in a b exponent slope of 6.4.

Use Eq. 5.1 and Fig. 5.4 to find the number of cycles to fail at the above stress level:

$$N_2 = N_1 \left(\frac{S_1}{S_2} \right)^b$$

where $N_1 = 10^8$, cycles to fail at 7500 lb/in.^{2} from Fig. 5.4
$S_1 = 7500$ lb/in.^{2}, stress to fail at 10^8 cycles
$S_2 = 12{,}933$ lb/in.^{2}, wire tensile stress from Eq. 5.9

$$N_2 = (10^8) \left(\frac{7500}{12{,}933} \right)^{6.4} = 3.058 \times 10^6 \text{ cycles to fail with 5 } G \text{ input} \tag{5.11}$$

The time to fail T_f can be obtained from the natural frequency and the cycles to fail.

$$T_f = \frac{N_2}{f_n} = \frac{3.058 \times 10^6 \text{ cycles}}{\left(\frac{175 \text{ cycles}}{\text{s}} \right) \left(\frac{3600 \text{ s}}{\text{h}} \right)} = 4.85 \text{ h to fail for a 5-}G \text{ input} \tag{5.12}$$

Next find the dynamic force acting on the four wires for a 6-G peak sine input where

$W = 0.6$ lb, weight of inductor
$G_{\text{in}} = 6.0$ peak input acceleration

$$Q = 0.5 \left[\frac{175}{(6.0)^{0.6}} \right]^{0.76} = 11.2 \text{ transmissibility (Ref. Eq. 4.9)} \tag{5.13}$$

$$F = (0.6)(6.0)(11.2) = 40.32 \text{ lb, dynamic force} \tag{5.14}$$

Next find the dynamic tensile stress S_t on the four copper wires:

$$S_t = \frac{F}{A_w} = \frac{40.32}{0.00283} = 14{,}247 \text{ lb/in.}^2 \text{ tensile stress on 4 wires} \tag{5.15}$$

Use Eq. 5.1 and Fig. 5.4 to find the number of cycles to fail at the above stress level:

$$N_2 = (10^8) \left(\frac{7500}{14{,}247} \right)^{6.4} = 1.646 \times 10^6 \text{ cycles to fail with a 6-}G \text{ input} \tag{5.16}$$

The time to fail T_f can be obtained from the natural frequency and the cycles to fail:

$$T_f = \frac{N_2}{f_n} = \frac{1.646 \times 10^6 \text{ cycles}}{\left(\frac{175 \text{ cycles}}{\text{s}} \right) \left(\frac{3600 \text{ s}}{\text{h}} \right)} = 2.61 \text{ h to fail for a 6-}G \text{ input} \tag{5.17}$$

Solution 2 Use Eq. 5.2 to find the change in the fatigue life from a 5-G peak input to a 6-G peak input using the number of fatigue cycles N and the dynamic displacements Z:

$$Z_1 = \frac{9.8 G_{\text{in}} Q_1}{f_n^2} \qquad \text{(Ref. Eq. 3.17)}$$

where $G_{\text{in}} = 5\text{-}G$ peak input acceleration
$Q_1 = 12.2$ transmissibility (Ref. Eq. 5.7)
$f_n = 175\,\text{Hz}$, natural frequency

$$Z_1 = \frac{(9.8)(5.0)(12.2)}{(175)^2} = 0.0195 \text{ in. dynamic displacement for a 5-}G\text{ input} \quad (5.18)$$

$$Z_2 = \frac{9.8 G_{\text{in}} Q_2}{f_n^2}$$

where $G_{\text{in}} = 6\text{-}G$ input peak acceleration
$Q_2 = 11.2$ transmissibility (Ref. Eq. 5.13)
$f_n = 175\,\text{Hz}$, natural frequency

$$Z_2 = \frac{(9.8)(6.0)(11.2)}{(175)^2} = 0.0215 \text{ in. dynamic displacement for a 6-}G\text{ input} \quad (5.19)$$

Substitute into Eq. 5.2 to find the change in the fatigue life from a 5-G to a 6-G input where

$N_1 = 3.058 \times 10^6$ cycles to fail (Ref. Eq. 5.11)
$Z_1 = 0.0195$ in. displacement (Ref. Eq. 5.18)
$Z_2 = 0.0215$ in. displacement (Ref. Eq. 5.19)

$$N_2 = (3.05 \times 10^6)\left(\frac{0.0195}{0.0215}\right)^{6.4} = 1.63 \times 10^6 \text{ cycles to fail for a 6-}G\text{ input} \quad (5.20)$$

The fatigue life in Eq. 5.16 compares well with the fatigue life in Eq. 5.20.

Solution 3 Use Eq. 5.4 to find the change in the fatigue life from a 5-G peak input to a 6-G peak input using the fatigue life time T and the response acceleration G level where

$T_1 = 4.85\,\text{h}$ life for a 5-G peak input (Ref. Eq. 5.12)
$G_{\text{out}} = G_{\text{in}} Q = (5.0)(12.2) = 61.0\ G$ peak response (Ref. Eq. 5.7 for 5 G)

$G_{out} = G_{in}Q = (6.0)(11.2) = 67.2\ G$ peak response (Ref. Eq. 5.13 for 6 G)

$$T_2 = T_1\left(\frac{G_1}{G_2}\right)^b = (4.85)\left(\frac{61.0}{67.2}\right)^{6.4} = 2.61 \text{ h to fail for 6-}G\text{ input} \tag{5.21}$$

The time to fail shown in Eq. 5.21 agrees with the time to fail shown in Eq. 5.17.

Sample Problem: Effects of Changing the Test Time and Acceleration Level

A customer is interested in an electronic box that just completed a qualification test with a 3.0-G peak input and a resonant dwell period of 4.0 h. The customer requires the box to pass a 4.5-G peak input for a period of 0.75 h. Will the present box be able to pass the new requirements?

Solution Use Eq. 5.4 with the following data:

$T_1 = 4.0$-h vibration test passed
$G_1 = 3.0$-G peak input acceleration level passed
$G_2 = 4.5$-G peak input acceleration customer requirement

$$T_2 = T_1\left(\frac{G_1}{G_2}\right)^b = (4.0)\left(\frac{3.0}{4.5}\right)^{6.4} = 0.298 \text{ h to fail} \tag{5.22}$$

The customer requires a vibration time capability of 0.75 h with an acceleration input level of 4.5 G so the expected fatigue life of only 0.298 h is not acceptable. Some design changes will have to be made to improve the fatigue life of the box.

5.3 THERMAL CYCLE FATIGUE EXPONENT SLOPE *b* FOR SOLDER

Equation 5.5 can also be used to find the fatigue exponent slope b for solder joint thermal cycling conditions when the solder joint stress is the driving condition, based on thermal cycling test data using Fig. 5.3:

$N_1 = 8000$ cycles to fail
$N_2 = 500$ cycles to fail
$S_1 = 500\ \text{lb/in.}^2$ stress to fail
$S_2 = 1520\ \text{lb/in.}^2$ stress to fail

$$\frac{8000}{500} = \left(\frac{1520}{500}\right)^b \quad \text{so} \quad 16.0 = (3.04)^b \quad \text{so} \quad b = \frac{\log_{10} 16}{\log_{10} 3.04} = \frac{1.204}{0.482} = 2.5 \tag{5.23}$$

The slope b of the solder fatigue curve can be applied when the temperature change is the driving force in thermal cycling fatigue conditions. Here N represents the number of thermal cycles to fail and Δt represents the temperature change, which becomes the driving force:

$$N_1 \, \Delta t_1^b = N_2 \, \Delta t_2^b \tag{5.24}$$

Sample Problem: Establishing an Automotive Proof of Life Test Program

A manufacturer of automobile electronic systems wishes to establish a reliability proof of life test program to cover a warranty period of 10 years. The PCB solder joints under the hood are the major concern. Because there is a large amount of scatter in fatigue life testing, the electronic system will be designed for a life of 20 years using a scatter factor (safety factor) of 2.0. The average underhood temperature rise for a certain model is expected to be about 55°C. The average use is expected to have 7 engine starts per day, 6 days per week, 52 weeks per year, for the assumed 20 years. The upper management people are willing to fund an accelerated life test program for a working period of about 9 months using a double work shift. Recommendations were made for a thermal cycling test program from −40 to +110°C, with 30-min dwell periods at temperature extremes. The solder will have a chance to relieve the stresses at the high-temperature area so the analysis was based on the full temperature range of 150°C. Is the proposed test program acceptable?

Solution Assume every engine start is one thermal cycle. The number of thermal cycles expected over the 20 year period will be as follows:

$$N_1 = (7 \text{ starts/day})(6 \text{ days/week})(52 \text{ weeks/year})(20 \text{ years}) = 43{,}680 \text{ cycles} \tag{5.25}$$

$\Delta t_1 = 55°\text{C}$ average underhood temperature rise expected
$\Delta t_2 = 150°\text{C}$ recommended temperature cycle from −40 to +110°C with a ramp rate of 20°C/min and a 30-min dwell period at each temperature extreme.

Substitute into Eq. 5.24 to find the number of thermal cycles required for the test:

$$N_2 = N_1 \left(\frac{\Delta t_1}{\Delta t_2}\right)^b = (43{,}680)\left(\frac{55}{150}\right)^{2.5} = 3556 \text{ test cycles} \tag{5.26}$$

Each thermal cycle will take about 75 min. Each 8-h work shift will have 480 min of test time. A double shift will have 960 min of test time per day. The approximate number of working months required to run the accelerated life test program is

$$\text{Time} = \frac{(3556) \text{ cycles}(75 \text{ min/cycle})}{(960 \text{ min/day})(30 \text{ days/month})} = 9.26 \text{ months} \tag{5.27}$$

This accelerated life test program comes very close to the working time proposed by the upper management group; so the proposal should be acceptable.

Sample Problem: Thermal Forces, Stresses, and Fatigue Life of a Component

High temperatures from the wave soldering process causes heat to wick up the lead wires into the sensitive silicon chip causing degraded electrical performance. Spacers 0.060 high × 0.090 wide × 0.50 in. long epoxy glass will be mounted under the ceramic component to increase the length of the four 0.030-in. diameter kovar lead wire by raising the component above the 0.080-in.-thick PCB, as shown in Fig. 5.5 to solve the heat wicking problem. Find the approximate thermal induced forces, stresses, and expected fatigue life of the lead wires and solder joints for rapid temperature cycles from −50 to +80°C. Recommend design and material changes that may be used to improve the fatigue life of the component solder joints and lead wires.

Solution Set up a thermal expansion equilibrium equation where the component, PCB, and the spacer expansion along the Z axis are resisted by the kovar lead wires, which imposes a tensile force on the wires and a compressive reaction on the other members. The subscripts w, e, s, and c refer to the wire, epoxy PCB, spacer, and ceramic component. Since the wires are loaded in tension, the wire length does not stop at the component or PCB interface. The effective wire length will extend about two diameters into the PCB and two diameters into the component body [5]:

$$\alpha_w L_w \,\Delta t + \frac{P_w L_w}{A_w E_w} = \alpha_e L_e \,\Delta t - \frac{P_e L_e}{A_e E_e} + \alpha_s L_s \,\Delta t - \frac{P_s L_s}{A_s E_s} + \alpha_c L_c \,\Delta t - \frac{P_c L_c}{A_c E_c} \tag{5.28}$$

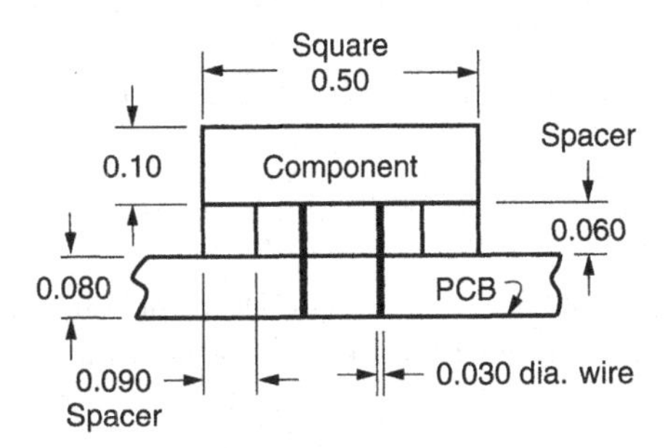

FIGURE 5.5 Spacers are placed under a heat-sensitive component to raise it above the PCB so less heat will wick up the longer wires during the wave soldering operation.

where $\alpha_w = 6.0 \times 10^{-6}$ in./in./°C, thermal coefficient of expansion (TCE) of kovar lead wire

$L_w = 0.060 + 4(0.030) = 0.180$ in., effective length for 4 wires

$\Delta t = \dfrac{80 - (-50)}{2} = 65$°C, for a rapid thermal cycle with no solder creep

$A_w = (4 \text{ wires})\left(\frac{\pi}{4}\right)(0.030)^2 = 0.00283$ in.^{2}, area of 4 wires

$E_w = 20 \times 10^6$ lb/in.^{2}, kovar wire modulus of elasticity

$\alpha_e = 98 \times 10^{-6}$ in./in./°C, TCE of epoxy fiberglass PCB in Z axis at 80°C

$L_e = 0.080/2 = 0.040$ in., effective epoxy fiberglass PCB thickness

$A_e = (2 \text{ spacers})(0.50)(0.090) = 0.090$ in.^{2}, area of PCB under 2 spacers

$E_e = 0.35 \times 10^6$ lb/in.^{2}, epoxy fiberglass PCB modulus along Z axis at 80°C

$\alpha_s = 98 \times 10^{-6}$ in./in./°C, TCE of epoxy fiberglass spacer in Z axis at 80°C

$L_s = 0.060$ in., height of epoxy fiberglass spacer

$E_s = 0.35 \times 10^6$ lb/in.^{2}, modulus of epoxy spacer same as PCB

$\alpha_c = 6.0 \times 10^{-6}$ in./in./°C, TCE of ceramic component

$L_c = 0.100/2 = 0.050$ in., effective height of component

$A_c = (2 \text{ spacers})(0.50)(0.090) = 0.090$ in.^{2} area of component on 2 spacers

$E_c = 42 \times 10^6$ lb/in.^{2}, ceramic component modulus of elasticity

Substitute into Eq. 5.28 for the tensile force in the wires and shear force in the solder:

$$\overbrace{(6.0 \times 10^{-6})(0.18)(65) + \frac{P_w(0.18)}{(0.00283)(20 \times 10^6)}}^{\text{wire}}$$

$$= \overbrace{(98 \times 10^{-6})(0.04)(65) - \frac{P_e(0.040)}{(0.090)(0.35 \times 10^6)}}^{\text{epoxy PCB}}$$

$$+ \overbrace{(98 \times 10^{-6})(0.06)(65) - \frac{P_s(0.06)}{(0.09)(0.35 \times 10^6)}}^{\text{epoxy spacer}}$$

$$+ \overbrace{(6 \times 10^{-6})(0.05)(65) - \frac{P_c(0.05)}{(0.09)(42 \times 10^6)}}^{\text{component}}$$

$$7.020 \times 10^{-5} + 3.180 \times 10^{-6} P_w$$

$$= 2.548 \times 10^{-4} - 1.270 \times 10^{-6} P_e + 3.822 \times 10^{-4} - 1.905 \times 10^{-6} P_s$$

$$+ 1.950 \times 10^{-5} - 1.323 \times 10^{-8} P_c$$

$$\textstyle\sum P = 0 \quad \text{so} \quad P_w = P_e = P_s = P_c$$

$$6.368 \times 10^{-6} P = 5.863 \times 10^{-4} \quad \text{so} \quad P = 92.07 \text{ lb force on 4 wires} \qquad (5.29)$$

Find the tensile stress S_t in each of the four-electrical lead wires:

$$S_t = \frac{P}{A_w} = \frac{92.07}{0.00283} = 32{,}533 \text{ lb/in.}^2 \text{ tensile stress in each wire} \tag{5.30}$$

Find the approximate fatigue life of the wires with the epoxy fiberglass spacer due to thermal cycling using Eq. 5.1 and Fig. 10.3 where

$N_2 = 1000$ cycles to fail at 84,000 lb/in.^{2} reference point
$S_2 = 84{,}000$ lb/in.^{2} stress to fail at 1000 cycles reference point
$S_1 = 32{,}533$ lb/in.^{2}, tensile stress in each lead wires (Ref. Eq. 5.30)
$b = 6.4$ slope of fatigue curve for lead wires with a stress concentration of 2

$$N_1 = N_2\left(\frac{S_2}{S_1}\right)^b = (1000)\left(\frac{84{,}000}{32{,}533}\right)^{6.4} = 4.33 \times 10^5 \text{ wire cycles to fail} \tag{5.31}$$

The shear stress S_s in the PCB lead wire solder joints, for a rapid thermal cycle, can be obtained from Eq. 5.29 using the average solder joint diameter with a plated through-hole diameter of 0.042 in. and a solder joint height assumed to be the same as the thickness of the PCB.

$$\text{Average solder joint diameter} \qquad d = (0.030 + 0.042)/2 = 0.036 \text{ in.}$$

$$A_s = (4 \text{ wires})(\pi d h) = (4)(\pi)(0.036)(0.080) = 0.0362 \text{ in.}^2 \text{ solder shear area}$$

$$\text{Solder shear stress } S_s = \frac{P}{A_s} = \left(\frac{92.07}{0.0362}\right) = 2543 \text{ lb/in.}^2 \tag{5.32}$$

The solder fatigue life can be estimated from Eq. 5.1 with the following information:

$N_1 = 80{,}000$ stress cycle life at 200 lb/in.^{2} reference point Fig. 5.6
$S_1 = 200$ lb/in.^{2} solder failure reference point Fig. 5.6
$S_2 = 2543$ lb/in.^{2} calculated solder stress

$$N_2 = (80{,}000)\left(\frac{200}{2543}\right)^{2.5} = 139 \text{ solder cycles to fail} \tag{5.33}$$

The solder joints are obviously far more critical than the lead wires. A reasonable fatigue life would be at least 1000 cycles; thus a design change is needed to improve the life.

Solution *Recommendations for Design Changes to Improve Solder Fatigue Life*

1. Change the spacer material under the component from epoxy fiberglass to aluminum, which has a much lower TCE but a higher modulus of elasticity.
2. Leave the epoxy fiberglass spacers under the component but add a 0.010-in.-thick layer of a very soft material such as a silicone pad or room temperature vulcanizing (RTV) under the spacers.
3. Put a small offset or a small kink in each lead wire to reduce the axial spring rate of the wire, to reduce the axial force produced in the wires.

Recommendation 1 Change spacer from epoxy to aluminum where

$\alpha_a = 23 \times 10^{-6}$ in./in./°C, TCE of aluminum spacer replacing epoxy fiberglass
$E_a = 10 \times 10^6$ lb/in.^{2}, aluminum modulus of elasticity
$L_a = 0.06$ in. length or height of aluminum spacer
$A_a = 0.090$ in.^{2}, area of 2 aluminum spacers (same as epoxy spacers)

Substituting into Eq. 5.28 using the aluminum spacers will result in the following:

$$\overset{\text{wire}}{(6 \times 10^{-6})(0.18)(65) + \frac{P_w(0.18)}{(0.00283)(20 \times 10^6)}}$$

$$\overset{\text{epoxy PCB}}{= (98 \times 10^{-6})(0.04)(65) - \frac{P_e(0.04)}{(0.09)(0.35 \times 10^6)}}$$

$$\overset{\text{aluminum spacer}}{+ (23 \times 10^{-6})(0.06)(65) - \frac{P_s(0.06)}{(0.09)(10 \times 10^6)}}$$

$$\overset{\text{component}}{+ (6 \times 10^{-6})(0.05)(65) - \frac{P_c(0.05)}{(0.09)(42 \times 10^6)}}$$

$$7.02 \times 10^{-5} + 3.180 \times 10^{-6} P_w$$

$$= 2.548 \times 10^{-4} - 1.270 \times 10^{-6} P_e + 8.970 \times 10^{-5} - 6.667 \times 10^{-8} P_s$$

$$+ 1.950 \times 10^{-5} - 1.323 \times 10^{-8} P_c$$

$$\sum P = 0 \quad \text{so} \quad P_w = P_e = P_s = P_c = P$$

$$4.530 \times 10^{-6} P = 2.938 \times 10^{-4} \quad \text{so} \quad P = 64.8 \text{ lb, force on 4 wires} \tag{5.34}$$

$$\text{Wire stress} = \frac{P}{A_w} = \frac{64.8}{0.00283} = 22{,}897 \text{ lb/in.}^2 \text{ on each of 4 wires} \tag{5.35}$$

Find the approximate fatigue life of the wires with the aluminum spacer due to thermal cycling using Eq. 5.1 and Fig. 10.3 where

$N_2 = 1000$ cycles to fail at 84,000 lb/in.^{2} reference point
$S_2 = 84{,}000$ lb/in.^{2} stress to fail at 1000 cycles reference point
$S_1 = 22{,}897$ lb/in.^{2} tensile stress in each lead wire (Ref. Eq. 5.35)
$b = 6.4$ slope of fatigue curve for lead wires with a stress concentration of 2

$$N_1 = (1000)\left(\frac{84{,}000}{22{,}897}\right)^{6.4} = 4.1 \times 10^6 \text{ wire cycles to fail} \tag{5.36}$$

The approximate solder joint shear stress can be obtained from the data in Eq. 5.32:

$$\text{Solder shear stress} = \frac{P}{A_s} = \left(\frac{64.8}{0.0362}\right) = 1790 \text{ lb/in.}^2 \tag{5.37}$$

The approximate fatigue life of the solder can be obtained from Eqs. 5.1. and 5.33:

$$N_2 = (80{,}000) = \left(\frac{200}{1790}\right)^{2.5} = 334 \text{ solder cycles to fail} \tag{5.38}$$

The solder fatigue life is still not adequate so another design change is required.

Recommendation 2 Add a 0.010-in.-thick soft silicone or RTV pad under the two spacers and change the two spacers back to epoxy fiberglass. The wire length increases by 0.010 in. and two more terms must be added in Eq. 5.28 for the soft pads:

$$+ \alpha_p L_p \, \Delta t - \frac{P_p L_p}{A_p E_p} \tag{5.39}$$

where $L_w = 0.180 + 0.010$ pad $= 0.19$ in., new wire length
$\alpha_p = 190 \times 10^{-6}$ in./in./°C, TCE of soft silicone or RTV pad at 80°C
$L_p = 0.010$ in., thickness of soft interface pad
$E_p = 1000$ lb/in.^{2}, modulus of elasticity of soft pad
$A_w = 0.00283$ in.^{2}, area of 4 wires
$A_p = 0.090$ in.^{2}, area of soft pad same as area of two epoxy spacers

$$\overset{\text{wire}}{(6 \times 10^{-6})(0.19)(65) + \frac{P_w(0.19)}{(0.00283)(20 \times 10^6)}}$$

$$\overset{\text{epoxy PCB}}{= (98 \times 10^{-6})(0.04)(65) - \frac{P_e(0.04)}{(0.09)(0.35 \times 10^6)}}$$

$$\overset{\text{epoxy spacer}}{+ (98 \times 10^{-6})(0.06)(65) - \frac{P_s(0.06)}{(0.09)(0.35 \times 10^6)}}$$

$$\overset{\text{component}}{+ (6 \times 10^{-6})(0.05)(65) - \frac{P_c(0.05)}{(0.09)(42 \times 10^6)}}$$

$$\overset{\text{silicone pad}}{+ (190 \times 10^{-6})(0.010)(65) - \frac{P_p(0.010)}{(0.09)(1000)}}$$

$$7.410 \times 10^{-5} + 3.357 \times 10^{-6} P_w$$

$$= 2.548 \times 10^{-4} - 1.270 \times 10^{-6} P_e + 3.822 \times 10^{-4} - 1.905 \times 10^{-6} P_s$$

$$+ 1.950 \times 10^{-5} - 1.323 \times 10^{-8} P_c + 1.235 \times 10^{-4} - 1.111 \times 10^{-4} P_p$$

$$\sum P_y = 0 \quad \text{so} \quad P_w = P_e = P_s = P_c = P_p = P$$

$$1.176 \times 10^{-4} P = 7.059 \times 10^{-4} \quad \text{so} \quad P = 6.002 \text{ lb force on 4 wires} \tag{5.40}$$

The solder shear stress can be obtained from the wire force and the solder shear area where

$P = 6.002$ lb, force in 4 wires
$A_s = 0.0362$ in.^{2}, shear area of 4 solder joints (Ref. Eq. 5.32)

$$S_s = \frac{P}{A_s} = \frac{6.002}{0.0362} = 166 \text{ lb/in.}^2, \text{ solder joint shear stress} \tag{5.41}$$

The approximate fatigue life of the solder joints for a rapid thermal cycle can be obtained from the general fatigue equation (5.1), and the *S–N* solder fatigue curve shown in Fig. 5.6 where

$N_1 = 80{,}000$ cycles to fail at 200 lb/in.^{2} reference point
$S_1 = 200$ lb/in.^{2} stress to fail reference point
$b = 2.5$ fatigue exponent slope for solder (Ref. Eq. 5.23)
$S_2 = 166$ lb/in.^{2}, solder shear stress for a rapid thermal cycle

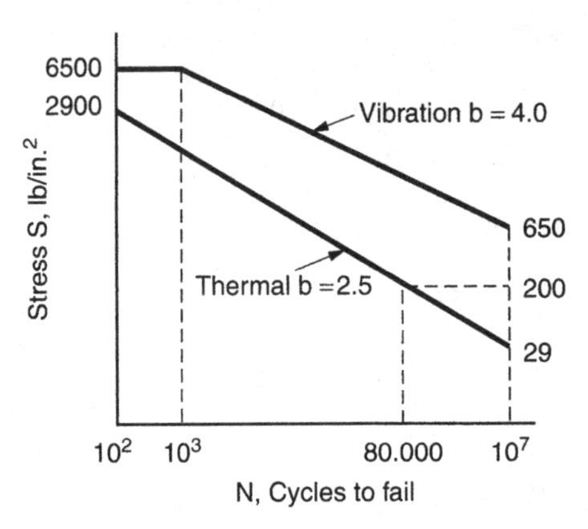

FIGURE 5.6 Typical log–log *S–N* fatigue life curve for 63/37 Sn/Pb solder showing the *b* exponent slope of 2.5 for a rapid thermal cycle and a *b* exponent slope of 4.0 for vibration.

Substitute into Eq. 5.1 to find the approximate fatigue life expected for the solder joints:

$$N_2 = (80{,}000)\left(\frac{200}{166}\right)^{2.5} = 1.27 \times 10^5 \text{ solder cycles to fail} \tag{5.42}$$

In a *worst-case* analysis with a very slow thermal cycle, the solder creep can double the solder shear stress, which would reduce the solder fatigue life to about 22,530 cycles.

Recommendation 3 Add a small kink or offset in the wire, Fig. 13.2. When there is room to add a small kink or a small offset in each lead wire it is often possible to substantially reduce the axial spring rate of the wire. This will reduce the forces and stresses in the wires and solder joints and increase the fatigue life.

CHAPTER 6

Octave Rule, Snubbers, Dampers, and Isolation for Preventing Vibration Damage in Electronic Systems

6.1 DYNAMIC COUPLING BETWEEN THE PCBs AND THEIR SUPPORT STRUCTURE

Vibration and shock-induced failures are often the result of careless design practices. The printed circuit boards (PCBs) are usually the heart of many different types of simple and complex electronic assemblies. Whether the electronic system is simple or complex, the location and the arrangement of the various PCBs within the assembly, and the properties of the structure supporting the PCBs, will often effect the response characteristics of the PCBs. These effects can damage the PCBs when they are operating in severe vibration and shock environments. There is also the feedback effects, where the dynamic response of the various PCBs can react back into the supporting structure and damage the supporting structure in a poorly designed electronic system.

Electronic assemblies with their various PCBs and support structures form a complex vibratory system that usually requires computer techniques to perform a dynamic analysis that can be very time consuming and expensive. The methods of analysis shown here were developed from computer parametric studies of two degree-of-freedom systems with springs, masses, and dampers. The energy transferred from the chassis to the PCB was the main concern. Vibration test data shows that most of the damage in electronic systems occurs in the PCBs. Therefore, the response of the PCB with respect to the chassis, which is the *coupled* response of the PCB, is required to obtain the acceleration G level acting on the PCB. The *coupled* response of the PCB was shown by Q_2. The uncoupled response of the chassis was shown as q_1 and the *uncoupled* response of the PCB was shown as q_2. The ratio R was used to show Q_2/q_1 values that were obtained from the computer study, based on the uncoupled natural frequency of the chassis f_1 and the uncoupled natural frequency of the PCB f_2 [16].

All of the computer and test data were based on plug-in PCBs with edge guides that firmly grip the edges of the PCB. The PCB response characteristics change very rapidly when the edges of the PCB are not gripped firmly. When the PCB has loose edge guides, the displacement amplitude of the PCB increases. This increases the period, which decreases the natural frequency of the PCB. Decreasing the PCB natural frequency brings the PCB natural frequency closer to the chassis natural frequency, which *increases* the dynamic coupling between the chassis and the PCB. This is a dangerous condition since it increases the acceleration levels acting on the PCB. Loose PCB edge guides should be avoided to ensure the reliability of the PCBs in severe vibration conditions.

The computer-generated response data for the chassis and the PCB were translated into a graphic form to cover a wide range of parameters in order to simplify analysis methods, which are normally quite complex. This approach was developed specifically for hand calculations, which will give quick but approximate answers to dynamic coupling effects between various PCBs and their support structures. These analysis methods can also be used as a sanity check for analyses performed on a computer. Computer users know that data input errors are easy to make. This condition is called garbage-in garbage-out, or GIGO. They also know that when a tired analyst has to enter a few thousand data points, it is very easy to make one or more data entry errors. One data entry error in a large program can ruin several weeks or months of hard work. Data entry errors are easy to make and very hard to find, especially on a large computer program. The approximate short-cut methods presented here can be used to check the results obtained from other computer-generated analyses. If there are large differences in the results of the two methods, further investigations should be made.

All machines in motion, from automobiles to washing machines, generate some form of vibration. As the speeds increase, the exciting frequencies and the acceleration forces also increase. One of the most important properties involved in dynamic motion is the transmissibility Q. This is a dimensionless ratio of the output (or response) divided by the input. The usual parameters are the displacements or the acceleration G levels, which can exhibit very high values depending on the damping in the structural members and the phase relations between the adjacent vibrating members. High displacements or high acceleration G levels can result in very rapid fatigue failures in improperly designed and manufactured products. When two adjacent structural elements, such as a chassis, considered to be mass 1, and a plug-in PCB, considered to be mass 2, have close natural frequencies with light damping, they will both have high transmissibility Q's. These conditions are extremely dangerous for a structural dynamic system. The response of the chassis becomes the input to the PCB. When the chassis has an input acceleration level of $10\,G$ and a transmissibility Q of 10, the chassis will see 10×10 or $100\,G$. The transmissibility Q's do not add; they multiply. The response of the chassis is the input to the PCB. When the PCB has a transmissibility Q of 10, the PCB response will be $10 \times 10 \times 10$ or $1000\,G$. These high acceleration levels can damage any electronic system.

The chassis receives the dynamic energy first; so the chassis is considered to be the first degree of freedom called mass 1. The PCBs receive their energy from the chassis; so the PCBs are considered to be the second degree of freedom called mass

2, as shown in Fig. 6.1. To prevent a rapid buildup of high destructive acceleration G levels, the natural frequencies of the chassis and the PCBs must be well separated. This is where the octave rule can be applied. Octave means to double. The natural frequency of the chassis must be examined by itself, without the PCBs, by considering it to be a simple spring–mass system. The natural frequency of the each PCB must be examined by itself with the proper edge supports, by also considering it to be a simple spring–mass system. The resulting natural frequencies of the chassis and the various PCBs should be separated by at least one octave or more to avoid severe dynamic coupling. When the chassis, mass 1, has a natural frequency of 100 Hz, each PCB, mass 2, should have a *minimum* natural frequency of 2×100, or a natural frequency greater than 200 Hz to avoid severe dynamic coupling with the chassis support structure. This is called the *forward* octave rule. The term *forward* is used to describe the direction of the load path from the chassis structure, forward into the PCB structure. Since the chassis receives the dynamic load first and then transfers the dynamic load to the PCB, the direction from the chassis to the PCB is called the forward direction. The *reverse* octave rule is used to describe the direction of the load path backward, from the PCB back to the chassis support structure. For the reverse octave rule the chassis, mass 1, would have a natural frequency of 200 Hz, and the PCB, mass 2, would have a natural frequency of 100 Hz [16].

An examination of Fig. 6.2 shows that when the frequency ratio f_2/f_1 is anywhere close to 1.0, and the weight ratio W_2/W_1 is very small, less than about 0.10, the coupling between the PCB and the chassis shown as $R = Q_2/q_1$ becomes very large. This means that the acceleration G levels acting on the PCB will be substantially increased above the chassis acceleration levels. These conditions can produce high accelerations and rapid PCB fatigue failures. When the frequency ratio $f_2/f_1 = 2.0$ or more, or when this frequency ratio is 0.5 or less on the horizontal axis, then the transmissibility ratio R on the vertical axis will be approximately 1.0. This means that the chassis, mass 1, and the PCB, mass 2, receive approximately the same acceleration G levels. There is virtually no increase in the PCB acceleration G level over and above the chassis acceleration G level. Also notice that *this condition only*

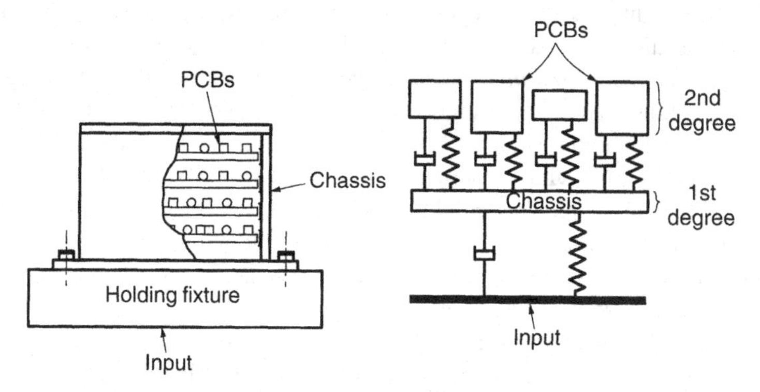

FIGURE 6.1 Mathematical model of a chassis with PCBs where the chassis is the first degree of freedom and the PCBs are the second degree of freedom.

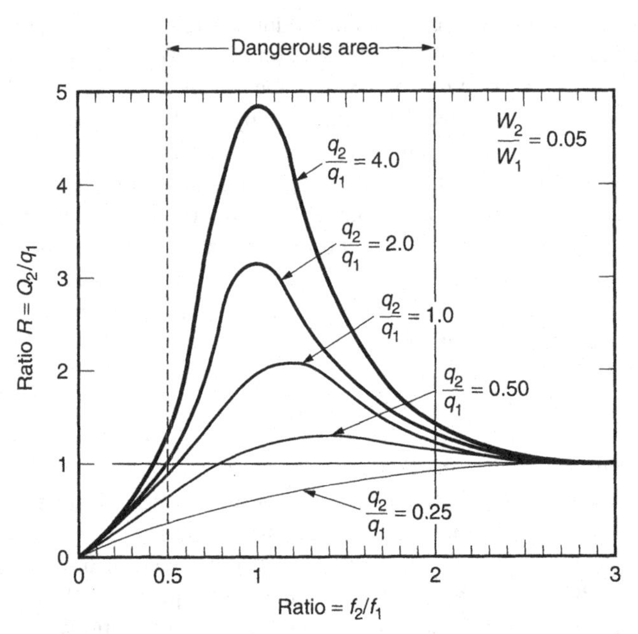

FIGURE 6.2 Dynamic coupling effects between the chassis and the PCB for various ratios of transmissibility when the weight of the chassis is 20 times the PCB weight. (Reprinted with permission from *Machine Design Magazine.*)

occurs when the weight ratio W_2/W_1, which is the ratio of the PCB weight divided by the chassis weight, is very small, less than about 10 to 1. Therefore the forward octave rule or the reverse octave rule can be used to reduce the dynamic coupling effects between the chassis and the PCB, only when the weight of each PCB is very small compared to the weight of the chassis.

Extreme care must be exercised in the design of any dynamic system when the weight of the second-degree-of-freedom member, mass 2, is substantially more than one tenth the weight of the first-degree-of-freedom member, mass 1. The use of the forward octave rule under these conditions will always work. *The use of the reverse octave rule under these conditions is not recommended.* An examination of Fig. 6.3*a* through 6.3*c*, shows that when weight W_2 (normally considered to be the PCB) is relatively large compared to weight W_1 (normally considered to be the chassis), then the reverse octave rule with a frequency ratio $f_2/f_1 = 0.5$ shows transmissibility ratios R of 4 to 5 or more. This condition will transfer acceleration G levels to the PCB that are 4 to 5 or more times greater than the acceleration G levels developed in the chassis. This condition can result in an extensive amount of damage and rapid PCB failures in electronic equipment.

A series of vibration tests were just completed as this is being written in May, 1999. The tests were run to examine the coupling between the chassis and the PCB when the weight of the PCB was one quarter the weight of the chassis and where the frequency ratio $f_2/f_1 = 0.5$. The vibration tests were run on real hardware samples,

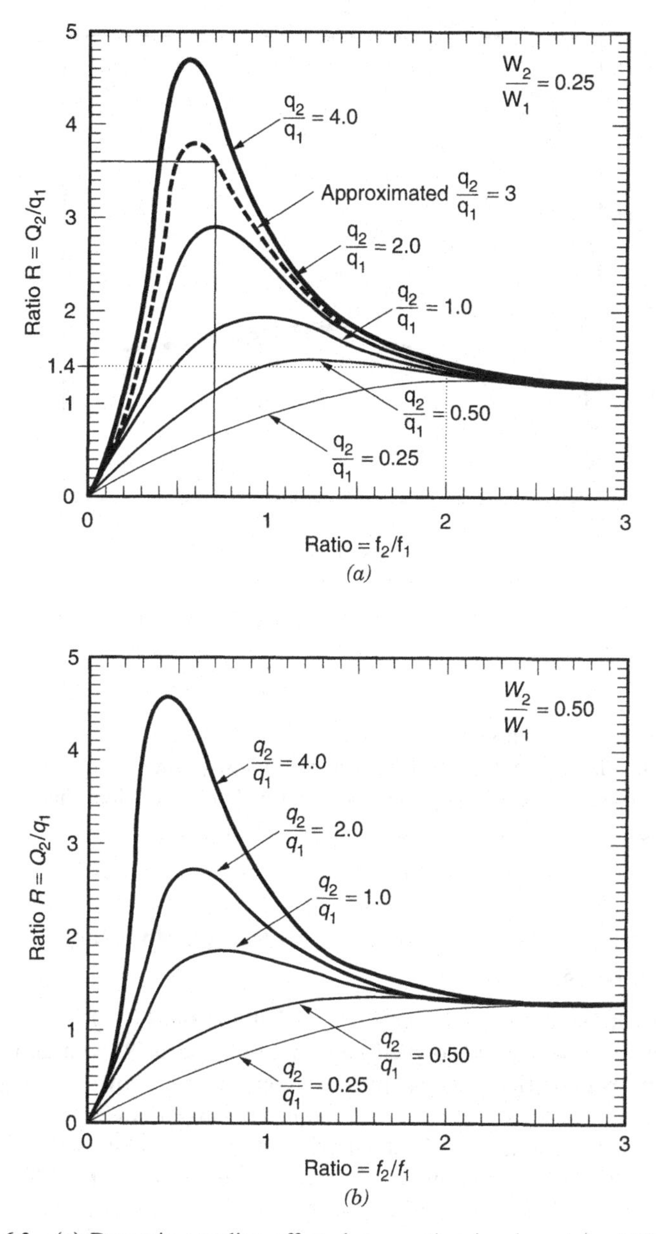

FIGURE 6.3 (*a*) Dynamic coupling effects between the chassis and the PCB for various ratios of transmissibility when the weight of the chassis is four times the PCB weight. (*b*) Dynamic coupling effects between the chassis and the PCB for various ratios of transmissibility when the weight of the chassis is two times the PCB weight. (*c*) Dynamic coupling effects between the chassis and the PCB for various ratios of transmissibility when the weight of the PCB is two times the chassis weight. (Reprinted with permission from *Machine Design Magazine*.)

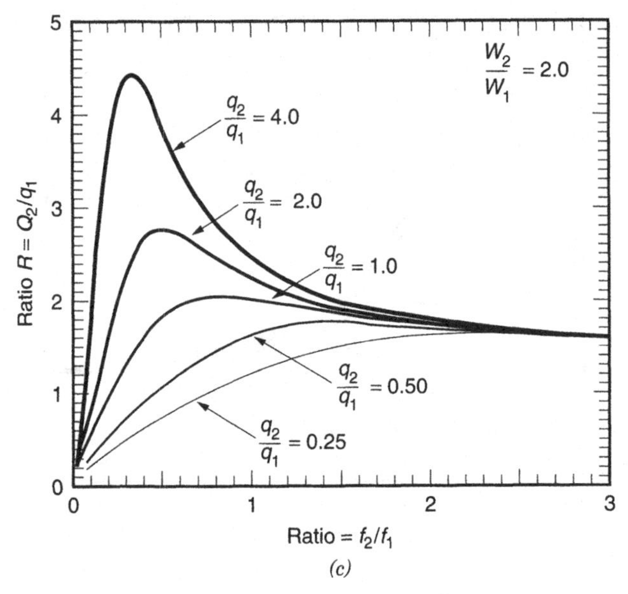

(c)

FIGURE 6.3 (*Continued*).

not mockups that were thrown together quickly for these tests. The tests showed acceleration levels on the PCB were about *five times greater* than the acceleration levels observed on the PCB where the weight ratio W_2/W_1 was less than about 0.10. This compares very well with the curves presented in Fig. 6.3.

Sample Problem: Effects of Vibration on Relays Mounted on Circuit Boards

Several relays are mounted at the center of a 2.5-lb plug-in PCB that was tested by itself and found to have an uncoupled transmissibility q_2 of 9, with an uncoupled natural frequency of 81 Hz. The relay manufacturer's catalog shows the relays have a 75-G peak rating for sine vibration, over a frequency range from 20 to 2000 Hz. The single PCB will be mounted in a 10-lb chassis that was tested by itself and found to have an uncoupled transmissibility q_1 of 3, with an uncoupled natural frequency of 120 Hz. The assembled system must be capable of passing a 5.0-G peak input (G_{in}) sine vibration qualification test, over a frequency range from 10 to 2000 Hz. Is the proposed design acceptable?

Solution The curves in Fig. 6.3*a* must be used to evaluate this system. These curves make use of several different ratios and a few equations, which are defined below:

f_2/f_1 = Uncoupled natural frequency for PCB weight W_2 to uncoupled natural frequency for chassis weight W_1

q_2/q_1 = Uncoupled transmissibility for PCB q_2 to uncoupled transmissibility for chassis q_1

$R = Q_2/q_1$ = Ratio for the *coupled* transmissibility of PCB Q_2 to the uncoupled transmissibility of chassis q_1

W_2/W_1 = Weight ratio of second-degree-of-freedom PCB W_2 to the first-degree-of-freedom chassis W_1

G_{in} = 5.0 peak acceleration

Uncoupled acceleration response of chassis: $$G_1 = G_{in}q_1 \tag{6.1}$$

Coupled acceleration response of PCB: $$G_2 = G_1Q_2 \tag{6.2}$$

Coupled acceleration response of PCB: $$G_2 = RG_{in}(q_1)^2 \tag{6.3}$$

Relay electrical contacts will chatter when the relay is exposed to vibration acceleration levels that exceed their rated value of 75-*G* peak. This will cause the relay to malfunction electrically. Relays are very seldom damaged when they are exposed to acceleration levels that exceed their rated value. When the vibration stops, or when the acceleration levels are reduced, the contacts stop chattering, and the relay works normally again. To find out if the relays will operate properly in the 5-*G* peak input acceleration, the response acceleration level expected on the PCB, G_2 must be evaluated to make sure it does not exceed 75-*G* peak. If the PCB response is determined to be greater than 75-*G* peak, then some corrective action has to be taken or the relays will fail electrically.

The peak acceleration response of the PCB, G_2 was shown by Eq. 6.3 to be related to the ratio R, which can be obtained from the various dynamic response curves shown in Fig. 6.3*a* to 6.3*c*. Each of these curves was derived for a different weight ratio. The ratio for this case of $W_2/W_1 = 2.5/10 = 0.25$ is shown in Fig. 6.3*a*.

$$\text{Frequency ratio } f_2/f_1 = 81/120 = 0.675 \tag{6.4}$$

$$\text{Uncoupled transmissibility ratio } q_2/q_1 = 9/3 = 3 \tag{6.5}$$

The next requirement is the ratio R at the points shown in Eqs. 6.4 and 6.5. An examination of Fig. 6.3*a* shows that there is no curve with a ratio $q_2/q_1 = 3$; so a curve with the ratio of 3 must be generated by following profiles between the uncoupled transmissibility ratios of 2 and 4. This new approximated curve was added to Fig. 6.3*a* as a dash line. The ratio R can now be obtained following Eqs. 6.4 and 6.5 in Fig. 6.3*a*.

$$\text{Ratio } R = Q_2/q_1 = 3.6 \quad \text{from Fig. 6.3}a \tag{6.6}$$

Next substitute into Eq. 6.3 for the PCB acceleration response:

$$G_2 = (3.6)(5.0)(3)^2 = 162 \text{ peak} \tag{6.7}$$

The PCB acceleration response of 162-G peak far exceeds the maximum allowable relay level of 75-G peak, so the relays for the expected environment are unacceptable. Some type of corrective action must be taken or the relays will fail electrically.

Proposed Corrective Action

1. Mount the relays at the edges of the PCB where the transmissibility Q is much lower than at the center of a plug-in PCB. This will require a new layout of the PCB with new interconnects on a multilayer PCB and new fabrication and assembly. This will delay delivery schedules and increase production costs.
2. Look for relays that have much higher acceleration G ratings above the 162-G peak. This may be a difficult job to find relays with the same size and same electrical functions with the same electrical contact locations.
3. Follow the octave rule by increasing the natural frequency of the PCB to a value two times the chassis natural frequency, or $2 \times 120 = 240$ Hz. This may be possible by adding several ribs to the PCB (if there is room available) to increase the stiffness. However, increasing the PCB natural frequency will also increase its uncoupled transmissibility q_2, which must be included in the analysis. Another potential problem may be the height of the ribs required to achieve the desired PCB frequency. High stiffening ribs may require additional space between the adjacent PCBs. This may require changing the location of the PCB edge guides and location of plug-in connectors on the interconnecting back plane.
4. Add snubbers to the relay PCB and to a couple of PCBs adjacent to the relay PCB to reduce the dynamic displacements during vibration. Snubbers are very effective in reducing dynamic displacements and often the resulting acceleration G levels. In this case, however, snubbers are not recommended. Snubbers will reduce the dynamic displacements in vibration. Snubbers will not work here because the impacting between the snubbers and adjacent PCBs will generate high-frequency and high acceleration level shock pulses on the relay PCB. Snubbers work very well on other PCBs with other types of semiconductor components because almost all semiconductor components can survive at least 5000 G of shock. The relays are only rated at 75 G acceleration. The shocks produced by impacting between the snubbers could easily exceed 75 G and cause other relay electrical problems; so snubbers are not recommended here.
5. Add damping to the relay PCB to reduce the transmissibility Q at its resonance. This will reduce the acceleration G levels developed in the PCB. The best and most efficient damping is with constrained layer dampers. These

members are usually in the shape of beams, with several alternate layers of a highly damped viscoelastic material with a thin foil of aluminum between each layer. Adding damping will reduce the transmissibility. Increasing the damping will work well for structures with low natural frequencies, below about 50 Hz, where the dynamic displacements are high. Damping is not very effective when the PCB natural frequency is higher, around 100 Hz or more, since the dynamic displacements are small; so the damping is small. However, when there is room on the PCB to add stiffening ribs to increase the natural frequency, to follow the octave rule, the ribs will almost always be a much better choice than damping.

Select option 3 above by adding stiffening ribs to increase the PCB natural frequency to $2 \times 120 = 240$ Hz, to follow the octave rule. The required new ratios are now

$$
\begin{aligned}
W_2/W_1 &= 2.5/10 = 0.25 && \text{weight ratio PCB/chassis} \\
f_2/f_1 &= 240/120 = 2.0 && \text{frequency ratio PCB/chassis} \\
q_2 &= \sqrt{240} = 15.5 && \text{uncoupled transmissibility of PCB} \\
q_2/q_1 &= 15.5/3 = 5.16 && \text{uncoupled transmissibility ratio PCB/chassis} \\
R &= 1.4 \text{ from Fig. 6.3}a \text{ for } Q_2/q_1
\end{aligned}
\tag{6.8}
$$

Substitute Eq. 6.8 into Eq. 6.3 to find the coupled acceleration response of the PCB:

$$G_2 = (1.4)(5.0)(3)^2 = 63\ G \text{ peak PCB response} \tag{6.9}$$

Since the peak acceleration level of 63 G for the PCB is well below the relay rating of 75 G peak, the modified PCB design should be acceptable.

6.2 USING SNUBBERS TO CALM PCBs IN SEVERE VIBRATION AND SHOCK

Large PCB displacements are often the cause of many different types of failures in shock and vibration environments, as outlined below:

1. High bending stresses and rapid fatigue failures in electrical lead wires
2. High acceleration problems in crystal oscillators, relays, and potentiometers
3. Short circuits caused by impacting between closely spaced PCBs
4. Solder joint cracking due to impacting between closely spaced PCBs
5. Component cracking due to impacting between closely spaced PCBs
6. Breaking pins on plug-in electrical connectors
7. Rapid fretting corrosion on electrical contacts causing intermittent failures

When the octave rule is followed, dynamic coupling effects are reduced so that large dynamic displacements on the PCBs are also reduced. However, when the octave rule cannot, or has not, been followed, then large displacements can often be reduced with the proper use of snubbers placed in critical areas on the PCB. Snubbers are small devices that can be attached to adjacent PCBs; they strike each other, reducing their displacements in severe vibration and shock environments. When the snubbers are properly placed, they can often be installed to have a slight interference fit between adjacent PCBs using a soft rubber-type material for *low-frequency* PCBs. Low usually means a natural frequency below about 50 Hz. More rigid materials, such as epoxy fiberglass, should be used for snubbers between *high-frequency* PCBs. High usually means a natural frequency above about 100 Hz. Since higher frequencies result in small dynamic displacements, small clearances should be used between the snubbers on adjacent PCBs when the PCB natural frequency is above 100 Hz. The clearances between the adjacent snubbers should be less than half of the dynamic PCB displacement expected in the particular environment. The smaller the spacing between the adjacent snubbers the better they will work. The best snubbers are ones that touch each other, with no space between them. This acts like a center support on the PCB, which can more than double the PCB natural frequency. Very small spacing between adjacent PCB snubbers may be desirable, but it can be difficult to achieve due to manufacturing and assembly tolerances. Normal manufacturing tolerances on locating plug-in PCBs is about ±0.015 in. With some increased costs this can be brought down to about ±0.010 in. The snubbers must impact against each other to reduce the dynamic displacements. This prevents the components and the solder joints on adjacent PCBs from striking each other. The impacting between adjacent snubbers will reduce the dynamic displacements on the PCBs. This will reduce bending deflections and stresses in the component lead wires and in the plug-in electrical connector pins and increase their fatigue life.

Snubbers can be made in different shapes using different materials. The cylindrical epoxy fiberglass shape about 0.18- to 0.25-in. diameter epoxy bonded to critical areas on the PCB works well. Different triangular, rectangular, or spherical shapes can be used depending upon the size, location, and material. Snubbers have been made from different grades of rubber, nylon, delrin, aluminum, and epoxy fiberglass. The harder materials such as epoxy fiberglass, delrin and aluminum should be used for high natural frequency PCBs that have smaller dynamic displacements. Softer materials such as rubber should be used for low natural frequency PCBs where the displacements are much larger. If a soft material such as rubber is used to reduce small dynamic displacements, the soft rubber will simply displace, which will allow the PCB to displace; thus it will not work.

The most critical area on a plug-in type of PCB is the center, which can 'oil can' during its resonant condition since the boards are usually very thin. Testing experience has shown that most of the damage on a plug-in PCB occurs at its fundamental (or lowest) natural frequency, where the dynamic displacement is usually the greatest. If the large dynamic displacements can be reduced, the failures will also be reduced. For new designs the PCB natural frequency can often be increased, which will decrease the dynamic displacements and often solve the problem. However, increasing the PCB natural frequency requires an increase in the stiffness. This usually means making the board thicker, or adding ribs to stiffen the board, or adding materials such as aluminum to stiffen the board. These items can add size, weight, and cost, which may cause problems. Instead of stiffening the PCBs to increase the natural frequency, it is often simpler to just add snubbers near the centers of a couple of critical PCBs to solve these problems as shown in Fig. 6.4 [17].

Solving the problem of large dynamic displacements on existing PCBs requires a slightly different approach. There may not be any free space available near the center of the PCBs to attach any type of snubber. Under these circumstances it will be necessary to attach snubbers in other areas that will allow the snubbers to either impact against other snubbers or to impact against the PCB in noncritical areas. These are areas away from solder joints, etched copper circuit traces, and other electronic components. One method that worked well was to make full-scale transparencies of both sides of each PCB that requires a snubber. All the transparencies are then placed together in their proper order, back to back, as they would be in the assembled chassis. Looking through the stacked transparent PCB layouts will often provide clues as to where different snubbers can be placed on adjacent PCBs to provide effective snubbing action.

There are many other ways that snubbing action can be used to reduce the dynamic displacements of PCBs in severe vibration and shock environments. When weight is not a real problem, and the PCBs are inside an enclosed box, with a bolted cover, it may be possible to snub the PCBs with the use of many small rubber balls. The small rubber balls can be added to the assembled box before the cover is bolted in place. The balls must be small enough to fit between the various PCBs. When the box is full, the cover is bolted in place. The balls between the various PCBs fill the space so the PCBs are held rigidly. An assembly of this type can withstand very high

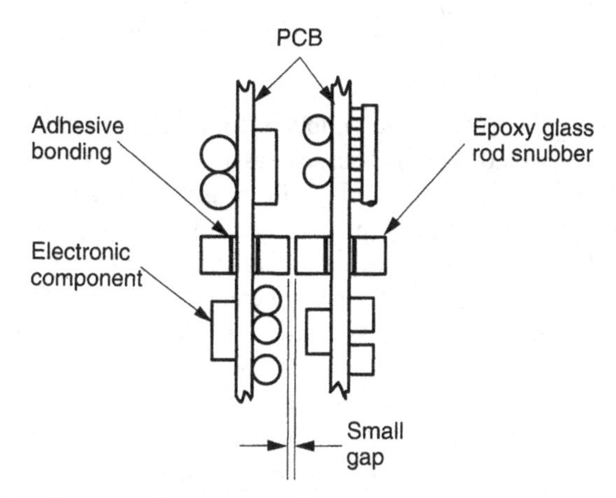

FIGURE 6.4 Small epoxy fiberglass snubbers are cemented near the center of adjacent PCBs with a small clearance between the snubbers to reduce the dynamic displacements and stresses in severe vibration and shock environments.

vibration and shock acceleration levels without failing. This box is easy to repair. The cover is removed and the balls are poured out and saved. Any PCB can easily be removed and replaced. The balls are then added again, the cover is replaced, and the box is ready for use. There is a slight probability of electrostatic voltage buildup during vibration. The rubber ball material used here should be checked for this possibility.

Sample Problem: Selecting Snubbers to Solve PCB Vibration Problems

Environmental stress screening (ESS) programs on production assemblies revealed some vibration problems on several PCBs within three different electronic chassis. The PCBs in the different chassis that presented the vibration problems are listed below:

1. PCBs with a natural frequency of 125 Hz and a 4-G peak sine vibration input
2. PCBs with a natural frequency of 45 Hz and a 2-G peak sine vibration input
3. PCBs with a natural frequency of 250 Hz and a 5-G sine peak vibration input

An examination of the problem PCBs shows there is room available near their centers for attaching some type of snubbers. The recommended course of action is to add epoxy fiberglass snubbers epoxy bonded to the PCBs using a clearance of 0.014 in. between the adjacent snubbers. Is this recommendation an acceptable solution?

Solution Find the expected PCB dynamic displacements using the information given in Table 6.1:

$$Z_0 = \frac{9.8 G_{\text{in}} Q}{f_n^2} \qquad \text{(Ref. Eq. 3.17)}$$

The recommended gap between the snubbers for all three chassis was 0.014 in. Chassis 1 needs a gap of 0.014 in. or less; so chassis 1 is good. Chassis 2 needs a gap of 0.032 in. or less; so the 0.014-in. gap will be good. Chassis 3 needs a gap of 0.006 in. or less, so the 0.014-in. gap will not work. A gap of 0.006 in. between snubbers will be difficult and expensive to obtain because of the manufacturing and assembly tolerances involved. If this is what the engineering and the manufacturing groups want to achieve, then they may have to go to matched assemblies. This is where each PCB, each set of snubbers, and each chassis are matched; so they cannot be interchanged with each other. If the matched sets are separated and assembled in different groups, the gap between the snubbers cannot be controlled to the desired 0.006-in. gap. Matched sets require an enormous amount of paperwork to keep track of the locations, problems, repairs, engineering changes, and any warranties on the reliability. It may be cheaper and easier to use stiffening ribs or a composite aluminum board for the PCB to increase the natural frequency.

TABLE 6.1 Gap Spacing Requirements Between Snubbers

Chassis	G_{in}	Q	f_n	Z_0 (in.)	Gap $= Z_0/2$ (in.)
1	4 G	$\sqrt{125} = 11.2$	125	0.0281	0.014
2	2 G	$\sqrt{45} = 6.7$	45	0.0648	0.032
3	5 G	$\sqrt{250} = 15.8$	250	0.0124	0.006

The natural frequency required to achieve a dynamic displacement of 0.006 in. can be obtained from a slight modification of Eq. 3.17, when $Q = \sqrt{f_n}$.

$$f_n = \left(\frac{9.8G_{in}}{Z_0}\right)^{2/3} \tag{6.10}$$

where $G_{in} = 5.0$-G peak, input acceleration level
$Z_0 = 0.006$ in., desired single amplitude displacement

$$f_n = \left[\frac{(9.8)(5.0)}{0.006}\right]^{2/3} = 405 \text{ Hz} \tag{6.11}$$

A natural frequency of 405 Hz will produce the desired displacement, but will it be worth the increase in the size, weight, and cost to achieve the results? It will be much easier to increase the natural frequency of a small PCB especially if the edges are clamped.

6.3 ADDING DAMPING TO PCBs TO REDUCE THE TRANSMISSIBILITY Q AT RESONANCE

Damping is often defined in a vibrating system as the conversion of kinetic energy into heat. When the energy conversion is contained within the material itself, due to internal friction or hysteresis in the molecular structure, it is called material damping. When the energy conversion is generated by friction, scraping, slapping, rubbing, or impacting at various interfaces and joints, it is called structural damping. The total damping in the system will be the sum of the material damping and the structural damping [18].

All real systems will generate some damping when they are vibrated. A system with very light damping will continue to oscillate for a long time after the exciting force has been removed. A system with heavy damping, like a good automobile shock absorber, may only oscillate once or twice after the exciting force has been removed. In all cases the system damping will eventually bring the system to rest after the exciting force has been removed. Since damping removes some of the energy within the system, there is less energy available to do work on the system.

Less energy means lower dynamic displacements and lower stress levels in a vibrating system, which will result in an increased fatigue life.

6.4 MATERIAL DAMPING PROPERTIES

Material damping relates to the internal energy that is lost when a structure is deformed. When a tensile axial force is applied to a bar, it does not instantly elongate to a new length. Instead there is a small time lag as the bar slowly creeps to its new stable length. When the external load is removed, the process is reversed and the bar slowly tries to shrink back to its original length. Since some of the strain energy has been converted into heat, there is less energy available so that the bar never quite reaches its original position. When an equal compressive load is now applied to the bar, the action described above takes place in the reverse direction. When the positive and negative strains are plotted with the positive and negative stress, it results in a hysteresis loop, which is slightly different for structural materials and viscoelastic materials, as shown in Fig. 6.5. The area enclosed within the loop is a measure of the energy lost with each stress cycle. The area then becomes a measure of the damping performance capability of the material. Higher stresses and strains in a material will increase the loop area, which indicates that higher stresses and strains will produce more damping.

Another way of estimating the material damping is to measure the amplitude change in a freely vibrating structure. When the beam structural element being tested is displaced and released, the beam will vibrate freely up and down with a decreasing amplitude because energy is being dissipated. Eventually the beam will come to rest.

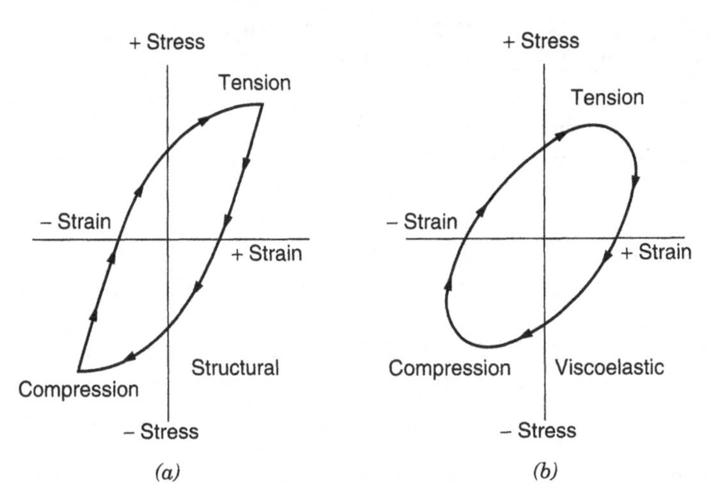

FIGURE 6.5 Stress–strain hysteresis loops: (*a*) structural materials and (*b*) viscoelastic materials.

Another way that material damping can be used is to cement strips of rubber on the inside surface of a bolted cover. The rubber strips are arranged so they press down on the unsupported top edges of the plug-in PCBs when the cover is bolted in place. The rubber strips provide two functions. They add some additional support to the PCBs and they increase the damping on the PCBs. This arrangement works well for fresh new systems. However, after these systems have been in service for a few years, vibration and shock environments will eventually wear down the rubber strips so that the extra support and the extra damping become less effective with time.

6.5 STRUCTURAL DAMPING PROPERTIES

Structural damping relates to the energy that is lost during vibration due to friction at bolted and riveted interfaces, friction at pivoting joints in built-up structures, and sliding, slapping, and impacting between various surfaces. Vibration tests on production built-up structures shows that no two identical appearing structures react in exactly the same way. There will always be small variations in their transmissibilities and natural frequencies. This is due to the differences in the manufacturing tolerances in sheet metal thickness, surface finish, interface pressures, surface flatness, material hardness, number of screws or rivets in an assembly, screw sizes, spacing and torque, dry or wet interfaces, humidity conditions, assembly sequence, and many other factors.

Vibration tests must be carefully planned to find the transmissibility and the natural frequency of any structure. It is always better to use the real structure instead of a mock up. When a mock up has to be used, it should closely resemble the real structure. Tests should be run using the required acceleration input levels where possible. When real hardware is used for the tests, there is always the fear that the real hardware will be damaged so lower test levels are often used. Plug-in types of PCBs are very nonlinear; so using a lower input acceleration test level will result in lower stresses and lower damping. Lower damping will result in a higher transmissibility and a slightly higher natural frequency. As long as the nonlinear properties of the structure are understood, there should be no problems with running vibration tests with reduced input acceleration levels.

The structural damping in a large system with many piece parts and bolted joints will always be greater than the material damping. Increasing the material stress will always increase the material damping and reduce the transmissibility. Increasing the material stress will also reduce the fatigue life of the material. This trade-off must be understood in order to design a cost-effective electronic system.

6.6 DAMPING PROPERTIES OF VISCOELASTIC MATERIALS

Damping converts kinetic energy, or mechanical energy, into heat, which is lost energy since it cannot be converted back into kinetic energy. Any time any energy is lost in a vibrating system, there is less energy available to do work on that system. This means that there will be less energy available to generate a high transmissibility when the natural frequency of the system is excited. Increasing the damping increases the energy dissipated in a vibrating system, which decreases the transmissibility. This reduces the dynamic displacements and stresses, which improves the fatigue life. There are many different ways to apply damping materials to electronic assemblies. Some methods work well and some methods do not. The best place to use damping materials is on the PCBs since they usually support most of the electronic components and cause most of the vibration problems.

Viscoelastic materials have properties that are very similar to rubber materials. They can stretch, compress, and deform through large displacements without failing. Many of these materials can dissipate large amounts of energy when they are deformed. Even more energy can be dissipated with the addition of materials such as asbestos. Natural rubber by itself does not have much strength or much damping. When carbon black is added, the strength and damping are increased.

Viscoelastic material properties in general are very sensitive to temperature and to frequency. Although these materials are generally very soft at room temperature, they become very hard at very low temperatures. They also behave like very hard materials when they are exposed to high frequencies. Their ability to deform under these conditions is sharply reduced. The modulus of elasticity of viscoelastic materials is difficult to measure because the modulus changes with the speed of the applied load. The modulus of elasticity for a rapidly applied load can be three times greater than the modulus obtained for a slowly applied load.

Fatigue failures can be reduced in vibrating beams and plates with the application of damping materials in the form of paints and tapes. These materials can be applied in single layers and multiple layers. Single layers of damping materials will not increase the damping very much. A single-layer damping material application is shown in Fig. 6.6. Multiple layers of damping materials will increase the damping. A multiple layer damping system may be made up of a single layer of damping material with a more rigid constraining strip added to the top of the viscoelastic damping material to increase the shearing strain, as shown in Fig. 6.7. More damping layers will increase the damping, but it will also increase the weight and make repairs on PCBs much more difficult. There are two popular viscoelastic application methods for increasing the damping. The first is to apply the damping material so it is forced to deform by stretching alone, as shown in Fig. 6.8. The second is by a combination of stretching and shearing, where a more rigid constraining strip is added to the top of the viscoelastic damping material to increase the shearing strain, as shown in Fig. 6.9. Several damping layers can be used to further increase the constrained layer damping properties. Each viscoelastic damping layer would be separated from the adjacent layer by using a thin aluminum foil, 0.005 in. thick to increase the shearing strain energy dissipated.

Multilayer viscoelastic dampers for plug-in types of PCBs are often fabricated in the shape of one or more tall narrow strips, laid across the center of the PCB. The center of the PCB has the greatest displacement, which produces more damping due to more shear energy dissipation. The damping strips have to be narrow since they take up some of the surface area that could be used for mounting electronic component parts.

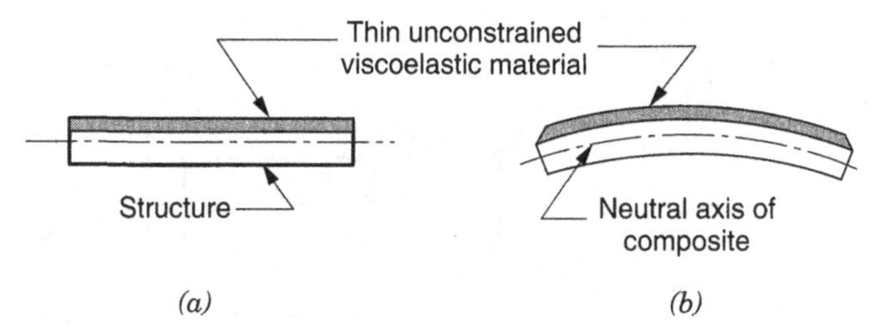

FIGURE 6.6 Two-layer composite system consisting of a thin unconstrained viscoelastic material applied to a structure: (*a*) at rest and (*b*) during vibration.

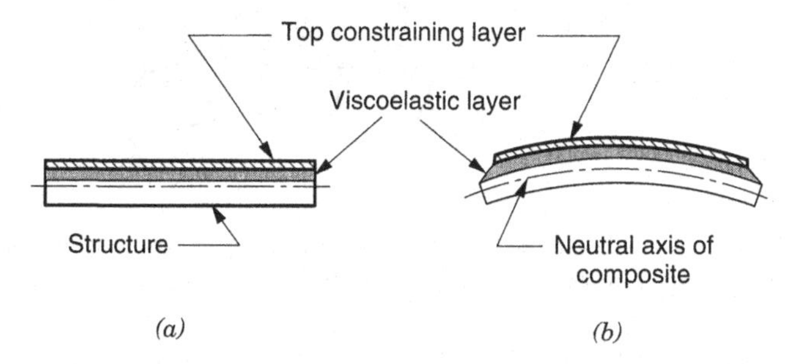

FIGURE 6.7 Three-layer composite system consisting of a thin viscoelastic material constrained between the structural elements: (*a*) at rest and (*b*) during vibration.

The damping strips provide the greatest amount of damping when the PCB displacements are large, since large displacements provide more energy dissipation. This is *contrary* to what is needed to provide a good fatigue life for the components mounted on the PCB. The dynamic displacements of the PCBs should be reduced, to reduce the stresses in the component lead wires and solder joints to increase the

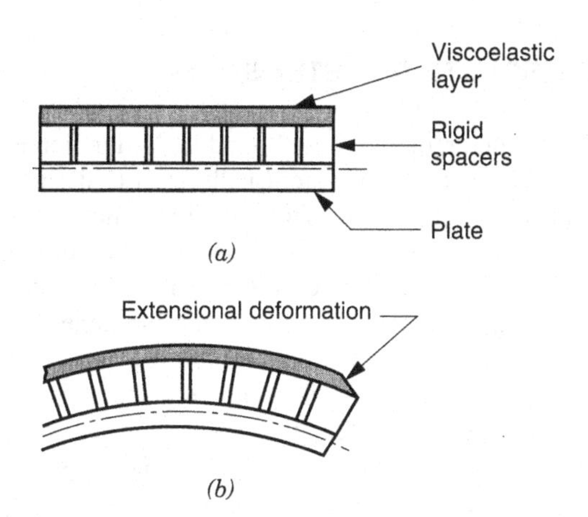

FIGURE 6.8 Using rigid spacers to improve the damping of viscoelastic materials without constraining layers: (*a*) at rest and (*b*) during vibration.

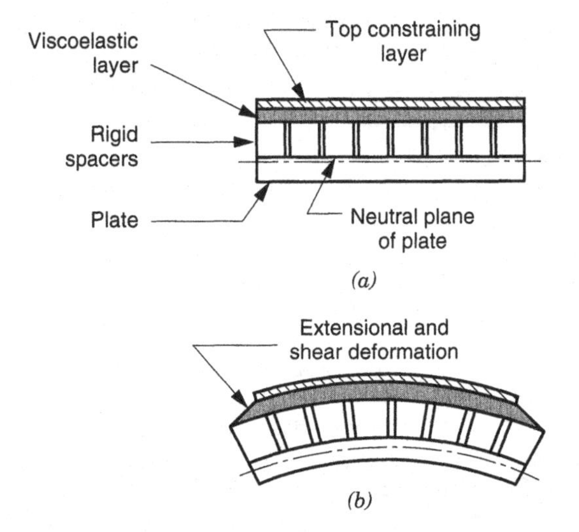

FIGURE 6.9 Using rigid spacers to improve the damping of viscoelastic materials with constraining layers: (*a*) at rest and (*b*) during vibration.

fatigue life. Instead of using constrained layer damping strips, it would be much better to use stiffening ribs rigidly attached to the PCBs to increase its natural frequency. Stiffening ribs will be much thinner than damping strips so ribs will take away less PCB surface area, which will allow more room for mounting more electronic components. Care must always be exercised when the desired PCB natural frequency is being established, to make sure the octave rule is being followed. Using the octave rule will avoid severe dynamic coupling between the chassis and the PCB, which will substantially improve the fatigue life of the electronic equipment [1].

6.7 VIBRATION ISOLATION SYSTEMS

Structural systems are often required to withstand high vibration acceleration levels without failing. Two different methods are usually used to design and manufacture reliable systems for severe vibration and shock environments. The first method is the brute force method, where a massive rugged structure is designed to have low dynamic stresses and displacements, to ensure a good fatigue life. This results in a large, heavy, and expensive structure, but it gets the job done. The second method examines the operating environment, looking for the most potential damaging forcing frequencies and produces a vibration and shock isolation system that significantly reduces the dynamic forces that are transferred to the structure. This results in a lighter and less expensive system that may not be smaller, when the size of the isolation system is included.

Many different types of materials are used in isolation systems, depending on the operating environments. Metal isolators must be used where very high temperatures are expected or where there is exposure to certain corrosive liquids. Rubbery-type materials are very popular since they can be molded into a wide variety of shapes and sizes to control the natural frequency, with different additives to control the damping. This allows the transmissibility to be controlled at the resonance. The isolator natural frequency must be much lower than the natural frequency of the structure it is trying to protect. Low natural frequencies work well for vibration isolation systems. Typical frequencies range from about 5 to 30 Hz, for many types of electronic assemblies. The typical transmissibility can often be controlled to 4 or less for most rubbery materials. Silicone rubber must be used for higher temperature applications. This results in slightly higher transmissibility values because silicone has lower damping properties than other rubbery materials [19].

Long dwell periods are often required for sine vibration tests at the isolator resonant frequency, where the isolator dynamic displacements are usually quite high. The high damping properties in the isolators produce a lot of heat at these conditions. This results in high temperatures, which may cause failures in the isolators after a few hours. Test conditions often combine many short-term vibration exposures in the actual operating environments into one long vibration qualification period. The lengthy extended test period, therefore, may not be realistic. When isolators fail due to overheating after a lengthy vibration dwell period, a test waiver should be requested. The test waiver might recommend a vibration dwell test program with 30 min of testing, followed by 15 min of rest, followed by 30 min of testing, repeated until the full testing period has been completed.

Isolation systems must often be capable of protecting sensitive equipment for both vibration and shock conditions. What is good for vibration isolation is not good for shock isolation. A good vibration isolation system would probably have a natural frequency of about 5 Hz. The structure that is being protected would then have a much higher natural frequency. When there is a large separation between the isolator natural frequency and the structure natural frequency, the isolation system can sharply reduce the vibration acceleration forces transferred into the structure. However, if a 5-Hz isolation system is used on a structure required to operate in a high shock condition, the shock displacement would be very high. High shock displacements will force the isolators to bottom out. This can produce very high acceleration forces that may rip the isolators or may damage the structure it is trying to protect. High shock conditions require an isolation system with higher natural frequencies to reduce the dynamic displacements. A compromise is usually made to provide an isolation system that can protect a sensitive structure against both vibration and shock. Isolation systems with a natural frequency of about 30 Hz have been found to work well for vibration and for shock in many different types of electronic assemblies.

6.8 MATCHED SETS OF ISOLATORS FOR PRECISION INSTRUMENTS

Precision instruments such as gyros for inertial navigation systems are very sensitive to vibration; thus they must be mounted on isolators when they are required to operate in harsh environments. The typical isolated gyro-stabilized platforms will have 6 degrees of freedom. These are translation along the x and y and z axes, and rotation about the x and y and z axes. This means the platform will have six natural frequencies. Pure translation motion along the x or y or z axes have very little effect on the accuracy of the gyros. Any rotation or rocking of the stable platform can introduce large navigation errors. Therefore, the isolators that support the gyro platform must be matched so they will allow translation only with virtually no rocking motion. This requires a close matching of the elastic center with the center of gravity. In addition the set of isolators for each gyro platform must be matched to have the same natural frequency and the same transmissibility. The natural frequency must be matched to ± 0.10 Hz and the transmissibility must be matched to ± 0.10 for the Q. Several hundred different isolators must be tested to obtain one set of four matched isolators. This process can run up the cost of four isolators from about \$16 to about \$5000.

6.9 APPLYING THE OCTAVE RULE TO ISOLATION SYSTEMS

The octave rule works very well when it is followed. Any time that natural frequencies in adjacent structural members are close to each other, severe coupling can occur, which can generate high acceleration G levels. Lightly damped structures will generate higher forces and stresses since they have a higher transmissibility. Highly damped structures may not generate high dynamic forces and stresses in themselves. However, when two highly damped structures couple, they can still produce large dynamic displacements, forces, and stresses. Severe coupling with light or heavy damping can cause rapid premature failures. The octave rule forces the equipment designer and the analyst to examine the structural load path through a dynamic system with the following questions in mind:

1. Where is the dynamic force coming from?
2. Where is the dynamic force going?
3. How does the dynamic force get there?

Once these questions are answered, the octave rule can be applied. The octave rule says that the natural frequency must be doubled for every extra degree of freedom to prevent severe dynamic coupling between adjacent structural members. It is not always easy to determine the load path through a structure. When there are several possible paths a load can take through a structure, the most probable path will be the path that is the stiffest. A stiff structure can carry a greater load than a soft structure for a given deflection.

Sample Problem: Recommended Frequencies for Use with Isolators

A vibration isolation system is being proposed for an electronic chassis that is expected to have a natural frequency of about 60 Hz. Several PCBs that support large surface-mounted fine-pitch semiconductor devices will be enclosed within the chassis. What are the recommended natural frequencies for the PCBs and the fine-pitch semiconductor devices to avoid vibration problems?

Solution Follow the octave rule. The octave rule requires the natural frequency to be doubled for each added degree of freedom. The chassis represents the first degree of freedom; so the chassis natural frequency should be two times greater than the isolators. Since the chassis natural frequency is about 60 Hz, the isolators should have a natural frequency less than half of the chassis natural frequency, or about 25 Hz. The natural frequency of the PCBs should be more than two times the natural frequency of the chassis; so the natural frequency of the PCBs should be about 125–130 Hz. The natural frequency of the fine-pitch components mounted on the PCB should be more than two times the PCB; so the natural frequency of the fine pitch components should be about 270–280 Hz.

Vibration test data for large surface-mounted fine-pitch semiconductor devices on PCBs has shown many lead wire failures. The fine-pitch lead wires act like low rate springs, and the component body acts like a mass. Test data and analyses have shown that the natural frequency is usually quite high, often above 800 Hz. This is well above the recommended natural frequency of about 280 Hz. The problem here is that PCBs can be excited at higher harmonics. The third or fourth harmonic on a PCB could easily reach 800 Hz, especially in a random vibration environment. Therefore, it is a good practice to bond the four corners of these devices to the PCB using a semiflexible adhesive similar to polyurethane. An epoxy adhesive is usually too brittle; so it may crack in thermal cycling conditions due to thermal expansion differences between the PCB and the component body.

CHAPTER 7

Displacements, Forces, and Stresses in Axial Leaded Component Wires Due to Thermal Expansions

7.1 ELECTRONIC COMPONENTS MOUNTED ON CIRCUIT BOARDS

Electronic components come in a wide variety of sizes and shapes, with and without electrical lead wires, for surface mounting and through-hole mounting on a wide variety of printed circuit boards (PCBs). Many of the old-type axial leaded components such as resistors, capacitors, transformers, diodes, and flat pack integrated circuits are still being used for modern electronic systems. Many of the newer components still use axial lead wires such as semiconductors with fine-pitch leads, microprocessors, and various types of hybrids. Some of these devices can be used for through-hole mounting or surface mounting with a small change in their lead wires. Epoxy fiberglass is still the most common material used for the PCBs. Some polyimide glass is also popular, along with kevlar, quartz, and ceramic.

The power densities are always increasing, while the package size always seems to be decreasing. High power densities usually result in high temperatures unless special attention is paid to proper cooling methods. Conduction cooling is often used for the PCBs. Materials such as aluminum, copper, or magnesium can be added to conduct the heat to some type of heat sink. Aluminum is very popular because it is a very good heat conductor, and it is not very expensive. Aluminum has a high thermal coefficient of expansion (TCE) and a high modulus of elasticity. When aluminum is added to a PCB to improve the heat removal, its high TCE along with its high modulus of elasticity will usually produce higher thermal expansions. Higher expansions will generate higher forces and stresses in the lead wires and solder joints, which can lead to more field failures in thermal cycling environments. Special precautions must be taken to reduce these high forces and stresses to ensure a reliable system with an adequate fatigue life.

7.2 EVALUATING COMPONENT LEAD WIRES AS FRAMES AND BENTS

Electronic assemblies that are exposed to thermal cycling environments will expand when the temperatures are increased, and they will contract when the temperatures are decreased. These expansions and contractions will occur in the PCBs, the components, and the electrical lead wires. The component bodies and the PCBs are typically much larger and much stiffer than the electrical lead wires. The lead wires on axial leaded components must be bent 90° so they can extend to the PCB where they are soldered to make electrical connections, as shown in Fig. 7.1. When there are expansion differences between the PCB and the component body, the expansion differences will force the vertical leg of the lead wire to bend. Since the bending stiffness of a thin wire is much less than the axial stiffness of the PCB or the component body, virtually all of the distortions due to thermal cycling will be reflected in the bending of the wire. Under these conditions the known bending displacement of the wire can be used to find the forces, moments, and stresses in the various lead wires and associated solder joints for through-hole or surface-mounted components [1, 2].

Two different methods can be used to find the displacements, forces, and bending moments in the lead wires. These methods are superposition and strain energy. The superposition method makes use of standard deflection equations that can be obtained from various handbooks. These standard equations are then combined in various ways to obtain the desired results. Strain energy methods make use of Castigliano's theory using partial differential equations. The superposition method will be used first.

7.3 SUPERPOSITION METHOD FOR OBTAINING LEAD WIRE DISPLACEMENTS AND MOMENTS

The component body is much stiffer than the lead wire so the wire, where it joins the component body, will be analyzed as a clamped, or fixed, end. The other end of the wire joins the PCB through a solder joint. Since the PCB is much stiffer than the

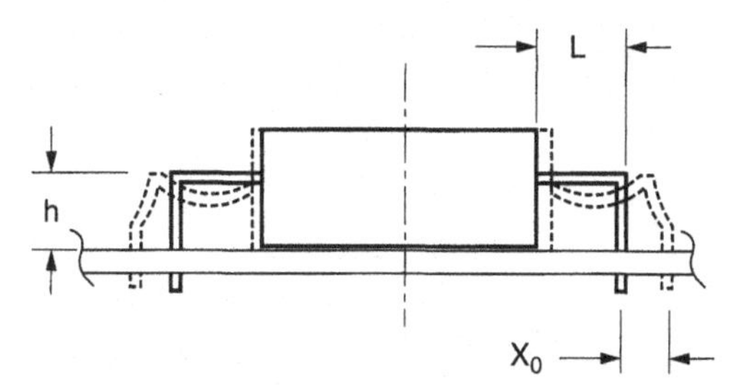

FIGURE 7.1 Differences in the TCEs of the component and the PCB will force the lead wires to bend and displace during thermal cycling conditions.

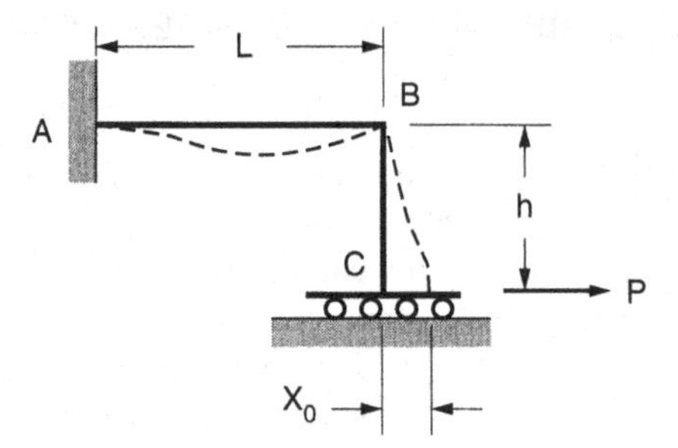

FIGURE 7.2 Structural model of the lead wire showing the bending and displacements.

wire, the wire at the solder joint will also be analyzed as a clamped end. When the PCB expands, it will force the vertical leg of the wire to bend, as shown in Fig. 7.2, but the wire will still have a zero slope at the PCB. Next, consider the horizontal leg of the wire *AB*, which is analyzed using superposition as shown in Fig. 7.3. The angular rotations of the wire at different points can be obtained as follows [2]:

$$\theta_2 = \frac{M_A L}{3EI} \tag{7.1}$$

$$\theta_4 = \frac{M_B}{6EI} \tag{7.2}$$

The slope at point *A* must be zero, so

$$\theta_2 - \theta_4 = 0; \quad \text{then} \quad \theta_2 = \theta_4 \tag{7.3}$$

$$\frac{M_A L}{3EI} = \frac{M_B L}{6EI} \quad \text{so} \quad M_A = \frac{M_B}{2} \tag{7.4}$$

At point *B*:

$$\theta_B = \theta_3 - \theta_1 = \frac{M_B L}{3EI} - \frac{M_A L}{6EI} \quad \text{but} \quad M_A = \frac{M_B}{2}$$

$$\theta_B = \frac{M_B L}{4EI} \tag{7.5}$$

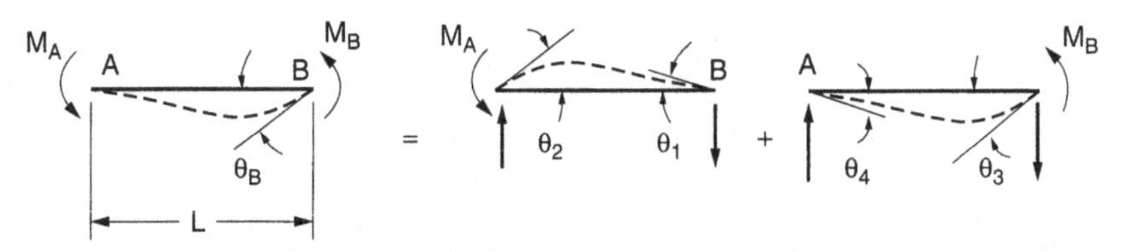

FIGURE 7.3 Superposition model for bending in the horizontal segment of the wire.

Next consider the vertical leg of the wire BC, as shown in Fig. 7.4:

$$\alpha_1 = \frac{Ph^2}{2EI} \tag{7.6}$$

$$\alpha_2 = \frac{M_B h}{EI} \tag{7.7}$$

$$\theta_B = \alpha_1 - \alpha_2 \quad \text{but} \quad \theta_B = \frac{M_B L}{4EI}$$

so

$$\frac{M_B L}{4EI} = \frac{Ph^2}{2EI} - \frac{M_B h}{EI} \quad \text{or} \quad M_B\left(h + \frac{L}{4}\right) = \frac{Ph^2}{2}$$

$$M_B = \frac{Ph^2}{2}\left(\frac{4}{4h + L}\right) = \frac{2Ph^2}{4h + L} \tag{7.8}$$

Now find the horizontal displacement X_0 for the vertical leg BC:

$$X_1 = \frac{Ph^3}{3EI} \tag{7.9}$$

$$X_2 = \frac{M_B h^2}{2EI} \tag{7.10}$$

$$X_0 = X_1 - X_2 = \frac{Ph^3}{3EI} - \frac{M_B h^2}{2EI} \quad \text{but} \quad M_B = \frac{2Ph^2}{4h + L} \tag{7.11}$$

$$X_0 = \frac{Ph^3}{3EI} - \left(\frac{2Ph^2}{4h + L}\right)\left(\frac{h^2}{2EI}\right)$$

$$X_0 = \frac{Ph^3}{EI}\left(\frac{1}{3} - \frac{h}{4h + L}\right) \quad \text{horizontal displacement} \tag{7.12}$$

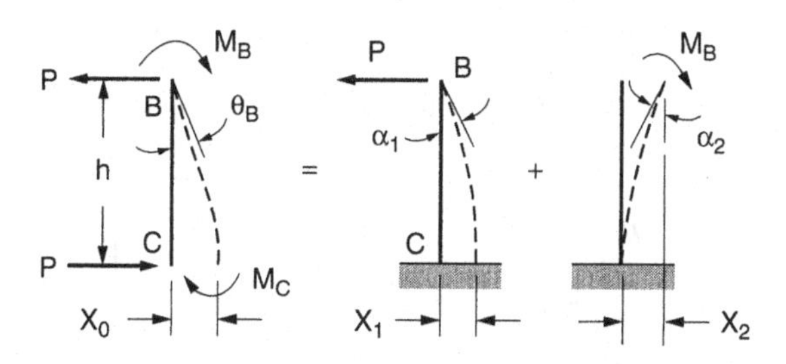

FIGURE 7.4 Superposition model for bending and displacement in the vertical segment of the wire.

Find horizontal displacement for the square frame when $h = L$; substitute into Eq. 7.12:

$$X_0 = \frac{Ph^2}{7.5EI} \tag{7.13}$$

Find the bending moment in the wire at point C from Fig. 7.4 when $h = L$. Point C is at the end of the wire in the PCB solder joint. This is the maximum bending moment in the lead wire. This is also the overturning moment that produces the shear tear-out stresses in the solder joint. To ensure a good solder joint fatigue life in thermal cycling environments, the solder shear tear-out stress should not exceed a value of about 400 psi:

$$M_B + M_C = Ph \quad \text{but} \quad M_B = \frac{2Ph^2}{4h + h} = \frac{2Ph}{5} \tag{7.14}$$

$$\frac{2Ph}{5} + M_C = Ph \quad \text{so} \quad M_C = Ph - \frac{2Ph}{5}$$

$$M_C = \frac{3Ph}{5} = 0.60Ph \quad \text{when} \quad h = L \tag{7.15}$$

Find horizontal displacement when the horizontal leg $L = 2h$; substitute into Eq. 7.12:

$$X_0 = \frac{Ph^3}{EI}\left(\frac{1}{3} - \frac{h}{4h + 2h}\right) = \frac{Ph^3}{6EI} \tag{7.16}$$

Find the bending moment in the wire at point C from Fig. 7.4 when $L = 2h$:

$$M_B + M_C = Ph \quad \text{but} \quad M_B = \frac{2Ph^2}{4h + 2h} = \frac{Ph}{3}$$

$$\frac{Ph}{3} + M_C = Ph \quad \text{so} \quad M_C = Ph - \frac{Ph}{3}$$

$$M_C = \frac{2Ph}{3} = 0.667Ph \tag{7.17}$$

Find the horizontal displacement when the vertical leg $h = 2L$; substitute into Eq. 7.12:

$$X_0 = \frac{Ph^3}{EI}\left[\frac{1}{3} - \frac{h}{4h + (h/2)}\right] = \frac{Ph^3}{EI}\left(\frac{1}{3} - \frac{2}{9}\right) = \frac{Ph^3}{9EI} \tag{7.18}$$

Find the bending moment in the wire at point C from Fig. 7.4 when $h = 2L$:

$$M_B + M_C = Ph \quad \text{but} \quad M_B = \frac{2Ph^2}{4h + (h/2)} = \frac{4Ph}{9}$$

$$\frac{4Ph}{9} + M_C = Ph \quad \text{so} \quad M_C = Ph - \frac{4Ph}{9}$$

$$M_C = \frac{5Ph}{9} = 0.556Ph \tag{7.19}$$

7.4 STRAIN ENERGY METHOD FOR OBTAINING LEAD WIRE DISPLACEMENTS AND MOMENTS

Castigliano's strain energy theorem can be used to find the displacements, moments, and forces in the lead wires. The theorem states that the partial derivative of the strain energy with respect to an applied force will give the deflection produced by that force in the direction of the force. Castigliano's theorem can also be used to find angular rotations in the lead wires. This theorem states that the partial derivative of the strain energy with respect to an applied moment will give the angular rotation produced by the moment in the direction of the moment. See Fig. 7.5 for reference points.

The total strain energy U in a system can usually be obtained by considering tension and compression, bending, torsion, and shear. Their individual properties are

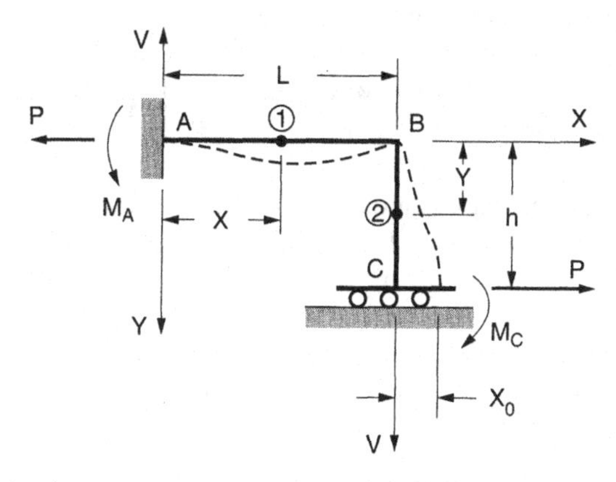

FIGURE 7.5 Strain energy model for the horizontal and vertical segments of the wire.

as follows:

Tension and compression: $$U_T = \int_a^b \frac{P^2 \, dX}{2EA} \tag{7.20}$$

Bending: $$U_B = \int_a^b \frac{M^2 \, dX}{2EI} \tag{7.21}$$

Torsion: $$U_T = \int_a^b \frac{T^2 \, dX}{2GJ} \tag{7.22}$$

Shear: $$U_S = \int_a^b \frac{V^2 \, dX}{2GA} \tag{7.23}$$

Axial displacements in the lead wires will be very small compared to the bending displacements; thus only the bending strain energy needs to be considered in the horizontal and vertical sections of the lead wires. The bending moment for the horizontal wire will be taken at point 1. The bending moment in the vertical wire will be taken at point 2. The counterclockwise bending moments will be considered as positive moments. The wire bent is statically indeterminate. This means there are more unknown forces then there are equations available to solve for the forces.

Find the bending moment equation at point 1 on the horizontal leg *AB* of the wire, shown in Fig. 7.5:

$$M_1 = M_A - VX \tag{7.24}$$

Find the bending moment equation at point 2 on the vertical leg *BC* of the wire:

$$M_2 = M_A - VL + PY \tag{7.25}$$

Use the strain energy of bending from Eq. 7.21 and take the partial derivative to find the displacement of the bending lead wire:

$$X_0 = \frac{\partial U_B}{\partial P} = \int_a^b \frac{2M(\partial M/\partial P) \, dX}{2EI} = \frac{1}{EI} \int_a^b M \frac{\partial M}{\partial P} \, dX \tag{7.26}$$

The above wire displacement equation cannot be used now because there are two unknown quantities in Eqs. 7.24 and 7.25, the bending moment M_A and the vertical force V. Therefore, two equations must be developed to find these two unknown quantities.

The first equation can be developed by knowing that the vertical displacement Y at point A, due to the vertical force V, is zero. This must be related to the strain energy in both legs of the wire for force V, as

$$Y = \frac{1}{EI_1} \int_0^L M_1 \frac{\partial M_1}{\partial V} \, dX + \frac{1}{EI_2} \int_0^h M_2 \frac{\partial M_2}{\partial V} \, dY = 0 \tag{7.27}$$

where $\dfrac{\partial M_1}{\partial V} = -X$ (Ref. Eq. 7.24)

$\dfrac{\partial M_2}{\partial V} = -L$ (Ref. Eq. 7.25)

$$Y = \frac{1}{EI_1}\int_0^L (M_A - VX)(-X)\,dX + \frac{1}{EI_2}\int_0^h (M_A - VL + PY)(-L)\,dY = 0$$

$$Y = \frac{1}{EI_1}\left(-\frac{M_A X^2}{2} + \frac{VX^3}{3}\right)_0^L + \frac{1}{EI_2}\left(-M_A LY + VL^2 Y - \frac{PLY^2}{2}\right)_0^h = 0$$

$$-\frac{M_A L^2}{2} + \frac{VL^3}{3} - M_A Lh + VL^2 h - \frac{PLh^2}{2} = 0$$

$$-M_A\left(h + \frac{L}{2}\right) + V\left(\frac{L^2}{3} + Lh\right) - \frac{Ph^2}{2} = 0 \tag{7.28}$$

This is the first of the two equations required to solve for the two unknowns.

The second required equation can be developed by knowing that the relative slope at point A, due to moment at point A, is zero. This must be related to the strain energy in both legs for the moment M_A as

$$\theta_A = \frac{1}{EI_1}\int_0^L M_1 \frac{\partial M_1}{\partial M_A}\,dX + \frac{1}{EI_2}\int_0^h M_2 \frac{\partial M_2}{\partial M_A}\,dY = 0 \tag{7.29}$$

where $\dfrac{\partial M_1}{\partial M_a} = 1$ (Ref. Eq. 7.24)

$\dfrac{\partial M_2}{\partial M_A} = 1$ (Ref. Eq. 7.25)

$$\theta_A = \frac{1}{EI_1}\int_0^L (M_A - VX)\,dX + \frac{1}{EI_2}\int_0^h (M_A - VL + PY)\,dY = 0$$

$$\theta_A = \frac{1}{EI_1}\left(M_A L - \frac{VL^2}{2}\right) + \frac{1}{EI_2}\left(M_A h - VLh + \frac{Ph^2}{2}\right) = 0$$

$$M_A(L + h) - V\left(\frac{L^2}{2} + Lh\right) + \frac{Ph^2}{2} = 0 \tag{7.30}$$

This is the second of the two equations required to solve for the two unknowns.

Equations 7.28 and 7.30 can be solved simultaneously for the two unknowns, M_A and V. To simplify the solution solve for the relations on a square frame where the wire length L is equal to the wire height h.

Substitute $L = h$ in Eq. 7.28:

$$-M_A\left(\frac{3L}{2}\right) + V\left(\frac{4L^2}{3}\right) - \frac{PL^2}{2} = 0 \tag{7.31}$$

Substitute $L = h$ in Eq. 7.30:

$$M_A(2L) - V\left(\frac{3L^2}{2}\right) + \frac{PL^2}{2} = 0 \tag{7.32}$$

Multiply Eq. 7.31 by $\frac{4}{3}$ so it will be easier to eliminate M_A:

$$-M_A(2L) + V\left(\frac{16L^2}{9}\right) - \frac{2PL^2}{3} = 0 \tag{7.33}$$

Add Eqs. 7.32 and 7.33 to eliminate M_A and solve for V:

$$VL^2(-\tfrac{3}{2} + \tfrac{16}{9}) + PL^2(\tfrac{1}{2} - \tfrac{2}{3}) = 0$$

$$V = \tfrac{3}{5}P \tag{7.34}$$

Substitute back into Eq. 7.32 to solve for M_A:

$$M_A(2L) - \left(\frac{3P}{5}\right)\left(\frac{3L^2}{2}\right) + \frac{PL^2}{2} = 0$$

$$2M_A - \frac{9}{10}PL + \frac{PL}{2} = 0$$

$$M_A = \frac{PL}{5} \tag{7.35}$$

Equation 7.26 can now be used with a slight modification to find the horizontal wire displacement X_0 based on the bending strain energy in the horizontal and vertical legs:

$$X_0 = \frac{1}{EI_1}\int_0^L M_1 \frac{\partial M_1}{\partial P}\, dX + \frac{1}{EI_2}\int_0^h M_2 \frac{\partial M_2}{\partial P}\, dY \tag{7.36}$$

Substitute Eqs. 7.34 and 7.35 into Eqs. 7.24 and 7.25:

$$M_1 = P\left(\frac{L}{5} - \frac{3X}{5}\right) \quad \text{and} \quad \frac{\partial M_1}{\partial P} = \left(\frac{L}{5} - \frac{3X}{5}\right) \tag{7.37}$$

$$M_2 = P\left(-\frac{2L}{5} + Y\right) \quad \text{and} \quad \frac{\partial M_2}{\partial P} = \left(-\frac{2L}{5} + Y\right) \tag{7.38}$$

Substitute Eqs. 7.37 and 7.38 into Eq. 7.36:

$$X_0 = \frac{1}{EI_1}\int_0^L P\left(\frac{L}{5} - \frac{3X}{5}\right)\left(\frac{L}{5} - \frac{3X}{5}\right) dX + \frac{1}{EI_2}\int_0^h P\left(\frac{-2L}{5} + Y\right)\left(\frac{-2L}{5} + Y\right) dY$$

Note that the area moments of inertia for the wire are the same for the horizontal and the vertical legs. Also note that the frame is square now so $L = h$:

$$X_0 = \frac{P}{EI}\left[\left(\frac{L^2X}{25} - \frac{6LX^2}{(25)(2)} + \frac{9X^3}{(25)(3)}\right)_0^L + \left(\frac{4L^2Y}{25} - \frac{4LY^2}{(5)(2)} + \frac{Y^3}{3}\right)_0^L\right]$$

$$X_0 = \frac{P}{EI}\left(\frac{L^3}{25} + \frac{4L^3}{25} - \frac{2L^3}{5} + \frac{L^3}{3}\right) = \frac{2PL^3}{15EI} \tag{7.39}$$

$$X_0 = \frac{PL^3}{7.5EI} \quad \text{for a square frame where } L = h$$

This agrees with Eq. 7.13, which was derived using superposition for a square wire frame where $L = h$.

7.5 EFFECTIVE LENGTH OF ELECTRICAL LEAD WIRES IN TENSION AND BENDING

Computer studies using finite element methods (FEM) of analysis on high-speed computers, combined with thermal cycling tests of different types of electronic components, have shown that lead wire lengths can be deceptive. For example, consider a transformer with lead wires extending from its bottom surface, flush mounted and soldered to a through-hole PCB. Since there is no gap between the bottom surface of the transformer and the top surface of the PCB, at first glance the length of the wire would appear to be zero. If the wire length is zero, the wire spring rate must be infinite. This is impossible. The wire spring rate cannot be infinite. Therefore, there must be some effective length to the wire. Using a combination of testing to failure and FEM analysis, it was shown that a good approximation for the effective length of a lead wire loaded in direct tension is that the wire will extend into the body of the component about two wire diameters. The effective length of the wire will also extend into the solder joint in the PCB another two wire diameters, or the thickness of the PCB (for plated through-holes) whichever is less. Lead wires loaded in bending appear to extend into the body of the component about one wire diameter, and another one wire diameter into the PCB.

Sample Problem: Thermal Expansion Displacements, Forces and Stresses in a ASIC with Axial Lead Wires for Surface Mounting and Through-Hole Applications

Determine the forces and stresses in the lead wires and solder joints of a 1.0-in.^{2} application-specific integrated circuit (ASIC) for surface-mounted and through-hole-mounted applications over a rapid temperature cycling range from -30 to $+100$°C. The forces are due to a mismatch in the TCE between the ceramic ASIC and the

0.070-in.-thick epoxy fiberglass PCB. Axial leaded kovar wires 0.020 in. diameter are located on two opposite sides of the hybrid, as shown in Fig. 7.6.

Solution Set up the continuity equation with the lead wire bending deflection. The same component with the same lead wires can be used for a surface-mounted or a through-hole-mounted application with a small adjustment to the lead wires for soldering to the PCB. In thermal cycling environments the PCB will expand more than the ceramic component body. This action will force the wires to bend. The stiffness of the wires is so much smaller than the stiffness of the PCB or the ASIC that almost all of the deflection difference will be taken up by the bending in the wire. The basic wire geometry will be the same for the surface mounting and the through-hole mounting. The displacements, forces, and, moments will be determined for the basic wire geometry first. The solder joint shear stresses can be obtained later from the solder joint geometry.

An examination of the structural elements shows that the expansion of the PCB must be equal to the diagonal expansion of the square ASIC body, plus expansion of the lead wires in the horizontal direction, plus the bending deflection of the vertical leg on the lead wire. This is shown in the following equation [5]:

$$X_P = X_C + X_H + X_W \tag{7.40}$$

where X_P = thermal expansion of the PCB in the horizontal XY plane
X_C = diagonal thermal expansion of the square ceramic ASIC in the XY plane
X_H = thermal expansion of the horizontal leg of the wire in the XY plane
X_W = bending of the vertical leg of the wire along the horizontal XY plane

The expansion differences between the PCB and the ASIC will produce a horizontal force in the system, which will force the vertical leg of the wire to bend. The expansion differences can be obtained from the TCEs of the various materials and the expected temperature change developed in the thermal cycling environment. The dwell period at the high-temperature part of the thermal cycle is important. A long dwell period at the high temperature will allow the solder joint to creep and strain

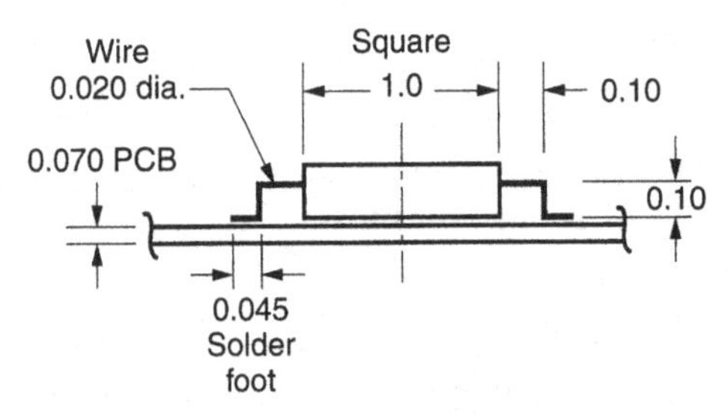

FIGURE 7.6 Ceramic ASIC with kovar lead wires is mounted on an epoxy fiberglass PCB.

relieve itself. This will change the slope of the vertical leg of the lead wire slightly, where it joins the PCB. The solder creep will reduce the magnitude of the bending moment and the bending stress in the wire very slightly, which will be ignored here with very little effect on the wire. The solder creep, however, will increase the fatigue strain and the fatigue damage in the solder joint and reduce the solder joint fatigue life. The solder joint stress levels and strain relief creep effects are very difficult to calculate. These properties are related to a rapid thermal cycle and to a slow thermal cycle. The extra damage occurs in the solder during a slow thermal cycle when high temperatures are involved. The slow thermal cycle allows more creep, which results in a reduced solder joint fatigue life, with very little effect on the lead wire forces, stresses, and fatigue life.

In a rapid thermal cycle the solder does not have a chance to strain relieve itself, so the forces and stresses are easier to calculate. In the actual operating environments the high temperatures are often maintained for a long period, perhaps for several hours as in an automobile or an airplane electronic system. For very long operating periods at high temperatures, the solder joints can strain relieve themselves down to a near zero stress condition. When this happens on both sides of the high- and low-temperature ends, the solder joint stresses can be doubled. This means that very slow thermal cycling conditions can double the stress levels in the solder joints but not in the lead wires. Doubling the stress levels in the solder joints will reduce the solder fatigue life by a factor of more than five times. Because of these conditions, solder joint stress levels should not be allowed to exceed a value of about 400 lb/in.^2, under any stress condition when a long fatigue life is required or desired.

In this analysis a rapid thermal cycle will be used; so the solder will not have a chance to creep and relieve the strain and stress to a near zero condition. Test data shows there is a wide scatter in the fatigue life of virtually all load-carrying structural elements. Therefore, it is a good practice to use safety factors, sometimes called factors of ignorance, in any analysis to ensure the desired fatigue life.

The bending displacement in the vertical leg of the wire can be obtained from Eq. 7.40, with some modifications as follows:

$$X_W = \alpha_P L_P \,\Delta t - \alpha_C L_C \,\Delta t - \alpha_H L_H \,\Delta t \tag{7.41}$$

where $\alpha_P = 15 \times 10^{-6}$ in./in./°C, TCE of the epoxy fiberglass PCB in the XY plane

$L_P =$ half the diagonal ASIC + wire $= \sqrt{2.0}/2 + 0.071 = 0.778$ in.

$\Delta t = 100 - (-30) = 130$°C full temperature range

$\Delta t = \frac{130}{2} = 65$°C neutral to peak for a rapid temperature cycle; no solder creep

$\alpha_C = 6 \times 10^{-6}$ in./in./°C, TCE of ceramic ASIC component

$L_C = \sqrt{2.0}/2 = 0.707$ in., half the diagonal length of the ASIC

$\alpha_H = 6 \times 10^{-6}$ in./in./°C, TCE of horizontal section of kovar wire

$L_H = 0.10 \cos 45° = 0.071$ in., projected horizontal wire length

Substitute into Eq. 7.41 to find the expected bending deflection of the vertical wire leg:

$$X_W = (15 \times 10^{-6})(0.778)(65) - (6 \times 10^{-6})(0.707)(65) - (6 \times 10^{-6})(0.071)(65)$$

$$X_W = 7.59 \times 10^{-4} - 2.76 \times 10^{-4} - 2.77 \times 10^{-5} = 4.55 \times 10^{-4} \text{ in. displacement} \tag{7.42}$$

Equations 7.13 and 7.39 show the bending displacement of the vertical leg of a square bent frame where the horizontal and vertical wire legs are equal, due to a horizontal load P acting on the wire. Since the horizontal displacement of the wire is now known, the horizontal force can be obtained:

$$P = \frac{7.5 E_W I_W X_W}{(L_W)^3} \tag{7.43}$$

where $E_W = 20 \times 10^6 \text{ lb/in.}^2$, kovar wire modulus of elasticity

$I_W = \frac{\pi d^4}{64} = \frac{\pi (0.02)^4}{64} = 7.85 \times 10^{-9} \text{ in.}^4$, wire area moment of inertia

$L_W = 0.10 + 0.02$ wire dia. $= 0.12$ in., effective length of wire

$X_W = 4.83 \times 10^{-4}$ in., wire bending displacement (Ref. Eq. 7.42)

$$P = \frac{(7.5)(20 \times 10^6)(7.85 \times 10^{-9})(4.55 \times 10^{-4})}{(0.12)^3} = 0.310 \text{ lb, wire force} \tag{7.44}$$

The bending moment in the wire is maximum at the PCB solder joint M_C:

$$M_C = 0.60 \text{ Ph} \quad \text{(Ref. Eq. 7.15)}$$

where $P = 0.310$ lb (Ref. Eq. 7.44)

$h = L = 0.12$ in., effective length of wire

$$M_C = (0.60)(0.310)(0.12) = 0.0223 \text{ lb} \cdot \text{in.} \tag{7.45}$$

7.6 THERMAL CYCLING BENDING STRESS IN THE ELECTRICAL LEAD WIRE

The bending stress in the lead wire can be obtained from the standard bending stress equation. A stress concentration factor k must always be considered when fatigue damage is expected at holes, notches, and deep scratches, along with many thousand stress cycles. A common stress concentration factor for lead wires is about 2.0. This value has already been factored into the lead wire fatigue curve shown in Fig. 5.2, which results in a fatigue exponent $b = 6.4$, which will be used here. The stress concentration k can be used in the bending stress equation, or in the slope of the fatigue curve, but not in both places. *The stress concentration factor must be used only once* [20]:

$$S_W = \frac{kM_C c}{I_W} = \frac{(1.0)(0.0223)(0.010)}{7.85 \times 10^{-9}} = 28{,}408 \text{ lb/in.}^2 \tag{7.46}$$

7.7 APPROXIMATE WIRE THERMAL CYCLING FATIGUE LIFE

The approximate fatigue life expected for the lead wire for the thermal cycling condition can be obtained from Eq. 5.1 along with Fig. 5.2, which shows the lead wire fatigue curve with a stress concentration $k = 2.0$ that gives a fatigue exponent b a value of 6.4. The kovar wire will also have a slope of 6.4, but the stress value at point 1 on the S–N fatigue curve will be 14,000 psi, and the stress value at point 2 on the S–N fatigue curve will be 84,000 psi [21]:

$$N_2 = N_1 \left(\frac{S_1}{S_2} \right)^b \quad \text{(Ref. Eq. 5.1)}$$

where $N_1 = 10^8$ wire cycles to fail at 14,000 lb/in.^{2}
$S_1 = 14{,}000$ lb/in.^{2}, wire stress at 10^8 cycles with $k = 2$
$S_2 = 28{,}408$ lb/in.^{2}, wire bending stress (Ref. Eq. 7.46)
$b = 6.4$ slope of wire fatigue curve

$$N_2 = (10^8) \left(\frac{14{,}000}{28{,}408} \right)^{6.4} = 1.08 \times 10^6 \text{ wire cycles to fail} \tag{7.47}$$

7.8 SOLDER JOINT THERMAL CYCLE STRESS AND APPROXIMATE FATIGUE LIFE

The ASIC component can be soldered to through-hole PCBs and to surface-mounted PCBs, with just a slight modification in the termination of the lead wire. Consider the through-hole mounting method first. The shear tear-out stress in the through-hole solder joint is based on the average solder joint diameter in the following equation:

$$S_{ST} = \frac{M}{hA} \quad \text{(Ref. Eq. 2.15)}$$

where $M = 0.0223$ lb · in., overturning moment (Ref. Eq. 7.45)
$h = 0.070$ in., solder joint height, assumed equal to PCB thickness
$d = \frac{0.020 + 0.032}{2} = 0.026$ in. diameter average solder joint diameter

$$A = \frac{\pi d^2}{4} = \frac{\pi (0.026)^2}{4} = 0.000531 \text{ in.}^2, \text{ average solder joint area}$$

$$S_{ST} = \frac{0.0223}{(0.070)(0.000531)} = 600 \text{ lb/in.}^2 \text{ solder shear stress} \tag{7.48}$$

The estimated solder joint fatigue life for a rapid thermal cycle, with no solder creep, can be obtained from the solder fatigue curve in Fig. 5.3 as follows:

$$N_2 = N_1 \left(\frac{S_1}{S_2}\right)^b = (80{,}000)\left(\frac{200}{600}\right)^{2.5} = 5132 \text{ solder cycles to fail} \tag{7.49}$$

Next consider only the average solder joint shear stress for the ASIC lead wires when they are surface mounted on the PCB. The solder footprint for the wire is expected to be about 0.020 × 0.045 in. The solder shear stress can be obtained from the standard shear stress equation:

$$S_S = \frac{P}{A_S} \tag{7.50}$$

where $P = 0.310$ lb, force in wire (Ref. Eq. 7.44)
$A_S = (0.020)(0.045) = 0.00090 \text{ in.}^2$, solder area

$$S_S = \frac{0.310}{0.00090} = 344 \text{ lb/in.}^2, \text{ solder shear stress} \tag{7.51}$$

The approximate solder joint fatigue life for a rapid thermal cycle, with no solder creep, can be obtained from the solder fatigue curve in Fig. 5.3:

$$N_2 = N_1 \left(\frac{S_1}{S_2}\right)^b = (80{,}000)\left(\frac{200}{344}\right)^{2.5} = 20{,}619 \text{ solder cycles to fail} \tag{7.52}$$

CHAPTER 8

Designing Electronic Equipment for Sinusoidal Vibration

8.1 BASIC FAILURE MODES IN SINUSOIDAL VIBRATION

Sinusoidal vibrations can produce very high acceleration G levels in lightly damped structures, when the natural frequency is excited. Transmissibility Q values can be greatly magnified, resulting in very high displacements, forces, accelerations, and stresses, which often result in electrical malfunctions and failures. High displacements often result in impacting between adjacent structural members such as circuit boards, resulting in cracked components, cracked solder joints, broken electrical lead wires, and broken connector pins. High forces can produce high stresses in load-carrying elements such as screws, rivets, and ribs, which may become loose or may fracture. High accelerations can cause relays to chatter, crystal oscillators to malfunction, and potentiometers to lose their calibration accuracy. High stresses typically result in very rapid fatigue failures in various electronic elements from aluminum housings to cables and harnesses [22, 23].

Failures that occur during vibration are often difficult to understand or to analyze. Sometimes the failures can be examined and evaluated using finite element methods (FEM) with a high-speed computer. As powerful as these computers may be, they often do not provide answers that can solve the problem. Sometimes it becomes necessary to test a real unit or to build a test model that can be vibration tested using a strobe light with sinusoidal vibration to examine the structural motion during its resonant condition. This type of examination is always dramatic, and it always reveals important insights related to the dynamic response characteristics of the structure [1, 2].

Vibration failures are usually very difficult to trace because there are so many different possible problem sources. This includes the design, manufacturing, assembly, trouble shooting, repair, burn-in, process change, stress screening with vibration and thermal cycling, handling, dropping, shipping, unauthorized attempts to repair, lying about how the product was used, and receiving inaccurate information. Many people have short memories, are poor observers, or often give information that is completely wrong, but they will insist that it is correct. This can cause extensive delays in solving a problem or the problem may never be solved.

A typical example involves the difficulty in determining the source of low-level vibration screening failures in the steps of a bearing shaft for a spinning gyro. A normal stress concentration can be expected at a change in the cross section of a load-carrying beam. However, this was included in the dynamic vibration stress analysis of the shaft, and the calculated stress levels were not high enough to produce the observed fatigue failures. The critical shaft only had a 0.062-in. diameter, and the shaft was being cut on a lathe. If only one cutter was being used to machine the shaft as a cantilever beam, the cutter might impose a bending moment high enough to produce minute cracks at the step of the shaft during the machining operation. The small initial crack would then propagate rapidly during vibration, until a complete fracture developed. When the machine shop foreman was asked if a single cutter was used, he became very angry and insisted that he knew more about fatigue failures than most engineers working in this area. He also insisted that all of his machinists were carefully instructed to use three cutters spaced 120° apart to avoid such a problem. The shop foreman went on to say that the design engineers should not blame their poor design practices on poor manufacturing practices. The investigation turned away from the manufacturing area for a while, but no other area seemed to offer an explanation for the failures. The manufacturing area was visited once again, but instead of talking to the foreman a visit was made directly to the machining section. A quick glance showed that a single cutter was being used to machine the critical shaft section. When the shop foreman was called over to verify that a single cutter was being used, his face turned white. He asked the machinist why he was using a single cutter when he was instructed to use three cutters. The machinist answered that it takes too much time to use three cutters so he cannot make his daily quota; but when he uses only one cutter the part turns out just as good and he saves a lot of time.

Obviously, the shop foreman never bothered to check back to see if his instructions were being followed. The shop foreman therefore passed on information that he thought was correct, but it was not. The gyros were being used on many different programs that were scheduled to continue for several more years. The source of the shaft failures could have continued for many years with more possible field failures. The message here is that many people will give very positive answers to questions, even when they really do not know all of the details. It is often very difficult to separate good information from bad information, especially in the investigation of a failure. To be a good investigator, you often have to see as much as you can personally, before any questions are asked.

People will often lie to save their jobs where failures are involved. An example of this involved a large 90-lb cast aluminum electronic enclosure with quarter inch thick walls. One of the walls had a V-shaped crack in the casting that was indented about half an inch, where each leg of the V was about 2 in. long. A team of engineers investigating the crack wanted to know if random vibration could cause this type of crack. The investigators were informed that vibration could not possibly produce such a crack, and that the cast aluminum enclosure was obviously dropped from a height of about 5 ft. The investigators said that they had positive proof that the casting was not dropped. Their proof was that they asked everyone associated with the cast aluminum enclosure if it was dropped, and everyone said no. When there are no witnesses, or when no one wants to see a friend get fired, no one will ever admit to dropping a system that may cost over $150,000. Probably the most disturbing part of the entire situation was that several graduate mechanical engineers could not recognize the obvious difference between a vibration fatigue fracture and a deep local fracture in a thick cast aluminum wall caused by a severe concentrated blow.

8.2 BEAM STRUCTURES WITH NONUNIFORM CROSS SECTIONS

Nonuniform structures are common in electronic assemblies since shapes often have to be changed to allow them to fit into cramped spaces in automobile instrument panels, near engines, in wing tips, and tail sections of aircraft. A lot of analysis work could be saved if some type of an average area moment of inertia was available that was easy to use and produced good results without too much error.

Consider a cantilever beam with two different cross sections that form a step as shown in Fig. 8.1. Castigliano's strain energy equation can be used to find the deflection X_0 produced at the end of the beam by the concentrated load as

$$X_0 = \frac{1}{EI_1}\int_0^a M_1 \frac{\partial M_1}{\partial P} dX_1 + \frac{1}{EI_2}\int_0^b M_2 \frac{\partial M_2}{\partial P} dX_2 \tag{8.1}$$

When both legs of the beam have equal lengths that are half the full length of the beam, the deflection at the end of the beam will have the value shown below [1]:

$$X_0 = \frac{PL^3}{24}\left(\frac{1}{EI_1} + \frac{7}{EI_2}\right) \tag{8.2}$$

The deflection of a *uniform* cantilever beam with a concentrated end load can be found in a structural handbook as shown below. An average moment of inertia was shown since this will be used to find an average moment of inertia for the stepped beam:

$$X_0 = \frac{PL^3}{3EI_{AV}} \tag{8.3}$$

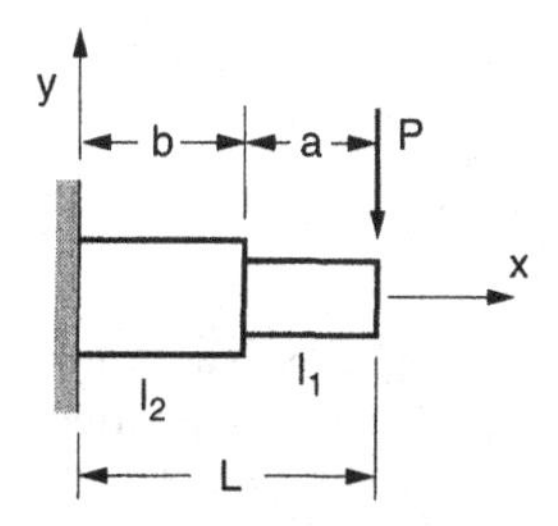

FIGURE 8.1 Two segment cantilever beam with two different lengths and moments of inertia.

Set Eq. 8.2 equal to Eq. 8.3 and solve for the average moment of inertia of the beam:

$$I_{AV} = \frac{8I_1 I_2}{7I_1 + I_2} \tag{8.4}$$

The average moment of inertia for the stepped beam can be obtained for a case where

$$I_1 = 3 \text{ in.}^4 \quad \text{and} \quad I_2 = 6 \text{ in.}^4 \tag{8.5}$$

Substitute Eq. 8.5 into Eq. 8.4 for the correct average moment of inertia:

$$I_{AV} = \frac{8(3)(6)}{7(3) + 6} = 5.33 \text{ in.}^4 \tag{8.6}$$

Now examine some other methods for finding the average moment of inertia without going through so much work. Consider the averaging method shown here:

$$I'_{AV} = \frac{aI_1 + bI_2}{a + b} \tag{8.7}$$

Substitute Eq. 8.5 into Eq. 8.7 to find the average moment of inertia:

$$I'_{AV} = \frac{(L/2)(3) + (L/2)(6)}{L/2 + L/2} = 4.5 \text{ in.}^4 \tag{8.8}$$

Comparing the approximate average moment of inertia in Eq. 8.8 with the correct value for the average moment of inertia shown in Eq. 8.6 shows that the approximate average moment of inertia is about 15.5% lower than the correct average moment of inertia. Since the natural frequency is related to the square root of the moment of inertia, the natural frequency will be about 8.1% too low. This results in a slight safety factor that is really desirable for designing electronic equipment for reliable vibration operation.

Sample Problem: Natural Frequency of a Chassis with Nonuniform Sections

Find the natural frequency of an electronic chassis with end supports where the length-to-height ratio permits the chassis to be evaluated as a beam simply supported at each end. The chassis has a total weight of 45.5 lb with four different length

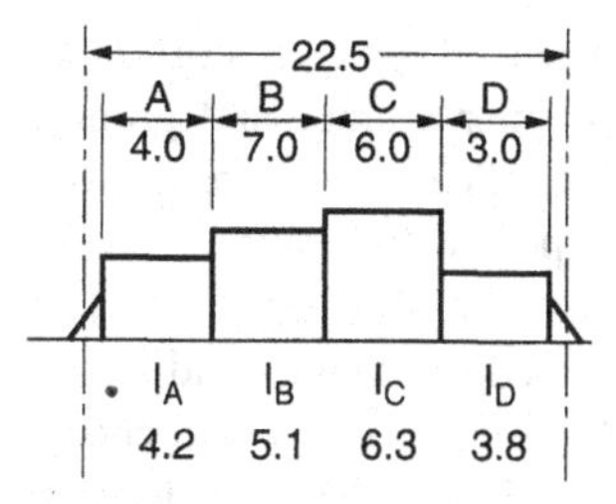

FIGURE 8.2 Simply supported chassis with four different lengths and moments of inertia.

segments and four different moments of inertia, as shown in Fig. 8.2. The chassis is restrained to vibrate along the vertical axis only:

$$I_A = 4.2 \text{ in.}^4 \qquad I_B = 5.1 \text{ in.}^4 \qquad I_C = 6.3 \text{ in.}^4 \qquad I_D = 3.8 \text{ in.}^4$$
$$A = 4.0 \text{ in.} \qquad B = 7.0 \text{ in.} \qquad C = 6.0 \text{ in.} \qquad D = 3.0 \text{ in.}$$

$$I'_{\text{AV}} = \frac{AI_A + BI_B + CI_C + DI_D}{A + B + C + D} \tag{8.9}$$

$$I'_{\text{AV}} = \frac{(4)(4.2) + (7)(5.1) + (6)(6.3) + (3)(3.8)}{4 + 7 + 6 + 3} = 5.08 \text{ in.}^4 \tag{8.10}$$

Solution Use the uniform beam equation to find the natural frequency:

$$f_n = \frac{\pi}{2}\sqrt{\frac{EIg}{WL^3}} \quad \text{(Ref. Eq. 3.29)}$$

where $E = 10.5 \times 10^6$ lb/in.^{2}, aluminum modulus of elasticity
$I = 5.08$ in.^{4}, average area moment of inertia
$g = 386$ in./s^2, acceleration of gravity
$W = 45.5$ lb, total chassis weight
$L = 22.5$ in., chassis length

$$f_n = \frac{\pi}{2}\sqrt{\frac{(10.5 \times 10^6)(5.08)(386)}{(45.5)(22.5)^3}} = 313 \text{ Hz} \tag{8.11}$$

8.3 COMPOSITE LAMINATED BUILT-UP STRUCTURES

Laminated structures with metals and plastics are used extensively in electronic systems to take advantage of their special mechanical and physical properties. Printed circuit boards (PCBs) are probably the most common composite laminated structures used. They are usually made with materials such as epoxy fiberglass or polyimide glass laminated to copper circuit traces and copper voltage and ground planes. Aluminum heat sinks may be laminated to the PCBs to improve the conduction heat transfer when the heat dissipation is expected to result in excessively high component temperatures. Electronic enclosures also make use of different combinations of various plastics and metals to reduce weight and costs. Two common materials often used in electronic structures are aluminum and epoxy fiberglass. These materials can be combined in many different ways to achieve special design features [1].

Combinations of epoxy impregnated with carbon graphite have become very popular for many different items such as tennis racquets, golf clubs, and other athletic equipment because of the light weight and high stiffness. These materials are being used extensively in virtually all of the newest military airplanes, tanks, and ships. Fuselage panels, wing panels, and tail sections make extensive use of these new composites.

Sample Problem: Natural Frequency of a Composite Laminated Beam

Find the natural frequency of the composite laminated uniform cantilever beam shown in Fig. 8.3, which is restricted to vibrate only in the vertical direction.

Solution Convert epoxy fiberglass section to an equivalent aluminum section. Since the laminations of the epoxy fiberglass and the aluminum sections are side by side, an equivalent uniform beam of one material, aluminum, can be utilized to find the natural frequency. The width of the epoxy fiberglass section can be reduced to make it equivalent to the *EI* bending stiffness of an aluminum section as shown below. Subscripts *A* and *E* refer to the aluminum and epoxy fiberglass respectively.

$$E_A I_A = E_E I_E$$
$$\frac{E_A b_A h_A^3}{12} = \frac{E_E b_E h_E^3}{12} \tag{8.12}$$

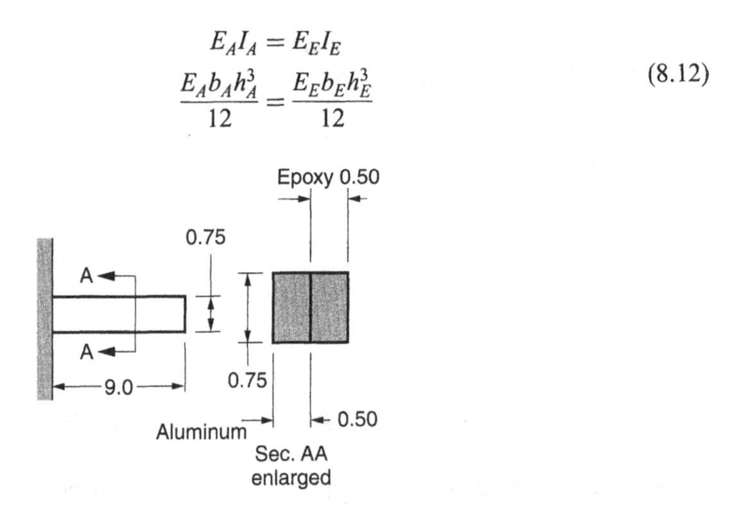

FIGURE 8.3 Composite cantilever beam with sections of aluminum bonded to epoxy fibreglass.

Since the heights h of the aluminum section and the epoxy fiberglass sections are the same, they cancel. The width b_A of the aluminum section that will give the same bending stiffness as the epoxy fiberglass section can be obtained as follows:

$$b_A = b_E \frac{E_E}{E_A} \tag{8.13}$$

where $b_E = 0.50$ in., width of epoxy fiberglass section
$E_E = 2.0 \times 10^6$ lb/in.^{2}, epoxy fiberglass modulus of elasticity
$E_A = 10.5 \times 10^6$ lb/in.^{2}, aluminum modulus of elasticity

Thus

$$b_A = (0.50)\left(\frac{2 \times 10^6}{10.5 \times 10^6}\right) = 0.0952 \text{ in.} \tag{8.14}$$

The equivalent width of the solid aluminum beam can be determined by adding the increment shown above to the original aluminum section as follows:

$$\mathrm{b_{EQ}} = 0.50 + 0.0952 = 0.5952 \text{ in.} \tag{8.15}$$

The natural frequency of the uniform solid aluminum beam can be obtained as follows:

$$f_n = \frac{3.52}{2\pi}\sqrt{\frac{EIg}{WL^3}} \tag{8.16}$$

where $E = 10.5 \times 10^6$ lb/in.^{2}, aluminum modulus of elasticity
$I = \dfrac{b_{\mathrm{EQ}}h^3}{12} = \dfrac{(0.5952)(0.75)^3}{12} = 0.0209$ in.^{4}
$g = 386$ in./s^2, acceleration of gravity
$L = 9.0$ in. length
$W_A = (9.0)(0.50)(0.75)(0.10 \text{ lb/in.}^3) = 0.338$ lb aluminum original
$W_E = (9.0)(0.50)(0.75)(0.065) \text{ lb/in.}^3) = 0.219$ lb epoxy fiberglass original
Original total weight $= 0.557$ lb weight

$$f_n = \frac{3.52}{2\pi}\sqrt{\frac{(10.5 \times 10^6)(0.0209)(386)}{(0.557)(9.0)^3}} = 256 \text{ Hz} \tag{8.17}$$

8.4 GEOMETRIC STRESS CONCENTRATION FACTORS

Anytime there is a change in the direction of the load path in a structural element, there will be an increase in the magnitude of the stress level. The areas where these higher stresses tend to accumulate are called stress concentration areas. They usually occur in areas where there are discontinuities, such as local impurities in metals or plastics, or in the geometric shape of the load-carrying element. They may be called geometric stress concentrations or stress risers. The geometric shapes that cause most of the problems are small holes, sharp notches, deep scratches, and sharp changes in the cross sections. Some typical geometric stress concentrations are shown in Fig. 8.4. These discontinuities, or stress risers, are most severe when they are exposed to vibration environments, where they experience thousands of stress reversals. With continued vibration exposure small cracks can develop. These cracks will grow slowly with each stress reversal, until there is a large change in the cross section, which then results in a sudden and complete fracture.

Highly ductile materials do not appear to have the same sensitivity to high static loads or to repeated alternating loads as low ductility materials. Geometric stress concentrations such as notches and small holes seem to have very little effect on ductile materials subjected to high static loads. This seems to be due to plastic creep and local yielding that occurs in the high stress areas, which reduces the magnitude of the maximum stress without causing any failures.

Alternating axial loads on notched test specimens appear to generate higher stress concentration factors than alternating bending loads. Stress concentration factors for notched flat plates and grooved shafts are shown in Figs. 8.5 through 8.8 [20].

Sample Problem: Fatigue Life of a Grooved Shaft in Bending

A gyro and bearing assembly with a weight of 0.50 lb is mounted on a small cantilevered shaft fabricated from 301 series stainless steel 3/4 hard, as shown in Fig. 8.9. The gyro shaft has a cut groove with a radius of 0.003 in. for a snap ring, which is required to retain the bearing. The system must operate in a 2.0-G peak sine vibration environment where the forcing frequency is expected to be close to the

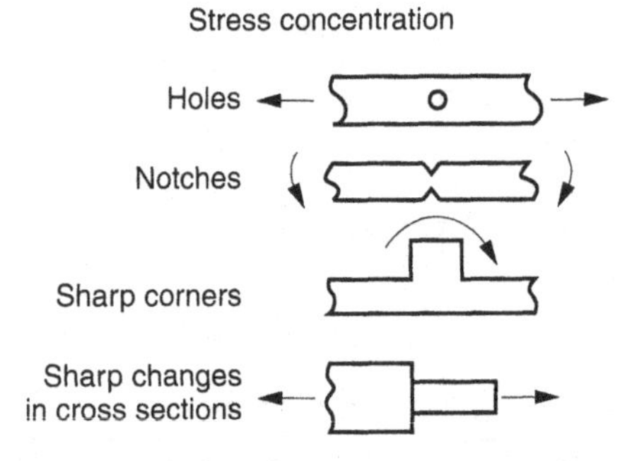

FIGURE 8.4 Different types of geometric stress concentrations in structural members.

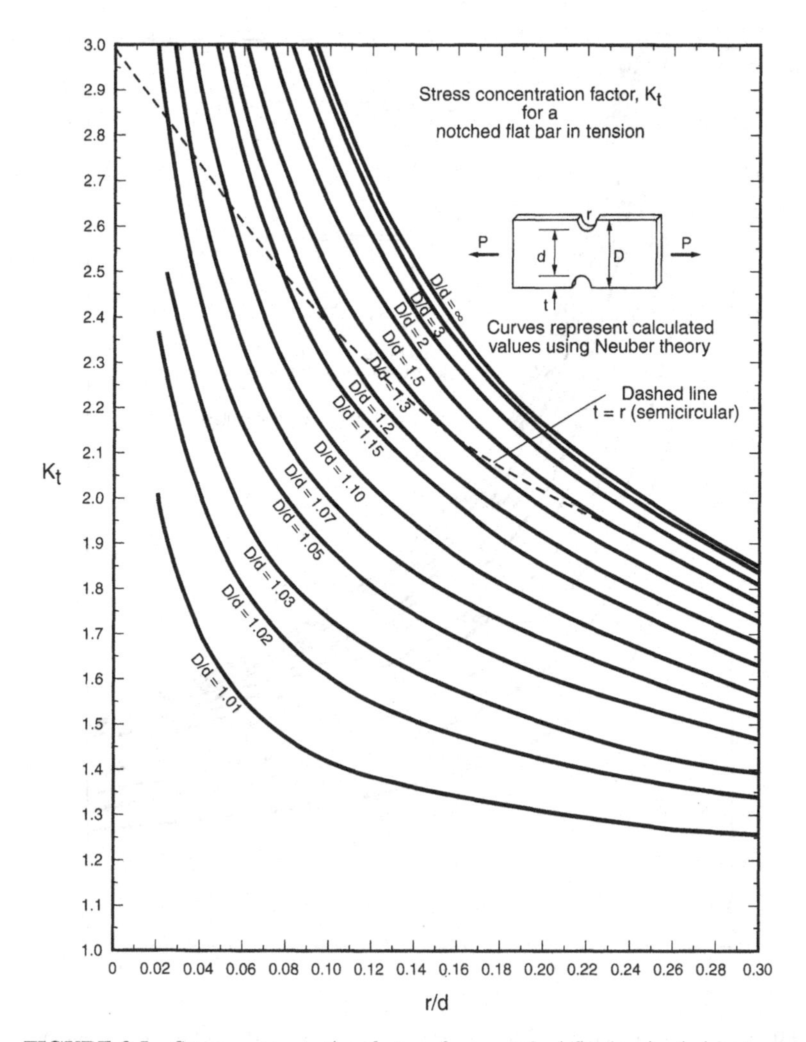

FIGURE 8.5 Stress concentration factors for a notched flat bar loaded in tension.

natural frequency of the system. Can this assembly be hard mounted or must it be isolated?

Solution Find the natural frequency of the gyro assembly, the transmissibility Q, the dynamic force acting on the gyro shaft, the stress concentration factor generated by the snap ring groove in the shaft, the dynamic bending moment at the groove, and the dynamic bending moment at the base of the shaft. Select the larger dynamic moment and calculate the dynamic bending stress. Calculate the b fatigue exponent slope for the 301 stainless steel shaft, then calculate the fatigue life for the most critical bending moment.

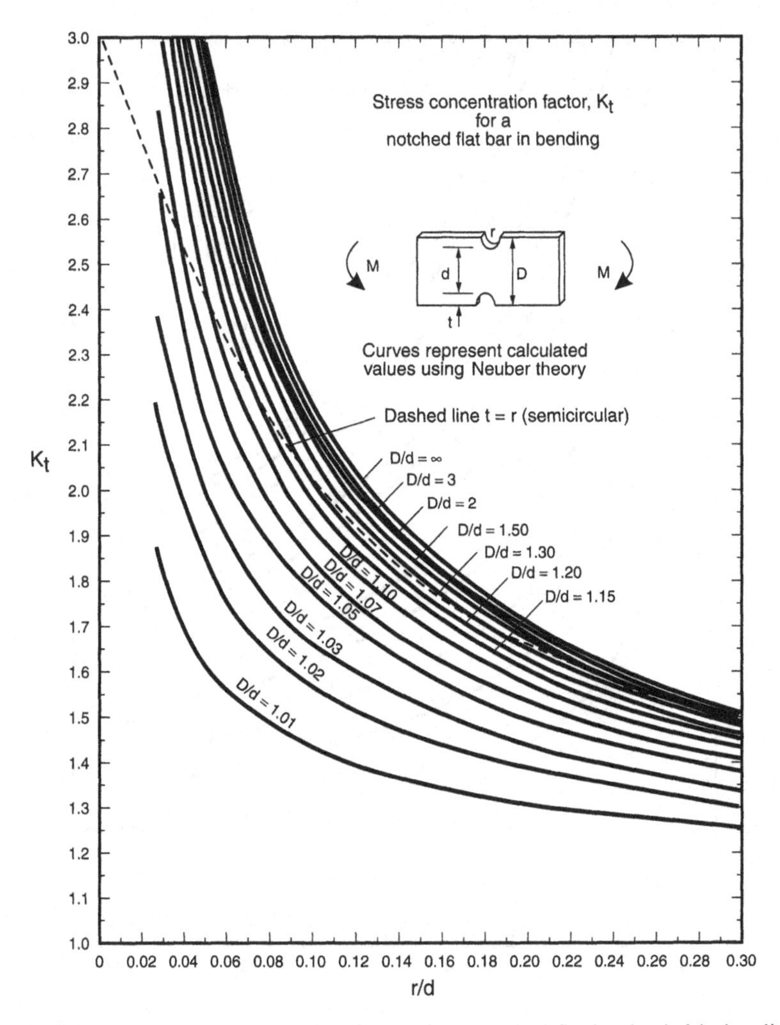

FIGURE 8.6 Stress concentration factors for a notched flat bar loaded in bending.

A. Natural Frequency of Gyro Assembly as a Cantilever Beam with an End Load. Where

$E = 29 \times 10^6$ lb/in.^{2}, stainless modulus of elasticity

$d_S = 0.180$ in., outer diameter of shaft

$d_N = 0.170$ in., diameter of shaft at notch for snap ring

$$I_S = \frac{\pi(d_S)^4}{64} = \frac{\pi(0.180)^4}{64} = 5.15 \times 10^{-5} \text{ in.}^4\text{, outer shaft moment of inertia}$$

$$I_N = \frac{\pi(d_N)^4}{64} = \frac{\pi(0.170)^4}{64} = 4.10 \times 10^{-5} \text{ in.}^4\text{, notch moment of inertia}$$

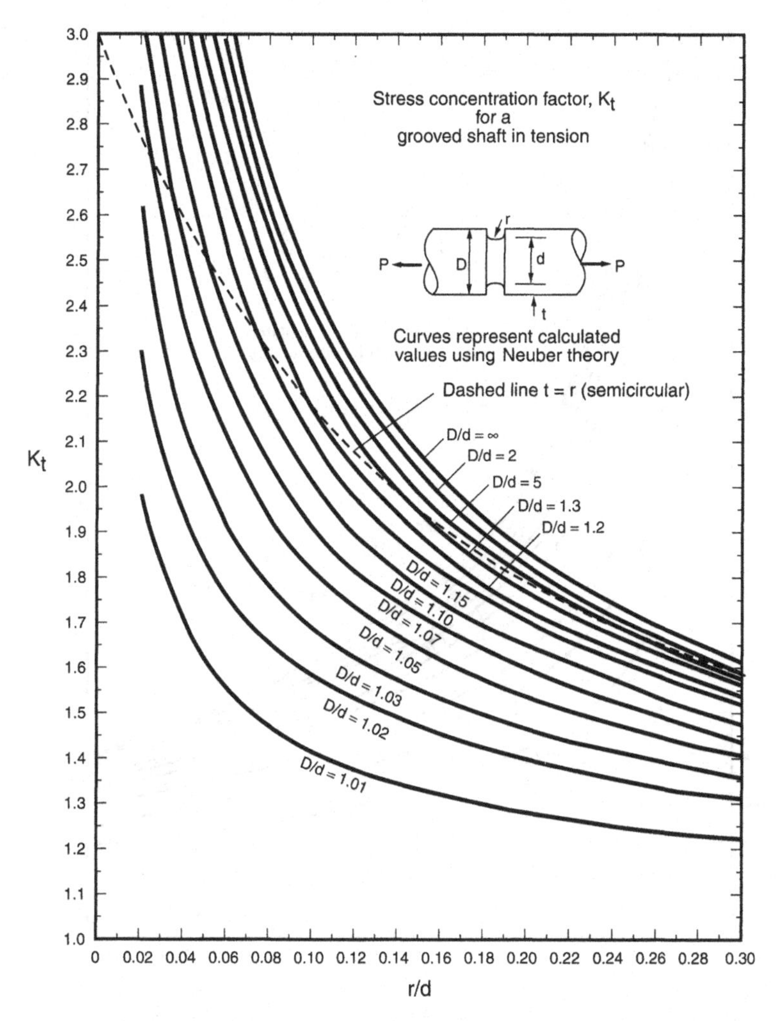

FIGURE 8.7 Stress concentration factors for a grooved shaft loaded in tension.

$g = 386$ in./s^2, acceleration of gravity
$W = 0.50$ lb, weight of gyro bearing assembly
$L_S = 0.90 + 0.50 = 1.40$ in., effective length of gyro shaft
$L_N = 0.50$ in., effective length of the gyro shaft at the snap ring notch

$$Y_{\text{ST}} = \frac{W(L_S)^3}{3EI_S} = \frac{(0.50)(1.40)^3}{(3)(29 \times 10^6)(5.15 \times 10^{-5})} = 3.062 \times 10^{-4} \text{ in static deflection} \tag{8.18}$$

$$f_n = \frac{1}{2\pi}\sqrt{\frac{g}{Y_{\text{ST}}}} = \frac{1}{2\pi}\sqrt{\frac{386}{3.062 \times 10^{-4}}} = 179 \text{ Hz natural frequency} \tag{8.19}$$

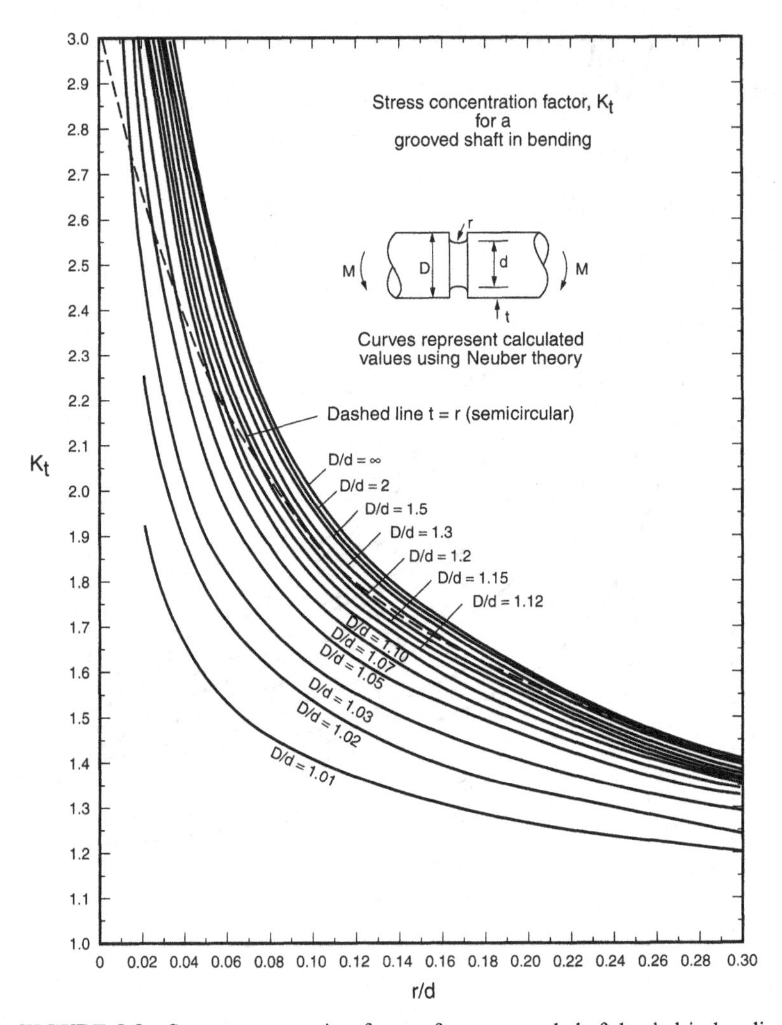

FIGURE 8.8 Stress concentration factors for a grooved shaft loaded in bending.

B. Approximate Transmissibility Q at the Natural Frequency for a Beam

$$Q = 2\sqrt{f_n} = 2\sqrt{179} = 26.7 \text{ dimensionless} \tag{8.20}$$

C. Dynamic Force Acting on the Gyro Shaft for a 2-G Peak Sine Input

$$P_d = WG_{\text{in}}Q = (0.50)(2.0)(26.7) = 26.7 \text{ lb} \tag{8.21}$$

D. Stress Concentration Factor k at Snap Ring Notch in Shaft from Fig. 8.8

$$\frac{r}{d} = \frac{0.003}{0.170} = 0.0176 \quad \text{and} \quad \frac{D}{d} = \frac{0.180}{0.170} = 1.059 \quad \text{so} \quad k = 2.85 \tag{8.22}$$

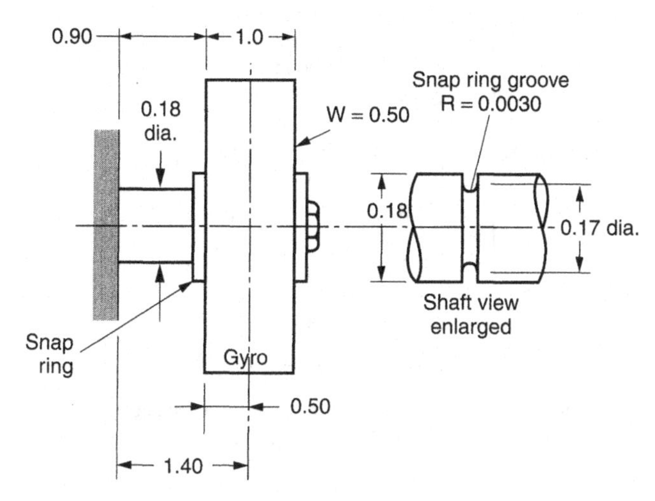

FIGURE 8.9 Gyro mounted on a cantilever shaft is retained by a snap ring in a groove.

E. Dynamic Bending Moment Acting at the Snap Ring Notch in the Shaft

$$M_N = P_d L_N = (26.7)(0.50) = 13.35 \text{ lb} \cdot \text{in.} \tag{8.23}$$

F. Dynamic Bending Stress Acting at the Snap Ring Notch in the Shaft
The stress concentration factor k must be considered in the stress evaluation. The stress concentration factor can be used in the bending stress equation or in the slope b of the fatigue curve in the S–N plot of the material fatigue life. Either place is acceptable, but it *must be used only once*, so it will be used only in the bending stress equation:

$$S_N = \frac{kM_N c_N}{I_N} = \frac{(2.85)(13.35)(0.170/2)}{4.10 \times 10^{-5}} = 78{,}879 \text{ lb/in.}^2 \tag{8.24}$$

G. Dynamic Bending Moment Acting at the Support Base of the Shaft

$$M_S = P_d L_S = (26.7)(1.40) = 37.38 \text{ lb} \cdot \text{in.} \tag{8.25}$$

H. Dynamic Bending Stress at the Support Base of the Smooth Shaft

$$S_S = \frac{M_S c_S}{I_S} = \frac{(37.38)(0.180/2)}{5.15 \times 10^{-5}} = 65{,}324 \text{ lb/in.}^2 \tag{8.26}$$

The highest shaft stress occurs at the snap ring notch in Eq. 8.24 because of the high stress concentration. This stress level will therefore be used to determine the minimum fatigue life of the shaft. This requires the slope of the fatigue curve b for the steel shaft, which can be obtained using the material properties shown in Fig. 8.10.

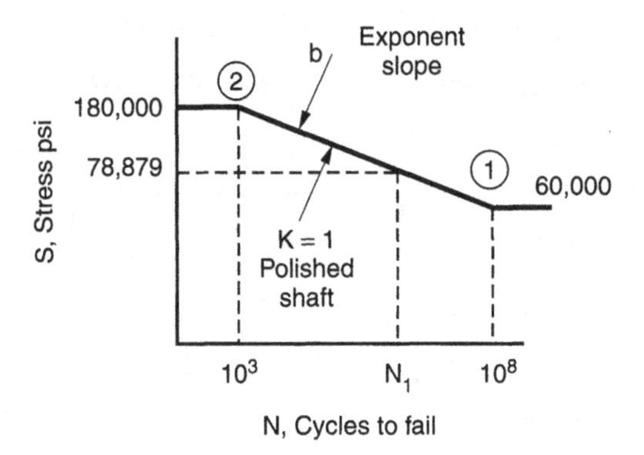

FIGURE 8.10 Typical log–log *S–N* fatigue life curve for a polished 301 stainless steel shaft 3/4 hard. with no stress concentration ($k = 1.0$), which results in a b exponent slope of 10.48.

The stress concentration factor k must be considered, but it can only be used once. Since it was previously used in the bending stress (Eq. 8.24), it must not be used here to find slope b:

$$\frac{N_1}{N_2} = \left(\frac{S_2}{S_1}\right)^b \quad \text{(Ref. Eq. 5.5)}$$

where $N_1 = 10^8$ cycles to fail at 60,000 lb/in.^{2} reference point 301 stainless 3/4 hard

$N_2 = 10^3$ cycles to fail at 180,000 lb/in.^{2} reference point 301 stainless steel

$S_1 = 60{,}000$ lb/in.^{2} reference point

$S_2 = 180{,}000$ lb/in.^{2} reference point

$$\frac{10^8}{10^3} = \left(\frac{180{,}000}{60{,}000}\right)^b \quad \text{so} \quad 10^5 = (3)^b \quad \text{then} \quad b = \frac{\log_{10} 10^5}{\log_{10} 3} = \frac{5}{0.477} = 10.48 \tag{8.27}$$

The fatigue life of the shaft will be determined by the stress at the snap ring notch:

$$N_1 = N_2\left(\frac{S_2}{S_1}\right)^b = (10^3)\left(\frac{180{,}000}{78{,}879}\right)^{10.48} = 5.69 \times 10^6 \text{ cycles to fail} \tag{8.28}$$

$$\text{Life} = \frac{5.69 \times 10^6 \text{ cycles to fail}}{(179 \text{ cycles/s})3600 \text{ s/h}} = 8.83 \text{ h to fail} \tag{8.29}$$

If this gyro system will be used in a military environment where the operational life is usually expected to be about 10,000 h, the gyro cannot be hard mounted because it will experience an early failure. The gyro assembly must be mounted on isolators to reduce the response acceleration levels and associated stresses to improve its life.

Sample Problem: Finding Input Vibration Level to Achieve a 10,000-h Life

Use the previous sample problem to find the input sine vibration acceleration G level that will increase the fatigue life to about 10,000 h.

Solution

$$\left(\frac{T_1}{T_2}\right)^{1/b} = \frac{G_2}{G_1} \quad \text{(Ref. Eq. 5.4)}$$

where $G_2 = 2.0$ original input sine acceleration level
$G_1 =$ input acceleration G level to achieve a 10,000-h fatigue life
$T_1 =$ 10,000-h desired new life
$T_2 =$ 8.83-h original life with 2.0 G sine input acceleration
$b = 10.49$ slope of fatigue curve for 301 stainless steel shaft (Ref. Eq. 8.27)

$$G_1 = \frac{G_2}{(T_1/T_2)^{1/b}} = \frac{2.0}{(10{,}000/8.83)^{1/10.48}} = 1.02\ G \text{ new input acceleration level} \tag{8.30}$$

8.5 HOW PCB COMPONENT SIZE, LOCATION, AND ORIENTATION EFFECT THE FATIGUE LIFE

Vibration testing experience has shown that large components, longer than about 1 in., cause most of the problems in the lead wires and solder joints. This includes leaded surface-mounted or through-hole-mounted devices and leadless surface-mounted devices. Many other factors must also be considered such as the lead wire material, geometry, and size, as well as component location and component orientation on the PCB. As the component size increases, the problems also increase. Low PCB natural frequencies, with high vibration acceleration G levels and large components of any type are an invitation to a disaster.

The fundamental natural frequency of a plug-in PCB will typically cause most of the vibration problems. The largest displacement and the most rapid change in the curvature is at the center of the PCB that is supported on three or four sides. Therefore, this is the most critical location for large components, as shown in Fig. 8.11. Try to mount large components near the edges of the PCB where the relative dynamic displacements and the curvature are reduced. The electrical circuit functions and the cooling requirements typically drive the locations of the components. When severe vibration levels are expected, the initial PCB layouts on plug-in types of PCBs should attempt to place the largest components near the edges of the PCB to improve their fatigue life.

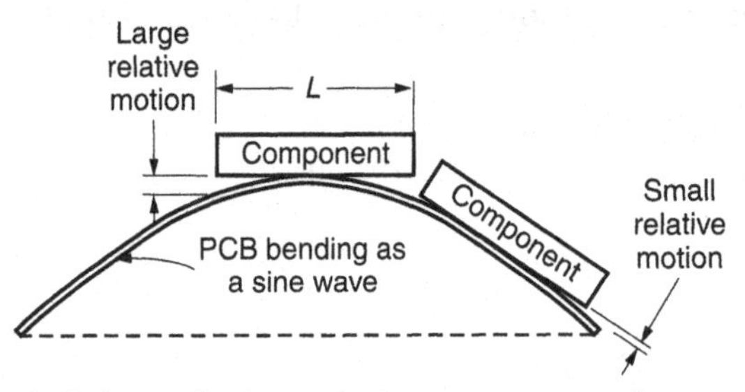

FIGURE 8.11 Relative motion between a long component and a supported PCB, bending in vibration, is reduced when the component is mounted near the PCB edge so the lead wire stresses are also reduced.

Dynamic displacements and the relative curvature in the PCBs can be used to find the approximate forces and stresses generated in the lead wires and solder joints of the components mounted on the PCBs. The relative curvature for different sections of the PCB can then be used to estimate the fatigue life for different types of components mounted in the different PCB sections. When the edges of a typical plug-in PCB are simply supported (or hinged), then a trigonometric expression can be used [1, 2, 24]:

$$Z = Z_0 \sin\frac{\pi X}{a} \sin\frac{\pi Y}{b} \quad \text{(Ref. Eq. 4.2)}$$

When the component is located at the center of the PCB, the displacement will be

$$X = \frac{a}{2} \quad \text{and} \quad Y = \frac{b}{2}$$
$$Z = Z_0 \sin\frac{\pi}{2} \sin\frac{\pi}{2} = Z_0 \tag{8.31}$$

The PCB displacement can be obtained for a component mounting position off center:

$$X = \frac{a}{2} \quad \text{and} \quad Y = \frac{b}{4}$$
$$Z = Z_0 \sin\frac{\pi}{2} \sin\frac{\pi}{4} = 0.707Z_0 \tag{8.32}$$

The PCB displacement can be obtained for a component at the quarter mounting points:

$$X = \frac{a}{4} \quad \text{and} \quad Y = \frac{b}{4}$$
$$Z = Z_0 \sin\frac{\pi}{4} \sin\frac{\pi}{4} = 0.50Z_0 \tag{8.33}$$

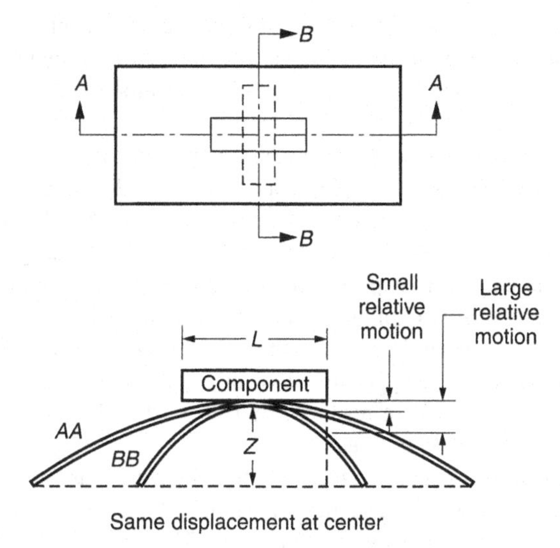

FIGURE 8.12 Vibration relative motion between the component and PCB and lead wire strain can be reduced if a long component is mounted parallel to the long edge of a rectangular PCB.

The orientation of long rectangular electronic components on a rectangular plug-in PCB can have a strong influence on the fatigue life of the lead wires and solder joints. If long rectangular components are mounted parallel to the short edge of the PCB, the fatigue life of the component can be significantly reduced. There is a more rapid change of curvature for the short side than there is for the long side of the PCB, since the displacement must be the same. This produces more relative motion between the PCB and the component, which will result in higher lead wire stresses, as shown in Fig. 8.12. The fatigue life of long components can be increased by mounting them parallel to the long side of a rectangular PCB.

8.6 DESIRED PCB NATURAL FREQUENCY FOR SINUSOIDAL VIBRATION

Plug-in types of PCBs are very popular in almost every type of commercial, industrial, and military electronic assembly. A PCB can be easily removed and repaired or replaced very quickly. PCBs are typically mounted within some type of enclosure, or chassis, for easy handling and to protect them from hostile environments. When this type of assembly is exposed to vibration, the outer chassis structure receives the energy first, so it is considered to be the first degree of freedom. The response of the outer chassis structure represents the input stimulus to the PCBs when the PCBs are attached to the chassis. So the PCBs represents the second degree of freedom, as shown in Fig. 6.1.

Sinusoidal vibration can excite the natural frequency of the outer chassis structure as well as the natural frequencies of the various plug-in PCBs enclosed within the chassis. When the chassis natural frequency is close to the natural frequency of any of the PCBs within the chassis, and when there is very little damping in the chassis and in the PCBs, severe dynamic coupling will be produced between the chassis and the PCBs. Individual PCBs within the chassis are usually much lighter in weight than the chassis. Under these conditions it is possible for certain PCBs to experience very high acceleration levels that will lead to very rapid fatigue failures. For example, when the chassis is exposed to a 10-G peak sine vibration input and the chassis has a transmissibility Q of 10, the chassis will generate a response of 10×10 or 100 G. This will represent the input to the PCBs within the chassis. When one of the PCBs has a transmissibility Q of 10, with the same natural frequency as the chassis, the PCB will generate a response of 100×10 or 1000 G. Levels this high can produce very rapid PCB failures.

One way to avoid this problem is to use the octave rule, as shown in Chapter 6. This rule requires the natural frequency of any one PCB to be two or more times greater than the natural frequency of the chassis. This is called the forward octave rule. When the weight of the chassis is 10 or more times greater than the weight of any one PCB, the reverse octave rule can also be applied. This is where the natural frequency of the chassis can be two or more times greater than the natural frequency of any one PCB. When the natural frequencies of the chassis and the various PCBs are well separated, the dynamic coupling between the chassis and the PCBs are dramatically reduced. The reverse octave rule must not be used when the weight of any one PCB approaches 25% of the chassis weight. This condition can produce very severe dynamic coupling between the chassis and the PCB.

Extensive PCB vibration tests using increasing acceleration levels to produce fatigue failures, combined with finite element modeling of the test specimens, have shown that the PCB dynamic displacement can be related to the approximate fatigue life of the PCB. The studies have shown that various components can achieve a fatigue life of about 10 million stress reversals in a sinusoidal vibration environment, when the PCB peak single-amplitude dynamic displacement is limited to the value shown below. The displacement represents the *maximum* allowable value. The equation includes a safety factor of 1.3 based on the lead wire stresses. The slope of the fatigue curve for lead wires is 6.4. Then $(1.3)^{6.4} = 5.36$ should provide additional fatigue life to complete 5.36 test programs. These programs usually have to be repeated over and over several times because so many things can go wrong. If the test unit fails before the qualification test is completed, a new test unit will be required to finish the test program. Upper management people get very upset when they have to scrap a system because of excessive fatigue damage accumulation. All PCB designs must be used with the octave rule to make sure the chassis natural frequency is well separated from the PCB natural frequency to reduce dynamic coupling:

$$Z = \frac{0.00022B}{Chr\sqrt{L}} \text{ inch,} \quad \text{maximum desired PCB displacement [1]} \tag{8.34}$$

where B = length of PCB edge parallel to the long edge of the component, in.
L = length of the rectangular or square component, in.
h = height or thickness of the PCB, in.
C = constant for different types of electronic components:
1.0 for a standard dual inline package (DIP)
1.26 for a DIP with side brazed lead wires
1.26 for any component with two parallel rows of wires extending from the bottom surface, hybrid, PGA, very large scale integrated (VLSI), application specific integrated circuit (ASIC), very high scale integrated circuit (VHSIC), and multichip module (MCM)
1.0 for a through-hole pin grid array (PGA) with many wires extending from the bottom surface of the component
2.25 for surface-mounted leadless ceramic chip carriers (LCCC)
1.26 for surface-mounted leaded ceramic chip carriers with thermal compression bonded J wires or gull wing wires
1.75 for surface-mounted ball grid array (BGA)
0.75 for axial leaded through-hole or surface-mounted components, resistors, capacitors, diodes
0.75 for fine-pitch surface-mounted axial leads around perimeter of component with four corners bonded to PCB to prevent bouncing
r = relative position factor for location of components on the PCB:
1.0 for component at the center of the PCB at
$X = a/2$ and $Y = b/2$ (Ref. Eq. 8.31)
0.707 for component offset from center of the PCB at
$X = a/2$ and $Y = b/4$ (Ref. Eq. 8.32)
0.50 for component at the quarter mounting points on the PCB at
$X = a/4$ and $Y = b/4$ (Ref. Eq. 8.33)

The approximate peak single-amplitude displacement Z at the center of the PCB can be obtained by assuming the PCB acts like a single-degree-of-freedom system when it is vibrating at its natural frequency:

$$Z = \frac{9.8G}{f^2} = \frac{9.8G_{\text{in}}Q}{f_n^2} \tag{8.35}$$

Dynamic displacements depend on the transmissibility Q of the PCB, which is a relatively complex function related to the damping of the vibrating structure. To avoid adding more complexity to the equations, approximations can be used for the transmissibility values. Equation 8.35 shows that high frequencies produce small displacements. Small displacements have less damping, which increases the Q value. Low frequencies produce high displacements with high damping. This decreases the Q value. A good approximation for the transmissibility Q for plug-in PCBs can be obtained from

$$Q = \sqrt{f_n} \tag{8.36}$$

The minimum desired PCB natural frequency, which will provide a fatigue life of about 10 million stress reversals for the component lead wires and solder joints, can be obtained by combining Eqs 8.34, 8.35 and 8.36 [1]:

$$f_d = \left[\frac{9.8G_{\text{in}}Chr\sqrt{L}}{0.00022B}\right]^{2/3} \quad \text{minimum desired PCB natural frequency} \tag{8.37}$$

Sample Problem: Finding the Minimum Desired PCB Natural Frequency

A surface-mounted leaded ceramic chip carrier measuring 2.0 × 1.0 in., with J wires is to be surface mounted at the center of a plug-in epoxy fiberglass PCB measuring 8.0 × 7.0 × 0.062 in. thick, parallel to the 7.0-in. edge. The PCB must be capable of operating in a 6.0-*G* peak sinusoidal vibration environment. Find the following:

A. Minimum desired PCB natural frequency
B. Expected fatigue life for a resonant dwell condition
C. Minimum desired PCB natural frequency if the component is moved to the quarter mounting points where X is $a/4$ and Y is $b/4$
D. Expected fatigue life for a resonant dwell condition at the new location

Solution A Minimum desired PCB natural frequency, substitute into Eq. 8.37 using the following information:

$G_{\text{in}} = 6.0$-G peak sine vibration input
$C = 1.26$ component type, leaded chip carrier with J leads
$h = 0.062$ in. PCB thickness
$r = 1.0$ relative position factor of component on PCB
$L = 2.0$ in., length of component
$B = 7.0$ in., length of PCB edge parallel to length of component

$$f_d = \left[\frac{(9.8)(6.0)(1.26)(0.062)(1.0)\sqrt{2.0}}{(0.00022)(7.0)}\right]^{2/3} = 261 \text{ Hz, minimum} \tag{8.38}$$

A 261-Hz natural frequency for a large PCB will be difficult to obtain with a 0.062-in. thick PCB. Some type of stiffening will have to be added in the form of ribs or by adding stiffer materials in multiple layers using materials such as aluminum or copper. Stiffening ribs will take up valuable room that will leave less room for mounting components. Buss bars soldered to the PCB can act as stiffening ribs when the buss bars are reinforced with a good adhesive. Using stiffer materials will add extra weight and cost to the PCB. Designers might consider adding snubbers to the local PCBs to reduce dynamic displacements if problems occur. Moving large components away from the center of the PCB will help to reduce the desired natural frequency requirements.

Solution B Expected fatigue life for a resonant dwell condition:

$$\text{Life} = \frac{10 \text{ million cycles to fail}}{(261 \text{ cycles/s})(3600 \text{ s/h})} = 10.64 \text{ h to fail} \tag{8.39}$$

Solution C Desired natural frequency when component is moved to $X = a/4$ and $Y = b/4$ where relative position factor r is 0.50:

$$f_d = \left[\frac{(9.8)(6.0)(1.26)(0.062)(0.50)\sqrt{2.0}}{(0.00022)(7.0)}\right]^{2/3} = 164 \text{ Hz minimum} \tag{8.40}$$

This lower desired natural frequency will be much easier to achieve.

Solution D Expected fatigue life for a resonant dwell condition:

$$\text{Life} = \frac{10 \text{ million cycles to fail}}{(164 \text{ cycles/s})(3600 \text{ s/h})} = 16.94 \text{ h to fail} \tag{8.41}$$

8.7 DAMAGE DEVELOPED DURING A SINE SWEEP THROUGH A RESONANCE

Sinusoidal vibration has been used extensively for determining fundamental natural frequencies along with various higher harmonics for a wide variety of very simple to very complex structures. Sinusoidal vibration is very useful for finding the transmissibility Q for different types of linear and nonlinear systems. Dynamic coupling characteristics between adjacent structural elements are easy to evaluate and to observe using a strobe light. Observations made with a strobe light are being used in physics of failure analyses to diagnose the reasons for field failures in vibration environments. Strobe lights often reveal dynamic relative motions and mode shapes that do not show up in finite element model evaluations. Many qualification test programs make extensive use of sinusoidal vibration to prove the structural integrity of equipment involving the public safety. The typical test will usually start at a low forcing frequency of 5–10 Hz. The frequency is usually increased slowly in the lower frequency ranges and increased more rapidly in the higher frequency ranges. This is typically a logarithmic sweep where the time to sweep is constant for each octave. Octave means to double. The time to sweep from 10 to 20 Hz is the same as the time to sweep from 100 to 200 Hz. Qualification tests used by some automobile radio manufacturers run from 10 to 500 Hz. A constant single-amplitude input displacement of about 0.050 in. may be used from 10 to about 31.3 Hz. This gives an input acceleration level of about 5.0 G peak at 31.3 Hz. The input level of 5.0 G peak is then held constant while the vibration sweeps from 31.3 to 500 Hz, at a sweep rate of one octave per minute. The process is usually reversed at this point, and the sweep is made in the reverse direction, from 500 to 10 Hz. It is a good practice to sweep up first, followed by a sweep down through the same starting and finishing points. The response should be plotted automatically for the sweep up and the sweep down and then examined to see if the resonant peak is symmetrical. If the resonant peak is not symmetrical, it is a sign of a nonlinear system.

Most of the damage produced in a vibration environment will occur when the natural frequency of the structure is excited. A high transmissibility Q can often be expected at the natural frequency. Most of the damage accumulated during a sinusoidal sweep test will occur as the frequency sweeps through the peak response area. This reference area is the half power points, used extensively by electrical engineers to characterize resonant peaks in electrical circuits. These are the points where the power that can be absorbed by damping is proportional to the square of the amplitude at a given frequency. For a lightly damped linear system, where the transmissibility is greater than about 10, the peak curve in the area of the resonance is approximately symmetrical. The half power points are then used to define the bandwidth of the system as shown in Fig. 8.13.

The time it takes to sweep through the half power points using a logarithmic sinusoidal input can be obtained from

$$t = \frac{\log_e \dfrac{1 + \frac{1}{2Q}}{1 - \frac{1}{2Q}}}{R \log_e 2} \tag{8.42}$$

where t = time in minutes to sweep through the half power points
R = sweep rate in octaves per minute
Q = transmissibility at resonance, dimensionless

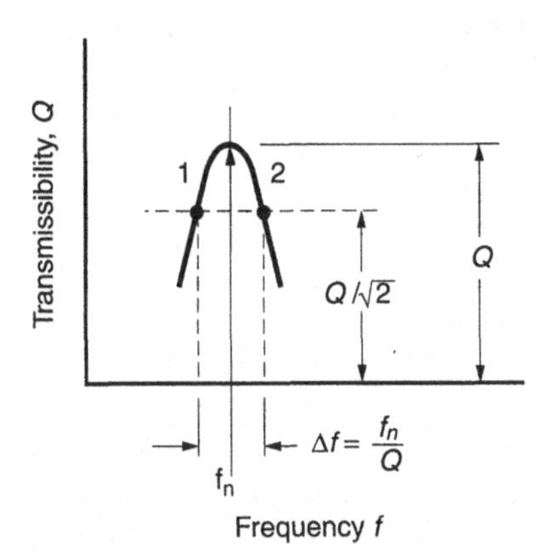

FIGURE 8.13 Properties of the half power points for a transmissibility curve.

The time it takes for a logarithmic sinusoidal sweep between two different frequencies can be obtained from

$$t = \frac{\log_e(f_2/f_1)}{R \log_e 2} \tag{8.43}$$

where f_2 = upper frequency, Hz
f_1 = lower frequency, Hz
R = sweep rate in octaves per minute

Sample Problem: Finding the Percent of Life Used Up by Sine Sweeps

Determine the approximate amount of life used up by 100 single sinusoidal vibration sweeps through the half power points for the 261-Hz natural frequency from the previous sample problem in Eq. 8.38, using the same input acceleration G level and a logarithmic sweep rate of 0.25 octaves per minute.

Solution Find the fatigue cycles for one sweep through the half power points. The approximate $Q = \sqrt{261} = 16.2$, and the sweep rate $R = 0.25$ octaves per minute. Substitute into Eq. 8.42.

$$t = \frac{\log_e \dfrac{1 + \frac{1}{32.4}}{1 - \frac{1}{32.4}}}{(0.25)\log_e 2} = \frac{0.0617}{0.1733} = 0.357 \text{ min} \tag{8.44}$$

$$\text{Life used up} = \frac{(0.357 \text{ min})(60 \text{ s/min})(261 \text{ cycles/s})(100 \text{ sweeps})}{10 \times 10^6 \text{ cycles life}} = 0.0559 \tag{8.45}$$

This shows that about 5.59% of the life will be used up by 100 sweeps.

Sample Problem: Time to Sweep Between Two Frequencies

Find the time it will take to sweep from 10 to 2000 Hz using a sweep rate of 0.50 octaves per minute.

Solution Substitute into Eq. 8.43 using a sweep rate $R = 0.50$ octaves per minute:

$$t = \frac{\log_e(2000/10)}{(0.50)\log_e 2} = \frac{5.298}{(0.50)(0.693)} = 15.29 \text{ min} \tag{8.46}$$

8.8 AVOID LOOSE EDGE GUIDES ON PLUG-IN PCBs

Plug-in types of PCBs must have some way of aligning blind mated electrical connectors at the bottom edge of the PCB. A wide variety of edge guides are available from spring clips to wedge clamps. Edge guides can also be cast or machined into enclosures that hold the plug-in PCBs. Vibration tests on electronic enclosures or chassis with plug-in PCBs that have loose edge guides have shown that these PCBs can experience very high acceleration levels due to severe coupling between the chassis and the PCBs. This happens when the natural frequency of the chassis is close to the natural frequency of one or more of the PCBs enclosed within

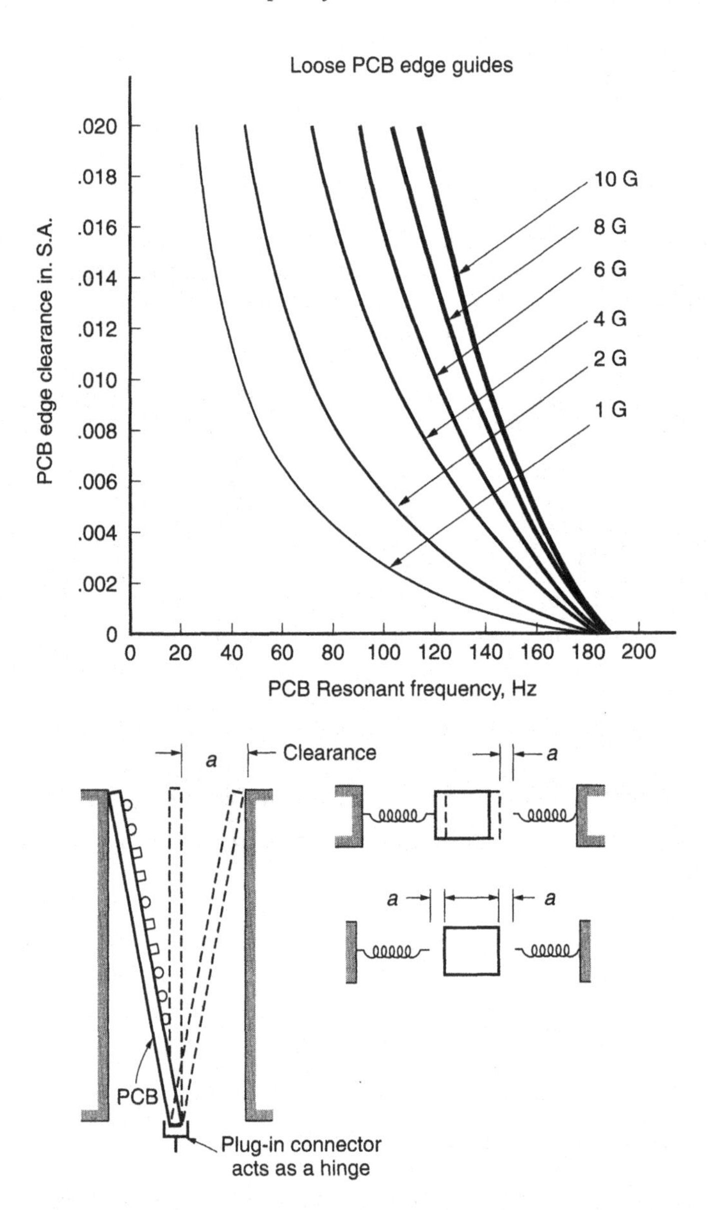

FIGURE 8.14 Natural frequency of a plug-in PCB with loose edge guides will be strongly influenced by the free sway space single amplitude and the input acceleration G level.

the chassis. The response of the chassis is the input to the PCB. When these structures are lightly damped, they will have high transmissibility *Q*'s. The *Q*'s will multiply and produce high PCB acceleration levels when their natural frequencies are close together. This can produce rapid PCB fatigue failures.

When the PCB has loose edge guides, its vibration period will be increased since the period depends on the displacement amplitude. The displacement amplitude of the PCB depends on the PCB bending stiffness plus any clearances. Loose edge guides generate a clearance, or a gap, that increases the total PCB displacement. This then increases the time it takes to complete one full alternating cycle. Therefore, the natural frequency of the PCB will be determined by the bending of the PCB, plus the edge guide clearance plus the velocity, which is controlled by the acceleration level. The introduction of a gap in the edge guide increases the PCB period, which results in a lower PCB natural frequency. A lower PCB natural frequency may bring it closer to the natural frequency of the chassis, which can increase the dynamic coupling if this condition is not corrected.

The octave rule discussed in Chapter 6 was recommended to avoid severe dynamic coupling between the outer chassis and the enclosed PCBs. Octave means to double. When the PCB natural frequency is two or more times greater than the outer chassis natural frequency, the dynamic coupling between these structures is greatly reduced. This separation of the natural frequencies between the chassis and the PCBs is easy to accomplish when the edges of the PCB are firmly supported. When the PCB edge guides are loose, it will take more time for the PCB to complete a full displacement cycle during vibration because of the increased displacement. This will increase the period and decrease the effective natural frequency of the PCB, which will bring the PCB natural frequency down closer to the chassis natural frequency. When the natural frequencies of the chassis and the PCBs are closer together, the dynamic coupling increases the acceleration *G* levels very rapidly. This is very dangerous because high acceleration levels can damage the PCBs.

Vibration test data on plug-in PCBs with loose edge guides showed that it was very difficult to prevent severe coupling between the chassis and the PCBs. Increasing the natural frequency of the chassis had very little effect on reducing the dynamic coupling between the chassis and the PCB. Increasing the natural frequency of the PCB also had very little effect on reducing the dynamic coupling between the chassis and the PCB. The only way that the severe dynamic coupling could be decreased between the chassis and the PCBs was to add clips that gripped the edges of the PCBs firmly. This removed the gaps between the chassis and the PCBs, which raised the PCB natural frequency. This increased the spread between the natural frequency of the chassis and the PCB, which reduced the dynamic coupling and improved the fatigue life of the PCBs. Figure 8.14 shows how the natural frequency of a plug-in PCB is affected by the size of the single-amplitude clearance between the PCB and the support and the input acceleration *G* level. A large single-amplitude gap and a low input acceleration *G* level will substantially reduce the effective natural frequency of the PCB [1].

CHAPTER 9

Assessment of Random Vibration on Electronic Design

9.1 INTRODUCTION

Random vibration continues to gain wide acceptance for use in many different testing programs associated with electronic equipment because random vibration has been shown to be more closely related to the actual operating environments. It has been found to be very effective for use in screening electronic equipment, to precipitate latent defects before the equipment is shipped to the customer. The defects are then repaired in the factory before the equipment is shipped, so the customer receives a very reliable piece of equipment. Random vibration is also very popular for performing accelerated life tests, to estimate the approximate fatigue life some equipment may have in a specific operating environment. Defective designs, defective materials, and defective manufacturing processes can be quickly uncovered and improved with the use of random vibration.

Displacement traces generated by random vibration never repeat themselves. This is probably the most obvious characteristic of random vibration. The past history of random vibration cannot be used to predict the precise magnitude at any specific instant of time, which is possible with sinusoidal vibration. A typical plot of a random displacement trace with respect to time is shown in Fig. 9.1. A close examination of this trace will reveal that it is made up of a very large number of varying individual sinusoidal traces superimposed upon each other, as shown in Fig. 9.2.

9.2 HOW RANDOM VIBRATION DIFFERS FROM SINUSOIDAL VIBRATION

Random vibration is characterized in terms of the bandwidth of the frequencies because it consists of many different sinusoidal frequencies superimposed upon one another. All of the frequencies within the bandwidth are present simultaneously at any instant of time for every frequency. When the frequency bandwidth is from 20 to 2000 Hz, every natural frequency of every structural member between 20 and 2000

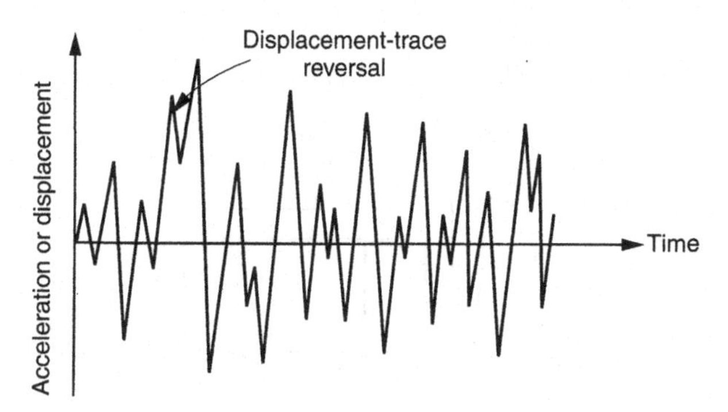

FIGURE 9.1 Typical plot of acceleration or displacement with respect to time for random vibration.

Hz will be excited at the same time. This includes every fundamental natural frequency, and every higher harmonic of every structural member within that bandwidth will be excited at the same time. For nonlinear systems it is possible to excite natural frequencies that are well above the highest frequency in the bandwidth. For example, test data using an oscilloscope shows that there is usually a substantial amount of vibration energy at frequencies as high as 100,000 Hz even when there is a sharp cutoff at 2000 Hz with a roll off at 90 dB per octave. Because there is no way of knowing the exact phase relationships of the various sinusoidal waves, it is impossible to know how, when, or where the various waves will add or subtract. It then becomes necessary to deal with probability distribution functions to evaluate random vibrations.

Electronic assemblies excited by sinusoidal vibration will respond differently than electronic assemblies excited by broadband random vibration. Failures generated by random vibration are very seldom similar to failures generated by sinusoidal vibration. Some types of random vibration failures cannot be reproduced by sinusoidal vibration. This can be demonstrated by considering two different spring–mass systems subjected to broadband random vibration and sinusoidal vibration, as shown

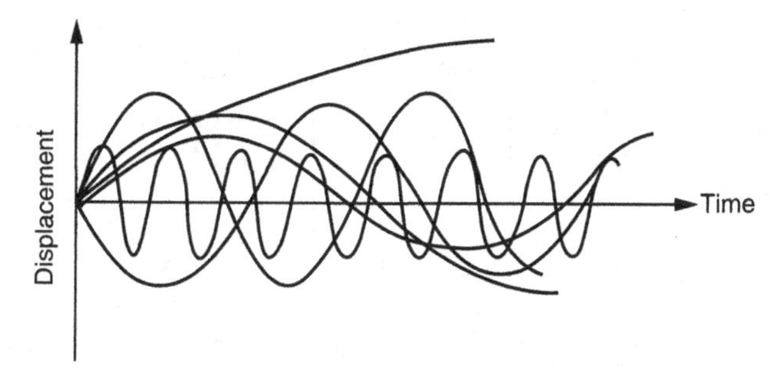

FIGURE 9.2 Random vibration is made up of many different overlapping sinusoidal curves.

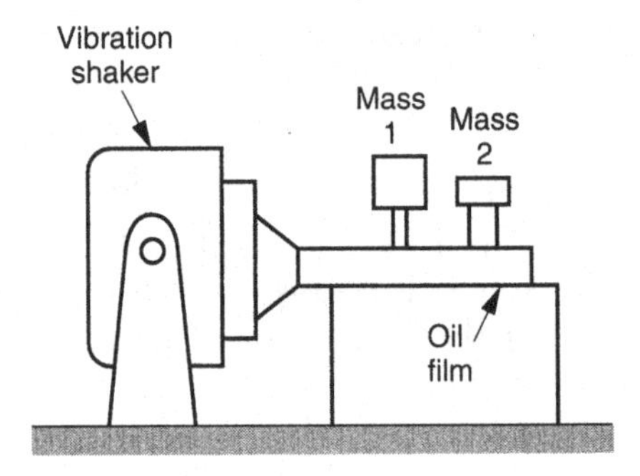

FIGURE 9.3 Two different cantilever spring–mass systems with two different natural frequencies are mounted on an oil film slider plate attached to a vibration shaker.

in Fig. 9.3. Mass 1 will be a cantilever beam with a low natural frequency, and mass 2 will be a cantilever beam with a much higher natural frequency. A failure will occur when the masses strike each other. This would be similar to the condition for closely spaced printed circuit boards (PCBs). If they strike each other during exposure to vibration, failures can occur due to cracked solder joints, component failures, and short circuits.

Start with a low sinusoidal vibration frequency and slowly sweep up to higher frequencies until the natural frequency of mass 1 becomes excited and responds with a high transmissibility Q and a high displacement, as shown in Fig. 9.4. Continue to sweep higher with sinusoidal vibration, and mass 1 will enter its isolation area where the Q response and the displacement are sharply reduced. Mass 2 will be quiet when mass 1 is excited because mass 2 has a much higher natural frequency. As the frequency is increased further, mass 2 will become excited with a high Q and a high displacement while mass 1 is quiet because it is in its isolation area. The sinusoidal vibration frequency can be swept up and down, but mass 1 and mass 2 can never strike each other. Since each mass is excited separately by the sinusoidal vibration, the displacement is not enough to produce impacting between the two masses. When

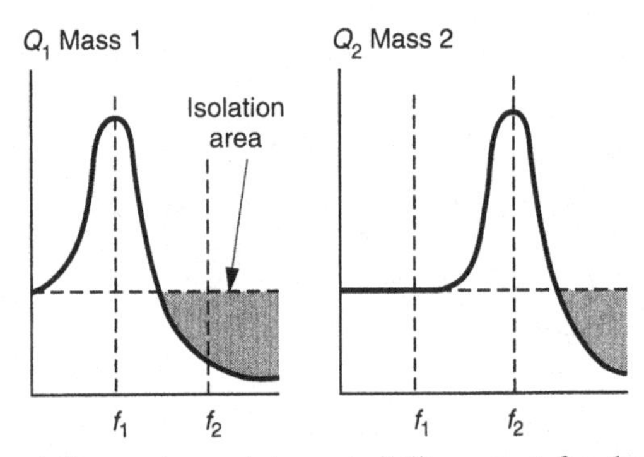

FIGURE 9.4 Natural frequencies and transmissibility curves for the two spring–mass systems mounted on the slider plate.

the same system is excited by broadband random vibration that contains both natural frequencies, both masses will be excited at the same time. Both masses will now produce their own large displacements that can combine to cause impacting between the two masses. This shows that some types of failures generated by random vibration cannot be duplicated with sinusoidal vibration.

9.3 RANDOM VIBRATION POWER SPECTRAL DENSITY CURVES

Random vibration input and response curves are typically presented in terms of a power spectral density (PSD) with acceleration in gravity units square per hertz, or G^2/Hz, which is plotted on the Y axis with the frequency in hertz (Hz) plotted on the X axis. Log–log plots that represent input curves are typically plotted with straight lines to represent a wide variety of shapes for different types of random vibration input environments, as shown in Fig. 9.5 [15].

The root mean square (rms) input acceleration G levels are obtained from the square root of the area under the input curves. The output or response rms acceleration G levels are obtained from the square root of the area under the response curves:

$$\sqrt{\text{area}} = \sqrt{\frac{G^2}{\text{Hz}} \times \text{Hz}} = \sqrt{G^2} = G \text{ rms} \tag{9.1}$$

Sample Problem: Obtaining the Input rms Acceleration Level

Find the input rms acceleration G level from the white noise (flat top) random vibration input specification shown in Fig. 9.6:

$$G \text{ rms} = \sqrt{(\text{PSD})(\Delta f)} = \sqrt{(0.10)(1000 - 5)} = 9.97\ G \text{ rms input} \tag{9.2}$$

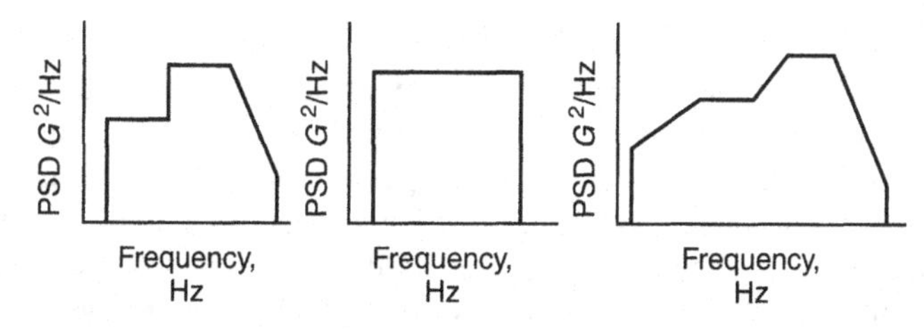

FIGURE 9.5 Random vibration PSD levels are typically plotted on log–log curves using straight lines to outline different shapes.

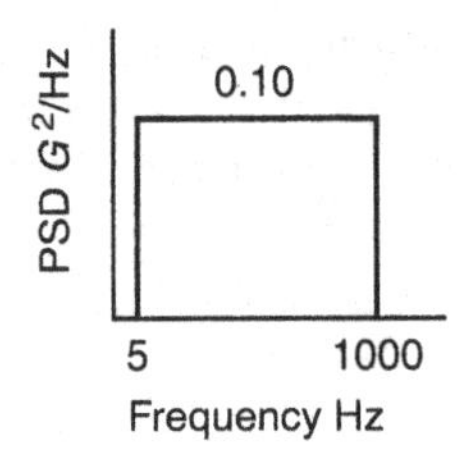

FIGURE 9.6 Typical log–log plot for white noise random PSD that has a flat horizontal top.

9.4 EXPRESSING THE SLOPE IN TERMS OF DECIBELS

The mathematical definition of decibel appears to be different for different functions related to sinusoidal vibration, random vibration, and shock as follows:

Sinusoidal vibration	Random vibration	Shock
$\text{dB} = 20 \log_{10} \frac{G_2}{G_1}$	$\text{dB} = 10 \log_{10} \frac{\text{PSD}_2}{\text{PSD}_1}$	$\text{dB} = 20 \log_{10} \frac{G_2}{G_1}$

(9.3)

In the above equations the acceleration G term is linear for sinusoidal vibration and shock so the constant in front of the equation is 20. For random vibration the PSD term is a square term where $\text{PSD} = G^2/\text{Hz}$, so the constant in front of the equation is 10.

For random vibration when the ratio of $\text{PSD}_2/\text{PSD}_1$ is 2.0, and for sine vibration and shock when the ratio of G_2/G_1 is 2.0, then in terms of dB this ratio would be expressed as follows:

$$\begin{aligned} &\text{Random vibration} && \text{dB} = 10 \log_{10}(2.0) = 10(0.301) = 3.01 \\ &\text{Sine vibration and shock} && \text{dB} = 20 \log_{10}(2.0) = 20(0.301) = 6.02 \end{aligned} \tag{9.4}$$

Another way of saying the same thing is that with random vibration 3 dB doubles and for sinusoidal vibration and shock 6 dB doubles.

9.5 FINDING THE AREA UNDER SHAPED RANDOM VIBRATION INPUT CURVES

Random vibration input curves are typically plotted on log–log paper using straight lines, which can be horizontal, vertical, or sloped. A straight line sloping up to the right is considered to be a positive (+) slope. A straight line sloping down to the right will have a negative (−) slope. Most input random vibration curves will have sloped lines, so special equations must be used to obtain the related area for a log–log plot. Only straight vertical lines must be used to break up segments under the sloped lines. Horizontal lines within the envelope must not be used to designate individual areas. The equations for finding the areas under the sloped lines are shown below. These equations cannot be used when the slope is negative (−) 3. The subscript 1 refers to the left side of an area segment and the subscript 2 refers to the right side of an area segment [1]:

$$A = \frac{3P_2}{3+S}\left[f_2 - \left(\frac{f_1}{f_2}\right)^{S/3}(f_1)\right] = (G\ \text{rms})^2 \tag{9.5}$$

or

$$A = \frac{3P_1}{3+S}\left[\left(\frac{f_2}{f_1}\right)^{S/3}(f_2) - f_1\right] = (G\ \text{rms})^2 \tag{9.5a}$$

where $P = G^2/\text{Hz}$, power spectral density (PSD)
$f = \text{Hz}$, frequency
$S = \text{dB}$ (dB per octave) slope of straight line

When the sloped section of the log–log plot has a value of −3, then the equations shown below can be used to find the area under the curve:

$$A = -f_2 P_2 \log_e \frac{f_1}{f_2} = (G\ \text{rms})^2 \tag{9.6}$$

or

$$A = f_1 P_1 \log_e \frac{f_2}{f_1} = (G\ \text{rms})^2 \tag{9.6a}$$

The area under the PSD curve with a flat top section that has a zero slope can be obtained from

$$A = P(f_2 - f_1) = (G\ \text{rms})^2 \tag{9.7}$$

Sample Problem: Finding the Input *G* rms from the Area under the PSD Curve

Find the input *G* rms acceleration level for the random vibration input curve shown in Fig. 9.7 when it is plotted on log–log paper.

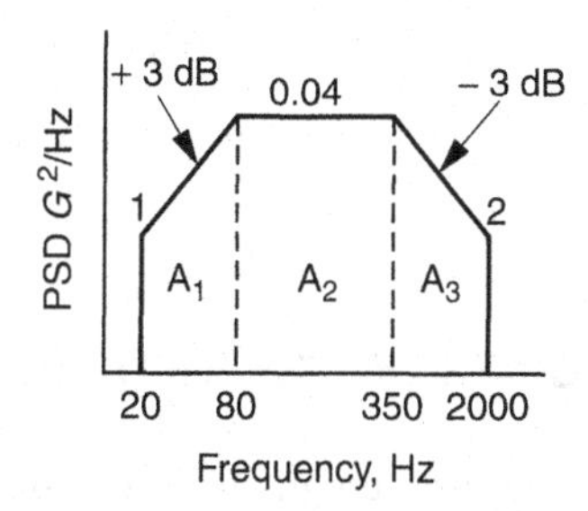

FIGURE 9.7 Breaking up a log–log PSD plot into three sections using only vertical lines.

Solution Substitute into Eqns. 9.5, 9.6a, and 9.7 using vertical lines only to break up the curve into three separate sections to find the individual areas where

$$f_1 = 20 \text{ Hz, frequency}$$
$$f_2 = 80 \text{ Hz, frequency}$$
$$P_2 = 0.040\ G^2/\text{Hz, power spectral density (PSD)}$$
$$S = 3 \text{ dB (or dB/octave) slope of PSD curve}$$

Substitute into Eq. 9.5 to find the area under section 1 of the input PSD curve (area 1):

$$A_1 = \frac{3(0.040)}{3+3}\left[80 - \left(\frac{20}{80}\right)^{3/3}(20)\right] = 1.50\ G^2 \tag{9.8}$$

Using Eq. 9.7 (for area 2):

$$A_2 = (0.040)(350 - 80) = 10.80\ G^2 \tag{9.9}$$

Use the following values for area 3:

$$f_1 = 350 \text{ Hz, frequency}$$
$$f_2 = 2000 \text{ Hz, frequency}$$
$$P_1 = 0.040\ G^2/\text{Hz, power spectral density (PSD)}$$
$$S = -3 \text{ dB (or dB/octave) slope of PSD curve}$$

Substitute into Eq. 9.6a to find the area under section 3 of the input PSD curve:

$$A_3 = (350)(0.040)\log_e \frac{2000}{350} = 24.4\ G^2 \tag{9.10}$$

The total area under the input PSD curve will determine the input G rms acceleration, using Eqs. 9.8, 9.9, and 9.10 to find the total area A_T:

$$\begin{aligned} A_T &= A_1 + A_2 + A_3 = 1.50 + 10.8 + 24.4 = 36.7\ G^2 \\ G \text{ rms} &= \sqrt{36.7} = 6.06\ G \text{ (rms input acceleration)} \end{aligned} \tag{9.11}$$

9.6 FINDING BREAK POINTS ON THE INPUT PSD CURVE

Random vibration input curves use straight lines on log–log plots to define the profile. This results in break points where the sloped lines meet the horizontal lines or the vertical lines. The curves are often defined by the break points and the slope of the line in dB. The PSD values at the break points can be obtained from

$$P_1 = P_2\left(\frac{f_1}{f_2}\right)^{S/3} \tag{9.12}$$

where $P = G^2/\text{Hz}$, power spectral density (PSD)
$f = \text{Hz}$, frequency
$S = \text{dB}$ (or dB/octave), slope of the inclined line

Sample Problem: Finding the PSD Values at Break Points

Find the PSD values at break points 1 and 2, which are shown in Fig. 9.7.

Solution Use Eq. 9.12 to find the PSD values at break points 1 and 2. At break point 1

$f_1 = 20$ Hz
$f_2 = 80$ Hz
$P_2 = 0.04\ G^2/\text{Hz}$
$S = 3$ dB/octave, slope of PSD curve

$$P_1 = (0.040)(20/80)^{3/3} = 0.010\ G^2/\text{Hz} \tag{9.13}$$

At break point 2

$f_1 = 350$ Hz
$f_2 = 2000$ Hz
$P_1 = 0.040\ G^2/\text{Hz}$
$S = -3\text{dB/octave}$, slope of PSD curve

$$P_2 = \frac{0.040}{(350/2000)^{-3/3}} = 0.0070\ G^2/\text{Hz} \tag{9.14}$$

9.7 RANDOM VIBRATION GAUSSIAN PROBABILITY DISTRIBUTION FUNCTIONS

Random vibration is nonperiodic, so probability distribution functions based on past history are used to predict various acceleration and displacement amplitudes. The distribution most often used is the Gaussian (or normal) distribution, shown in Fig. 9.8. This represents the value of the *instantaneous* acceleration levels at any time.

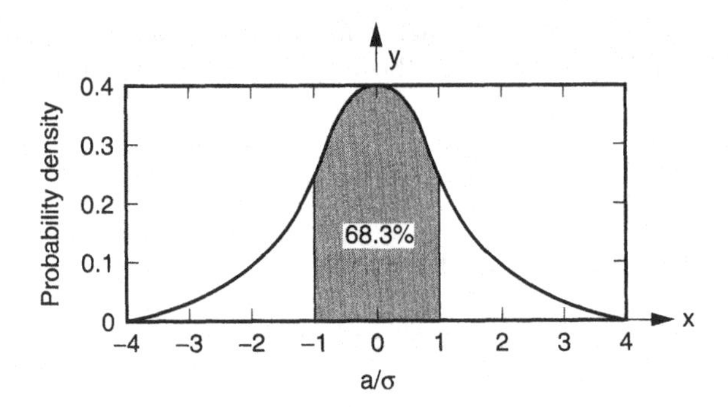

FIGURE 9.8 Gaussian probability distribution curve for random vibration where the probability of an occurrence is shown by the area under the curve.

The total area under the curve is unity. The area under the curve between any two points represents the probability that the accelerations will be between these two points. The shaded area in the figure shows the instantaneous accelerations that will be between the plus one sigma ($+1\sigma$) and the minus one sigma (-1σ) values about 68.3% of the time. The one sigma value here is the rms value. The instantaneous accelerations will be between the $+2\sigma$ and the -2σ value, which is two times the rms value, about 95.4% of the time. These accelerations will be between the $+3\sigma$ and the -3σ value, which is three times the rms level, about 99.73% of the time [1].

The maximum acceleration level considered for random vibration with the Gaussian function is the 3σ level. Much higher levels of 4σ and 5σ can occur in the real world, but they are almost always ignored. This is because the instantaneous accelerations are between $+3\sigma$ and -3σ levels about 99.73% of the time, which is almost 100% of the time. Also, all vibration machines electronically controlled are shipped with 3σ clippers, which clip off all input acceleration spikes greater than the 3σ level. Another factor was the MIL-STD 810 C military specification for random vibration, which stated that the maximum acceleration level may be limited to three times the rms level, which defines the Gaussian distribution.

Tests on many older electrodynamic shakers showed that the 3σ clippers were no longer working. Electronic equipment tested on these older shakers were often being subjected to acceleration peaks of 4σ, which are four times greater than the rms level instead of the maximum expected 3σ peaks, which are only three times greater than the rms acceleration levels. Higher acceleration levels will lead to more rapid fatigue failures. This makes it very difficult to determine if a random vibration test failure was due to a poor design or to a poor test. A poor vibration fixture design can also be the cause of premature vibration failures. The 3σ clippers on older vibration shakers should be checked to make sure they are still working properly to ensure a valid test.

The acceleration values discussed in most technical meetings refer to the rms levels. This means that when an acceleration level of 8 G is mentioned (it could be an input or a response), it refers to an rms acceleration level. The 8-G level must be multiplied by 2 to get the 2σ level of 16 G, which will act part of the time. The 8 G

must be multiplied by 3 to get the 3σ level of 24 G, which will act part of the time. Most of the damage will be produced by the 3σ acceleration levels since they are the highest levels expected. All of the above acceleration levels will be acting at the same time.

Random vibration is often used to perform qualification tests on electronic systems. Time is money. The random vibration tests are usually run at higher acceleration levels for a short period of time, to try to simulate the same amount of damage generated in the field by a lower vibration level for a longer period of time. Therefore qualification tests are often considered to be accelerated life tests.

9.8 RANDOM VIBRATION ANALYSIS USING THE THREE-BAND TECHNIQUE

Random vibration analysis at large companies is typically performed using finite element methods (FEM) with high-speed computers and a wide variety of software packages. The newer computers usually have a very large database, which is used to store data on the fatigue properties of many different materials. This makes it convenient to enter stress or strain data for a specific material, so the computer can evaluate the fatigue life. Many of the smaller companies do not have the time or the money available to support large sophisticated FEM analysis programs, so they rely more on hand calculations for their analyses. Hand calculations for random vibration analyses will normally be less accurate than computer analyses, but well documented hand calculations can still be a valuable tool for direct analysis or for verifying the accuracy of computer analyses. This is often called a sanity check for FEM. When a large computer program requires the input of more than 1000 data points, it is easy for a tired programmer to make an input error. Computer users have an expression: *garbage-in garbage-out*, or *GIGO*. One bad computer data point entry can ruin a month or more of hard work. Computer users also know that it is very difficult to check their own work, especially when they are tired. So it is often a common practice for them to seek some way to quickly obtain an approximate answer, which can be used as a rough check for their computer analysis. The hand calculation analysis methods shown here can be used for that purpose when the structures involved are not too complex.

The three-band technique method of analysis is based on the Gaussian distribution. The instantaneous accelerations *between* the $+1\sigma$ and the -1σ levels are assumed to act *at* the 1σ level 68.3% of the time. Instantaneous acceleration levels *between* the $+2\sigma$ and the -2σ levels are assumed to act 95.4 − 68.3 or 27.1% of the time. Instantaneous acceleration levels between $+3\sigma$ and -3σ are assumed to act 99.73 − 95.4 or 4.33% of the time. These three bands are shown for a quick reference [1]:

$$\begin{aligned} &1\sigma \text{ acceleration values occur } 68.3\% \text{ of the time} \\ &2\sigma \text{ acceleration values occur } 27.1\% \text{ of the time} \\ &3\sigma \text{ acceleration values occur } 4.33\% \text{ of the time} \end{aligned} \tag{9.15}$$

The three-band technique shown above is actually a Gaussian distribution for instantaneous accelerations skewed toward a Rayleigh distribution for peaks, which is discussed below.

9.9 RAYLEIGH PROBABILITY DISTRIBUTION FUNCTIONS

The Rayleigh distribution function shows the probability for the distribution of the *peak* accelerations developed in random vibration. The argument given here is that *peak* accelerations and *peak* stresses cause the fatigue damage, not the *instantaneous* stress levels. The Rayleigh distribution is shown in Fig. 9.9. The total area under the curve is also unity. The area under the curve between any two points is then used to represent the probability that a peak amplitude will be between these two points.

Extensive random vibration testing and analyses in the electronics industry has shown that structural fatigue failures appear to follow the Gaussian distribution more closely than the Rayleigh distribution. Therefore, the Gaussian distribution functions will be used to analyze random vibration in this book.

9.10 RESPONSE OF SINGLE-DEGREE-OF-FREEDOM SYSTEMS TO RANDOM VIBRATION

When a simple single-degree-of-freedom spring–mass system is excited by random vibration, the system can only respond by vibrating at its natural frequency. No other frequency is possible. The displacement versus time will have a varying amplitude, as shown in Fig. 9.10. The damage generated in any dynamic system is not determined by the input to the system, it is determined by the response (or output) of that system. The response of a lightly damped single-degree-of-freedom system to a broadband random vibration input can be obtained from Eq. 9.16. The response will

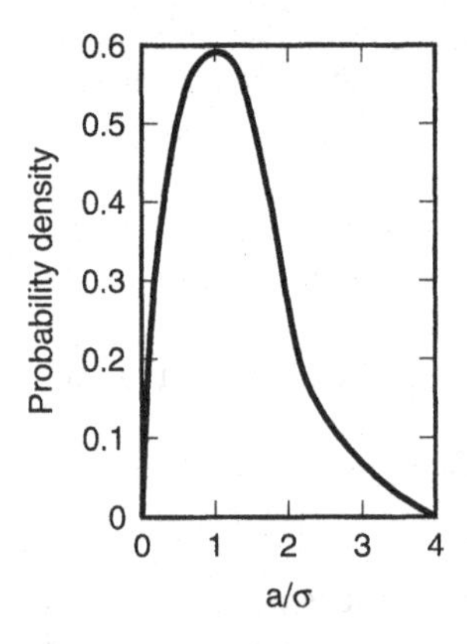

FIGURE 9.9 Rayleigh probability curve shows the distribution of the peak accelerations for random vibration.

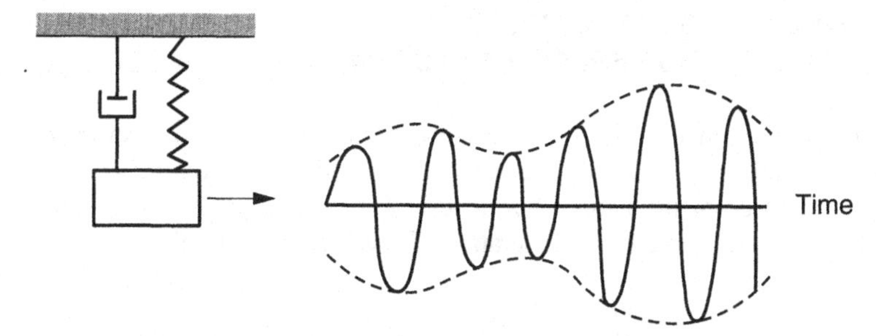

FIGURE 9.10 Displacement history for the response of a simple spring–mass system to a random vibration input.

be in terms of the rms. The equation is very accurate when the transmissibility Q is greater than about 10, and when the PSD curve is flat in the area of the resonant peak [1, 15]:

$$G \text{ rms} = \sqrt{\frac{\pi}{2} P f_n Q} \text{ response} \tag{9.16}$$

where $P = G^2/\text{Hz}$, PSD input at the natural frequency
$f_n = \text{Hz}$, natural frrequency
$Q =$ dimensionless, transmissibility at the natural frequency

9.11 NUMBER OF POSITIVE ZERO CROSSINGS FOR RANDOM VIBRATION

Every time a structure experiences a stress reversal, part of its life is used up. The number of stress cycles usually depends on the natural frequency of the structural element and the time. For a sinusoidal vibration dwell condition, this is simply the product of the natural frequency in cycles per second times the dwell time in seconds. It is not that simple for random vibration since it is broadband, so many natural frequencies are excited at the same time. Therefore, the number of positive crossings are used instead of the natural frequency. The number of positive zero crossings is defined as the number of times the displacement trace crosses the zero axis with a positive slope. This is basically the same definition of a sine wave. The exception is that with random vibration the displacement trace can have a positive slope, but it may reverse itself and not cross the zero axis, so it is not counted. The number of positive zero crossings is defined as [15]:

$$N_0^+ = \frac{1}{2\pi} \left(\frac{\sum \frac{P_i \, \Delta f_i \, Q_i^2}{\Omega_i^2}}{\sum \frac{P_i \, \Delta f_i \, Q_i^2}{\Omega_i^4}} \right)^{1/2} \tag{9.17}$$

9.12 DYNAMIC DISPLACEMENT OF A SINGLE-DEGREE-OF-FREEDOM SYSTEM

Dynamic displacements developed during random vibration can be evaluated using the method shown in Eq. 3.16. The number of positive zero crossings (N_0^+) must be used in the place of the natural frequency (f_n), and the peak acceleration G level must be replaced by the rms acceleration G level, as shown below:

$$Y_0 = \frac{9.8 \ G \ \text{rms}}{(N_0^+)^2} \ \text{single amplitude displacement, rms} \tag{9.18}$$

The number of positive zero crossings for random vibration requires a substantial amount of work using hand calculations. The amount of analysis time required to find the random vibration displacement can be reduced by using the fundamental natural frequency in the place of the number of positive zero crossing as shown below. This substitution will also reduce the accuracy of the dynamic analysis, so safety factors should be used to ensure the integrity of the analysis:

$$Y_0 = \frac{9.8 \ G \ \text{rms}}{f_n^2} \ \text{single amplitude displacement, rms} \tag{9.19}$$

The number of positive zero crossings is always greater than the fundamental natural frequency, so using the natural frequency substitution results in a larger dynamic displacement. This is equivalent to adding a safety factor.

9.13 RESPONSE OF PLUG-IN PCBs TO RANDOM VIBRATION

Broadband random vibration will excite the fundamental natural frequency and several higher harmonics of plug-in types of PCBs. The random vibration will also excite the various frequencies of the outer housing or chassis that supports the PCBs. The phase relations and coupling between the chassis and the PCBs are very difficult to calculate without a computer. Random vibration tests with the proper instrumentation will reveal the coupling properties if hardware, time, money, and test equipment are available.

When a computer is not available, and when hardware is not available for testing, the analysis must be done with hand calculations. The simplest approach is to work with the fundamental natural frequency of the plug-in PCB. The highest dynamic displacements and the most amount of damage occurs at the natural frequency of the PCB. When the octave rule is followed the natural frequency of the chassis and the PCB will be separated by a ratio of 2 to 1. This will reduce the coupling between the chassis and the PCB at the lowest natural frequency, but it will not reveal the coupling effects at the higher harmonics. Since hand calculations will not be as accurate as a computer, safety factors are recommended to compensate for the reduced accuracy of the hand calculations.

Sample Problem: Dynamic Response and Displacement of a Plug-in PCB

Find the dynamic acceleration response and the dynamic displacement of a plug-in PCB with a natural frequency of 225 Hz, during operation in a random vibration environment with an input PSD of 0.10 G^2/Hz.

Solution Substitute into Eqs. 9.16 and 9.19 for an approximate solution assuming the PCB acts like a single-degree-of-freedom structure where

$P = 0.10\ G^2/\text{Hz}$, PSD input

$f_n = 225$ Hz, natural frequency

$Q = \sqrt{225} = 15$ approximate transmissibility

The dynamic acceleration G response is

$$
\begin{aligned}
1\sigma G \text{ rms} &= \sqrt{\frac{\pi}{2}(0.10)(225)(15)} = 23.02 && 68.3\% \text{ of the time} \\
2\sigma &= 2(23.02) - 46.04 \text{ G} && 27.1\% \text{ of the time} \\
3\sigma &= 3(23.02) = 69.06\ G && 4.33\% \text{ of the time}
\end{aligned}
\tag{9.20}
$$

The dynamic single amplitude displacements using Eq. 9.19 are

$$
\begin{aligned}
1\sigma Y_0 &= \frac{(9.8)(23.02)}{(225)^2} = 0.00446 \text{ in. rms} && 68.3\% \text{ of the time} \\
2\sigma &= 2(0.00446) = 0.00892 \text{ in.} && 27.1\% \text{ of the time} \\
3\sigma &= 3(0.00446) = 0.0134 \text{ in.} && 4.33\% \text{ of the time}
\end{aligned}
\tag{9.21}
$$

Sample Problem: Estimating the Fatigue Life Using the Three-Band Technique with Miner's Cumulative Fatigue Damage Ratio

A 6061 T6 aluminum cantilevered beam 6.0 in. long × 0.50 in. wide × 0.30 in. thick with a concentrated end load of 0.50 lb, shown in Fig. 9.11, is restricted to move only in the vertical direction. The *S–N* fatigue curve for the beam includes a stress concentration factor of 2.0. This results in a fatigue exponent b of 6.4, which is shown in Fig. 9.12. The beam must operate in a white noise random vibration environment with an input PSD level of 0.30 G^2/Hz for 4.0 h. Use the three-band

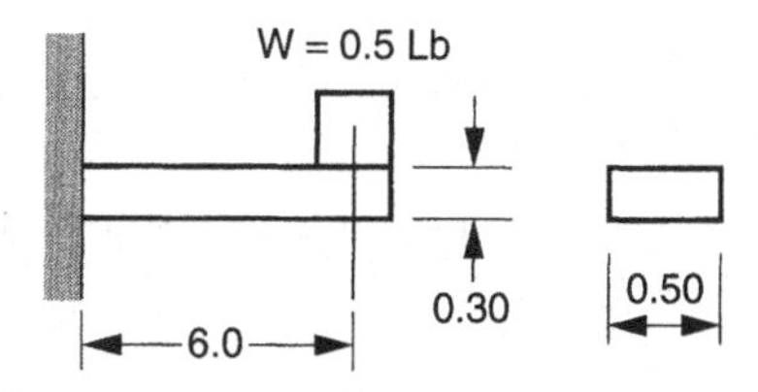

FIGURE 9.11 A cantilever beam with an end mass.

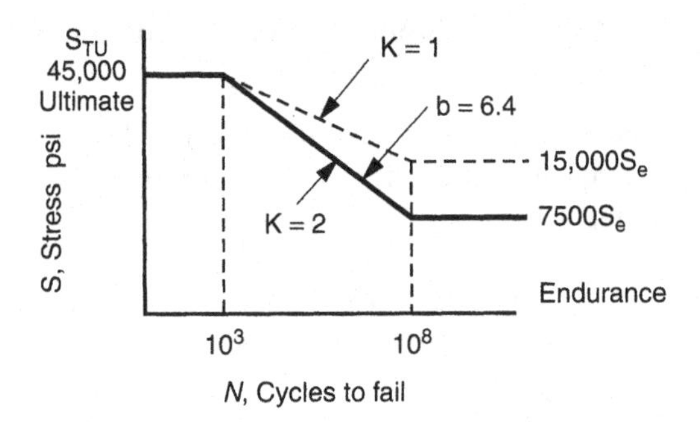

FIGURE 9.12 Typical log–log *S–N* fatigue life curve for 6061-T6 aluminum alloy with a stress concentration k factor of 2.0, which results in a b exponent slope of 6.4.

method to determine how much of the fatigue life is used up and to find the approximate fatigue life of the beam.

Solution Find the natural frequency, fatigue stress cycles, and the time to fail.

A. Cantilever Beam Natural Frequency Based on the Static Displacement

$$Y_{ST} = \frac{WL^3}{3EI} \qquad \text{static displacement, in.}$$

where $W = 0.50$ lb end load on beam
$L = 6.0$ in. beam length
$E = 10.5 \times 10^6$ lb/in.^{2}, aluminum modulus of elasticity
$I = \frac{(0.50)(0.30)^3}{12} = 0.00112$ in.^{4}, area moment of inertia

$$Y_{ST} = \frac{(0.50)(6.0)^3}{(3)(10.5 \times 10^6)(0.00112)} = 0.00306 \text{ in., static displacement} \qquad (9.22)$$

$$f_n = \frac{1}{2\pi}\sqrt{\frac{g}{Y_{ST}}} = \frac{1}{2\pi}\sqrt{\frac{386}{0.00306}} = 56.5 \text{ Hz, natural frequency} \qquad (9.23)$$

B. Response of Beam to Random Vibration Using Eq. 9.16

$P = 0.30\ G^2/\text{Hz}$, PSD input
$f_n = 56.5$ Hz, beam natural frequency
$Q = 2\sqrt{f_n} = 2\sqrt{56.5} = 15$ approximate beam transmissibility

$$G\text{ rms} = \sqrt{\frac{\pi}{2}(0.30)(56.5)(15)} = 20.0\ G \text{ rms acceleration response} \qquad (9.24)$$

C. Dynamic 1σ rms Bending Stress in Beam Acting 68.3% of the Time

$W = 0.50$ lb, weight on beam

$L = 6.0$ in., length of beam

$M = WGL = (0.50)(20)(6.0) = 60.0$ lb · in. rms, dynamic bending moment on beam

$k = 1.0$ smooth surface, no stress concentration used in the stress equation (a stress concentration of 2.0 will be used in the fatigue curve slope so $b = 6.4$)

$c = 0.30/2 = 0.15$ in., centroid neutral axis to outer beam surface

$I = 0.00112$ in.^{4}, area moment of inertia

$$1\sigma S_b = \frac{kMc}{I} = \frac{(1.0)(60.0)(0.15)}{0.00112} = 8036 \text{ lb/in.}^2 \text{rms, bending stress} \quad (9.25)$$

D. Number of Stress Cycles Required to Produce a Fatigue Failure, Three-Band Method The approximate number of stress cycles N_1 required to produce a fatigue failure in the beam for the 1σ, 2σ, and 3σ stresses can be obtained from Fig. 9.12 and

$$N_1 = N_2\left(\frac{S_2}{S_1}\right)^b \quad \text{(Ref. Eq. 5.1)}$$

where $N_2 = 1000$ cycles to fail at 45,000 lb/in.^{2}
$S_2 = 45{,}000$ lb/in.^{2}, stress at 1000 cycles
$S_1 = 8036$ lb/in.^{2} 1σ rms stress
$b = 6.4$ slope of fatigue line with the stress concentration $k = 2.0$ (be careful, do not use the stress concentration k more than one time, either in the bending stress equation or in the slope b of the fatigue line)

$$\text{rms } 1\sigma N_1 = (1000)\left(\frac{45{,}000}{8036}\right)^{6.4} = 6.14 \times 10^7 \text{ cycles to fail} \quad (9.26)$$

$$2\sigma N_2 = (1000)\left[\frac{45{,}000}{(2)(8036)}\right]^{6.4} = 7.27 \times 10^5 \text{ cycles to fail} \quad (9.27)$$

$$3\sigma N_3 = (1000)\left[\frac{45{,}000}{(3)(8036)}\right]^{6.4} = 5.43 \times 10^4 \text{ cycles to fail} \quad (9.28)$$

E. Actual Number of Stress Cycles Accumulated for 4 h of Vibration The actual number of stress cycles n accumulated during 4 h of vibration can be obtained for the 1σ, 2σ, and the 3σ conditions using the three-band technique with the natural frequency and the percent of time at each level, as shown in Eq. 9.20:

$$1\sigma n_1 = (56.5 \text{ cycle/s})(3600 \text{ s/h})(4 \text{ h})(0.683) = 5.55 \times 10^5 \text{ cycles} \quad (9.29)$$

$$2\sigma n_2 = (56.5 \text{ cycle/s})(3600 \text{ s/h})(4 \text{ h})(0.271) = 2.20 \times 10^5 \text{ cycles} \quad (9.30)$$

$$3\sigma n_3 = (56.5 \text{ cycle/s})(3600 \text{ s/h})(4 \text{ h})(0.0433) = 3.52 \times 10^4 \text{ cycles} \quad (9.31)$$

F. Use Miner's Cumulative Damage Ratio to Estimate the Beam Fatigue Life Miner's cumulative fatigue damage ratio R_n can be used to estimate the approximate fatigue life of the beam in the specified random vibration environment. Miner's ratio is based on the idea that every stress cycle uses up part of the fatigue life of a structure. It does not matter if the stress cycle is due to sine vibration, random vibration, thermal cycling, shock, or acoustic noise. A ratio n/N is used, where n is the actual number of stress cycles generated for a given condition. The number of stress cycles N, represents the number of cycles required to produce a fatigue failure at a given condition. When the ratios are all added together, a sum of 1.0 or greater means that all of the life has been used up so the part should fail. Miner's cumulative fatigue damage ratio is [11]

$$R_n = \frac{n_1}{N_1} + \frac{n_2}{N_2} + \frac{n_3}{N_3} + ++ = 1.0 \quad (9.32)$$

where $n_1 = 5.55 \times 10^5$ actual cycles accumulated for 1σ stress (Ref. Eq. 9.29)
$n_2 = 2.20 \times 10^5$ actual cycles accumulated for 2σ stress (Ref. Eq. 9.30)
$n_3 = 3.52 \times 10^4$ actual cycles accumulated for 3σ stress (Ref. Eq. 9.31)
$N_1 = 6.14 \times 10^7$ cycles to fail for 1σ stress (Ref. Eq. 9.26)
$N_2 = 7.27 \times 10^5$ cycles to fail for 2σ stress (Ref. Eq. 9.27)
$N_3 = 5.43 \times 10^4$ cycles to fail for 3σ stress (Ref. Eq. 9.28)

$$R_n = \frac{5.55 \times 10^5}{6.14 \times 10^7} + \frac{2.20 \times 10^5}{7.27 \times 10^5} + \frac{3.52 \times 10^4}{5.43 \times 10^4} = 0.960 \quad (9.33)$$

The above ratio shows that about 96% of the life is used up by the 4-h vibration test:

$$\text{Expected life} = 4.0 \text{ h} + (4.0)(1.0 - 0.960) = 4.0 + 0.16 = 4.16 \text{ h} \quad (9.34)$$

Since fatigue has a large amount of scatter, the proposed beam does not have a sufficient safety factor for the environment. The beam design should be modified to provide a fatigue life of about 8 h in the expected environment. This is equivalent to a safety factor of 2 on the fatigue life. The beam S–N fatigue curve already includes a safety factor of 2 based on the structure stress level. This resulted in a b exponent slope of 6.4. A safety factor of 2 on the life is equivalent to a structural stress safety factor of only 1.11:

$$\text{Structural safety factor} = (2.0)^{1/6.4} = 1.11 \tag{9.35}$$

9.14 SIMPLIFIED METHOD FOR ESTIMATING THE RANDOM VIBRATION FATIGUE LIFE

The three-band technique for random vibration was developed to permit the use of hand calculations for determining the approximate fatigue life of electronic equipment, without the use of a computer. The previous sample problem required a great deal of work to reach a final conclusion. A simplified method for analyzing random vibration was developed based on the total damage accumulated by the three-band technique for the percent of time exposure at the 1σ, 2σ, and 3σ levels as shown [1]:

$$\text{Damage} \qquad D = \sum NS^b = N_1 S_1^b + N_2 S_2^b + N_3 S_3^b$$

$$D = (0.683)(1.0)^{6.4} + (0.271)(2.0)^{6.4} + (0.0433)(3.0)^{6.4} = 72.55 \tag{9.36}$$

Now find one random σ acceleration level that generates the same total damage as the three band technique for the full Gaussian distribution shown above, assuming that it acts 100% of the time:

$$G^{6.4} = 72.55 \quad \text{so} \quad G = (72.55)^{1/6.4} = 1.95 \tag{9.37}$$

This means that when the 1.95σ acceleration value is assumed to act 100% of the time, it will generate the same damage as the previously combined 1σ, 2σ, and 3σ values.

The simplified relation shown above *only applies to the stress level* for finding the approximate fatigue life. This method *must not be used to determine the maximum acceleration G levels and the maximum displacements, which will still be the 3σ values*, based on the Gaussian distribution.

Sample Problem: Simplified Random Vibration Fatigue Life Analysis

Find the approximate fatigue life for the beam in the previous sample problem using the simplified random vibration analysis method.

Solution Use the 1.95σ acceleration level 100% of the time. The only item required here is the rms 1σ stress value of 8036 lb/in.^{2} as shown in Eq. 9.25, which can be substituted into Eq. 9.27 using the 1.95σ value instead of the 2σ value, to find the number of cycles to fail:

$$N_{1.95} = (1000)\left[\frac{45{,}000}{(1.95)(8036)}\right]^{6.4} = 8.55 \times 10^5 \text{ cycles to fail} \tag{9.38}$$

$$\text{Life} = \frac{8.55 \times 10^5 \text{ cycles to fail}}{(56.5 \text{ cycles/s})(3600 \text{ s/h})} = 4.20 \text{ h to fail} \tag{9.39}$$

Comparing the fatigue life of 4.16 h using the three-band technique in Eq. 9.34 with the fatigue life of 4.20 h using the short-cut method in Eq. 9.39 shows good correlation.

If the 1.95 value is rounded off to 2, the simplified analysis method will have a safety factor included, since it will result in a reduced fatigue life. This can be demonstrated using Eq. 9.27 with a 2σ fatigue life of 7.27×10^5 cycles as shown:

$$\text{Life} = \frac{7.27 \times 10^5 \text{ cycles to fail}}{(56.5 \text{ cycles/s})(3600 \text{ s/h})} = 3.57 \text{ h to fail} \tag{9.40}$$

The use of the 2σ value is equivalent to adding a very small additional safety factor on the fatigue life analysis of $4.20/3.57 = 1.176$. The safety factor on the structural stress will be much lower. With a fatigue exponent of $1/6.4$ the 1.176 value becomes only 1.025.

9.15 DESIRED PCB NATURAL FREQUENCY FOR RANDOM VIBRATION

Plug-in PCBs are usually mounted within an enclosure or chassis of some type. When this type of assembly is exposed to any dynamic environment, the chassis receives the energy first, so it represents the first degree-of-freedom. The PCBs represent the second degree-of-freedom because they receive their energy from the chassis. The response of the chassis then becomes the input to the PCBs. This is called coupling between the chassis and the PCBs, which can be very severe when the natural frequencies of the chassis and the PCBs are close together. Therefore, before any analysis can be performed on the PCBs, it is necessary to know how much coupling will occur between the chassis and the PCBs. Coupling problems can be avoided by following the octave rule. Octave means to double. When the natural frequency of each PCB is two or more times greater than the natural frequency of the chassis, the dynamic coupling between these two elements is substantially reduced. The chassis can then be ignored when the PCBs are analyzed. When the weight of any one PCB is less than 10% of the weight of the chassis, the reverse octave rule can also be applied. Here the natural frequency of the chassis is two or more times greater than the natural frequency of any PCB. When these conditions are followed, the dynamic coupling is again reduced, so the chassis can be ignored when the PCBs are analyzed. When the weight of any one PCB is much more than 10% of the chassis weight, the reverse octave rule is dangerous because of increased coupling from the chassis to the PCB. This can result in much higher PCB acceleration levels, which may result in more rapid PCB failures. The forward octave rule always works.

Combinations of extensive vibration testing and FEM analysis have shown that the approximate fatigue life of a PCB can be related to its dynamic displacement. The lead wires and solder joints of the electronic components are usually the most critical elements. A fatigue life of about 20 million stress reversals can be achieved in the critical elements when the maximum 3σ single-amplitude dynamic displacement Z of a perimeter-supported PCB is limited to the value shown below. This equation includes a structural safety factor of 1.3 to increase the probability of completing a qualification test when minor problems related to test procedures, instrumentation, or calibration require the test to be repeated a few times. The various terms are defined in Eq. 8.34:

$$Z = \frac{0.00022B}{CHr\sqrt{L}} \text{ in., } \textit{maximum} \text{ desired PCB displacement} \tag{9.41}$$

The real purpose of the above equation is to obtain the *minimum* desired PCB natural frequency that will provide a fatigue life of about 20 million stress reversals, for the most critical lead wires and solder joints of the various components mounted on the PCB. This can be accomplished by approximating the PCB as a single-degree-of-freedom system, so the number of positive zero crossings are the same as the natural frequency. Under these conditions several other equations shown below can be used for the evaluation:

$$\text{rms } G = \sqrt{\frac{\pi}{2} P f_n Q} \qquad \text{(Ref. Eq. 9.16)}$$

$$\text{rms } Z = \frac{9.8(G.\text{rms})}{(f_n)} \qquad \text{(Ref. Eq. 9.19)}$$

$$Q = \sqrt{f_n} \quad \text{approximate transmissibility}$$

Solving the above four equations simultaneously, and correcting for the 3σ value, will result in the *minimum* desired PCB natural frequency to achieve a PCB fatigue life of about 20 million stress reversals, as shown [1]:

$$f_d = \left[\frac{29.4Chr\sqrt{(\pi/2)PL}}{0.00022B}\right]^{0.8} \quad \textit{minimum} \text{ desired} \tag{9.42}$$

Sample Problem: Finding the Minimum Desired PCB Natural Frequency

A 2.0-in.-long hybrid with two rows of pins extending from the bottom surface can be surface mounted or through-hole mounted on a plug-in epoxy fiberglass PCB that measures 7.0 × 9.0 × 0.090 in. thick. The PSD input in the area of the natural frequency is expected to be flat with a value of 0.090 G^2/Hz, and the component will be mounted parallel to the 7.0-in, edge. Find the minimum desired PCB natural frequency and the approximate fatigue life of the lead wires and solder joints, for the two-component mounting positions shown below:

(a) Component mounted at the center of the PCB at $X = a/2$ and $Y = b/2$.
(b) Component mounted at the quarter points of the PCB at $X = a/4$ and $Y = b/4$.

Solution

Part A Substitute the following data into Eq. (9.42) where

$B = 7.0$ in., length of PCB parallel to the hybrid length
$h = 0.090$ in., PCB thickness
$L = 2.0$ in., length of hybrid component
$C = 1.26$ component constant for wires extending from bottom surface
$P = 0.090\ G^2/\text{Hz}$, PSD input to the PCB
$r = 1.0$ relative position factor for component at center of PCB

$$f_d = \left[\frac{(29.4)(1.26)(0.090)(1.0)\sqrt{(\pi/2)(0.090)(2.0)}}{(0.00022)(7.0)}\right]^{0.8} = 281 \text{ Hz, desired} \quad (9.43)$$

The approximate fatigue life can be obtained as

$$\text{Life} = \frac{20 \times 10^6 \text{ cycles to fail}}{(281 \text{ cycles/s})(3600 \text{ s/h})} = 19.77 \text{ h to fail} \quad (9.44)$$

A natural frequency of 281 Hz will be difficult to achieve with the PCB shown above.

Some type of stiffening will have to be added to the PCB. This could be adding stiffening ribs or perhaps an aluminum heat sink could be bonded to the PCB to also improve the thermal performance. Adding stiffening ribs may reduce the amount of PCB surface area available for mounting components. This could increase the number of PCBs required, which will make many electrical engineers and program managers very unhappy. Adding an aluminum heat sink to the PCB will increase the size, weight, and cost, which will make many management people very unhappy. Sometimes snubbers can be added near the center of the PCBs to solve severe dynamic problems. Ideas of this type should be carefully evaluated against the possibility of failing a qualification test or having a large number of field failures.

One easy solution may be just relocating the largest and most critical component away from the center of the PCB, which is the most critical location. This must be worked out with the electrical engineers and circuit designers.

Solution

Part B Substitute the following information into Eq. 9.42. The only change required is the component relative position factor, from 1.0 at the center of the PCB to 0.50 for the quarter mounting point. The results are

$$f_d = 161 \text{ Hz desired} \tag{9.45}$$

$$\text{Life} = \frac{20 \times 10^6 \text{ cycles to fail}}{161 \text{ cycles/s}(3600 \text{ s/h})} = 34.51 \text{ h to fail} \tag{9.46}$$

9.16 IGNORE THE INPUT rms *G* LEVEL AND CONCENTRATE ON THE INPUT PSD LEVEL

Input rms acceleration levels are often used to estimate the severity of random vibration environments. A high input rms acceleration level becomes cause for concern because it is taken to mean that a rugged, heavy, expensive electronic system must be developed. Past experience has shown that this usually involves a lot of time-consuming testing and analysis. Although a high input rms acceleration level can be of some importance in judging the severity of the random vibration environment, it is far more important to concentrate on the value of the input PSD level in the region of the structural natural frequency.

Random vibration acceleration levels are directly related to the square root of the areas under the input and the response curves, which are typically plotted on log–log paper. There are an infinite number of different input PSD profiles that will have the same areas that will result in exactly the same rms input acceleration levels. Some input random vibration profiles with exactly the same areas are shown in Fig. 9.13. The responses to these different input profiles can be substantially different, depending on the value of the PSD level at the natural frequency, which is demonstrated by Eq. 9.16. It shows that the response of the system is related to the

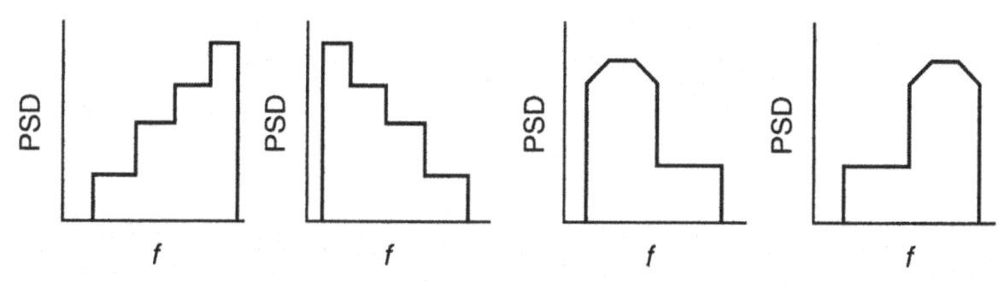

FIGURE 9.13 Various random vibration input curves with the same areas but with different PSD values so they will all have the same RMS input acceleration levels.

product of the PSD level and the natural frequency. For a given natural frequency, a low PSD input level will result in a low *G* rms response. For the same natural frequency, a high PSD input level will result in a high *G* rms response. Therefore, the *G* rms response levels can be different for each different PSD profile depending on the value of the PSD level at the natural frequency, even when all of the different profiles have the same areas and the same input rms *G* acceleration levels. A larger area under the input PSD profile can result in a larger input acceleration *G* level with a decreased *G* rms response, when the PSD level at the natural frequency is decreased. A high input acceleration *G* level does not mean that the system will have a high response acceleration *G* level. A low input acceleration *G* level does not mean that the system will have a low response acceleration *G* level.

A white noise input PSD profile is characterized by a flat-top horizontal PSD input curve. When the white noise input PSD level is increased, it will always increase the input acceleration *G* level and the response acceleration *G* level as long as the natural frequency of the system is within the frequency bandwidth.

CHAPTER 10

Combining Fatigue Damage for Random Vibration and Thermal Cycling

10.1 INTRODUCTION

Testing experience with many different thermal cycling and vibration environments has shown that the stress level and the number of alternating stress cycles have a very strong influence on the fatigue life of a structure. Miner [61] suggested the use of a ratio involving the actual number of stress cycles n from several environments, divided by the number of stress cycles required to produce a failure N in several environments. Fatigue life data for many different materials used in the manufacturing of electronic assemblies must be available to perform these analyses. This type of data is readily available for most of the more common structural materials. Many special materials are often used in electronic systems that have not been tested to failure; thus fatigue data are not available. These materials should be tested by experienced testing laboratories to establish the fatigue life data so that the data can be used to analyze the electronic systems.

Miner's cumulative fatigue damage ratio is a convenient way to find the approximate fatigue life of electronic systems that will experience fatigue damage due to operation in many different environments such as thermal cycling, sinusoidal vibration, random vibration, shock, and acoustic noise. This simple theory assumes that every stress reversal produces damage by using up a small part of the fatigue life in a structure. The damage accumulated in the various environments is assumed to be linear. This allows the damage from the individual environments to be simply added together to obtain the total damage. When the sum of all the ratios equals 1.0 or more, all of the life has been used up so the part will fail. This is an easy way to estimate the fatigue life. Miner's ratio is a relatively simple method for combining the damage accumulated in different environments that normally requires more complex computer analyses. Miner's cumulative damage ratio is

$$R_n = \frac{n_1}{N_1} + \frac{n_2}{N_2} + \frac{n_3}{N_3} + \frac{n_4}{N_4} + + + = 1.0 \quad \text{(Ref. Eq. 9.32)}$$

Sample Problem: Thermal Cycling and Vibration Fatigue Life of Electronics

An electronic manufacturer designs and builds control systems that monitor engine combustion performance. These systems are usually mounted in the engine compartments of delivery trucks. The outside ambient air is expected to be 30°C. Since engine areas tend to run hot, the manufacturer uses 8.0 × 6.0 × 0.060-in.-thick plug-in polyimide glass printed circuit boards (PCBs) because of their ability to withstand higher temperatures. Prior to shipping the products to the customers, the manufacturer plans to subject his electronics to an environmental stress screening (ESS) process. This will weed out latent manufacturing defects, which could cause premature failures in the field. The ESS consists of 40 rapid thermal cycles over a temperature range from −40 to +100°C, plus broadband random vibration with an input power spectral density (PSD) level of 0.045 G^2/Hz for a total of 1.5 h. The electronics manufacturer can also store his finished electronics systems for 10 years to ensure the availability of spare parts. The storage areas in the northern countries are expected to experience diurnal temperature changes of about 40°C.

A typical truck is expected to make 5 deliveries a day, 5 days per week, 48 weeks per year, for a period of 7 years. The engine is turned off every time the truck stops to make a delivery. The truck is expected to be used about 10 h a day. The average delivery time is expected to take about 48 min. For 5 stops this is a total of 4 h per day. The truck will be driven for 6 h per day, 5 days per week, 48 weeks per year, for a period of 7 years. The truck is expected to drive on city streets 40% of the time, where the average random vibration input level to the electronics is 0.020 G^2/Hz. The slow stop-and-go city traffic will result in a hot engine area where an average temperature rise of 70°C is expected for the electronics. The truck is expected to drive at higher speeds on highways 60% of the time, where the average random vibration input level to the electronics is 0.030 G^2/Hz. The higher speeds will provide better cooling for the engine area, where an average temperature rise of 60°C is expected for the electronics.

Find the stress levels generated in the hybrid lead wires and solder joints by the thermal cycling and random vibration conditions. Estimate the fatigue life of the hybrid shown in Fig. 10.1, for a surface-mounted and a through-hole-mounted condition, when the component is mounted off center, near the corners, parallel to the 8-in. edge.

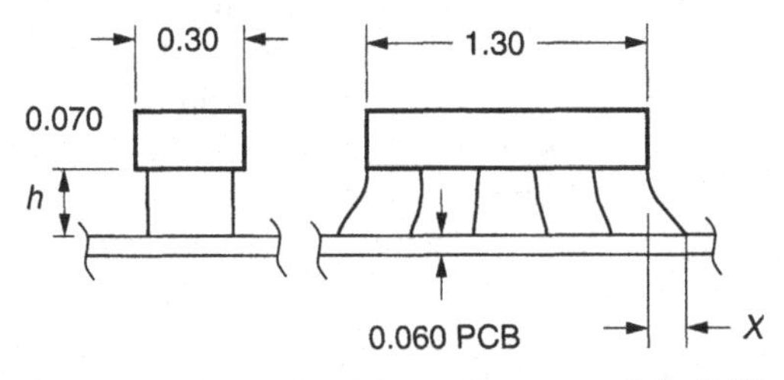

FIGURE 10.1 Differences in the TCEs of the component and the PCB will force long lead wires to bend and displace in thermal cycling conditions.

Solution

Use Miner's Damage Ratio to Find the Approximate Fatigue Life Random vibration response relations and thermal cycling equilibrium relations must be evaluated for the lead wires and solder joints in order to find the fatigue life. The investigations will be broken up into different random vibration and thermal cycling segments to make the analyses easier to follow. Start with the random vibration.

10.2 RANDOM VIBRATION FATIGUE DAMAGE IN LEAD WIRES AND SOLDER JOINTS

The total number of hours for driving is expected to be

$$(6.0\ \text{h/day})(5\ \text{days per week})(48\ \text{weeks/year})(7\ \text{years}) = 10{,}080\ \text{h} \qquad (10.1)$$

The number of hours of vibration and the PSD levels required are as follows:

A. Highway driving 60% $= (0.60)(10{,}080) = 6048$ h at 0.030 G^2/Hz (10.2)

B. City driving is 40% $= (0.40)(10{,}080) = 4032$ h at 0.020 G^2/Hz (10.3)

C. ESS production screen $= 1.5$ h at 0.045 G^2/Hz (10.4)

Solution A

Random Vibration for Lead Wires Highway Driving Evaluate random vibration condition A, highway driving 6048 h at 0.030 G^2/Hz. The minimum desired plug-in PCB natural frequency to achieve a component fatigue life of about 20 million stress cycles can be obtained from

$$f_d = \left[\frac{29.4Chr\sqrt{(\pi/2)PL}}{0.00022B}\right]^{0.8} \qquad \text{(Ref. Eq. 9.43)}$$

where $C = 1.26$ hybrid, with two rows of wires extending from the bottom surface
$h = 0.060$ in., thickness of polyimide glass PCB
$r = 0.50$ position factor for components mounted at $X = a/4$ and $Y = b/4$
$P = 0.030\ G^2/\text{Hz}$, PSD input to PCB for highway speed
$L = 1.3$ in., length of hybrid component
$B = 8.0$ in., length of PCB parallel to component length

$$f_d = \left[\frac{(29.4)(1.26)(0.060)(0.50)\sqrt{(\pi/2)(0.030)(1.30)}}{(0.00022)(8.0)}\right]^{0.8} = 56.9\ \text{Hz} \qquad (10.5)$$

The approximate fatigue life of the hybrid can be obtained from the estimated 20-million-cycle fatigue life for random vibration as shown below:

$$T_1 = \text{life} = \frac{20 \times 10^6 \text{ cycles to fail}}{(56.9 \text{ cycles/s})(3600 \text{ s/h})} = 97.6 \text{ h to fail} \tag{10.6}$$

The rms PCB acceleration G response to the random vibration input PSD level can be estimated from Eq. 9.16 using the following information:

$P = 0.030\ G^2\text{Hz}$, PSD input level for highway driving condition A
$f_n = 56.9$ Hz, minimum desired natural frequency
$Q = \sqrt{f_n} = \sqrt{56.9} = 7.5$ approximate PCB transmissibility

$$G = \sqrt{\frac{\pi}{2}(0.030)(56.9)(7.5)} = 4.48 \text{ rms, PCB response} \tag{10.7}$$

The approximate dynamic rms single-amplitude displacement of the PCB can be obtained from Eq. 9.19 using the following information:

$G = 4.48$ rms response (Ref. Eq. 10.7)
$f_n = 56.9$ Hz (Ref. Eq. 10.5)

$$Z_1 = \frac{9.8G}{f_n^2} = \frac{(9.8)(4.48)}{(56.9)^2} = 0.0136 \text{ in. rms} \tag{10.8}$$

The minimum required service life is 6048 h for highway driving, as shown by Eq. 10.2, so the expected life of 97.6 h shown by Eq. 10.6 for a natural frequency of 56.9 Hz is not adequate. A higher natural frequency is required to increase the fatigue life. An iterative process can be used where a natural frequency is assumed and the calculations are made. If the first assumption does not work, another iteration is made with another frequency to find one that will work. An adequate solution can usually be found with two or three iterations. Start with an assumed PCB natural frequency of 125 Hz and repeat the above analysis where

$P = 0.030\ G^2/\text{Hz}$, PSD input level for highway driving
$f_n = 125$ Hz, assumed PCB natural frequency
$Q = \sqrt{125} = 11.2$ approximate PCB transmissibility

$$G = \sqrt{\frac{\pi}{2}(0.030)(125)(11.2)} = 8.12\ G \text{ rms, PCB response} \tag{10.9}$$

The dynamic rms single-amplitude displacement of the PCB for the 125-Hz natural frequency can be obtained from Eq. 9.19 as follows:

$$Z_2 = \frac{(9.8)(8.12)}{(125)^2} = 0.0051 \text{ in. rms} \tag{10.10}$$

The approximate PCB fatigue life or time to fail (T_F) for the 125-Hz natural frequency for condition A, highway driving Eq. 10.2, can be obtained from the equation shown below, using Eq. 5.4 as a reference:

$$T_F = T_1 \left(\frac{Z_1}{Z_2}\right)^b \tag{10.11}$$

where $T_1 = 97.6$ h, time to fail for 56.9 Hz (Ref. Eq. 10.6)
$Z_1 = 0.0136$ in. rms PCB displacement for 56.9 Hz (Ref. Eq. 10.8)
$Z_2 = 0.0051$ in. rms PCB displacement for 125 Hz (Ref. Eq. 10.10)
$b = 6.4$ fatigue exponent, S–N curve for kovar wire, which includes a stress concentration $k = 2$ for lead wires with small cuts, nicks, and scratches

$$T_{\text{FA}} = (97.6)\left(\frac{0.0136}{0.0051}\right)^{6.4} = 5.2 \times 10^4 \text{ hours to fail} \tag{10.12}$$

Condition A: Wire Random Vibration Fatigue Cycle Ratio Highway Driving The fatigue cycle ratio showing the amount of life used up by condition A highway driving can be obtained from Eqs. 10.2 and 10.12 using Miner's damage method:

$$R_A = \frac{T_A}{T_{\text{FA}}} = \frac{6048}{5.2 \times 10^4} = 0.11631 \quad \text{wire vibration} \tag{10.13}$$

Solution B

Random Vibration for Component Lead Wires, City Driving Evaluate random vibration condition B city driving 4032 h at 0.020 G^2/Hz. The minimum desired plug-in PCB natural frequency to achieve a fatigue life of about 20 million stress reversals can be obtained using Eqs. 9.27 and 10.5:

$$f_d = \left[\frac{(29.4)(1.26)(0.060)(0.50)\sqrt{(\pi/2)(0.020)(1.30)}}{(0.00022)(8.0)}\right]^{0.8} = 48.4 \text{ Hz} \tag{10.14}$$

The approximate fatigue life of the hybrid can be obtained from the estimated 20-million-cycle fatigue life for random vibration as shown below:

$$T_1 = \text{life} = \frac{20 \times 10^6 \text{ cycles to fail}}{(48.4 \text{ cycles/s})(3600 \text{ s/h})} = 114.8 \text{ h to fail} \tag{10.15}$$

The approximate rms PCB acceleration G response to random vibration can be obtained using Eq. 9.16 with the following information:

$P = 0.020\ G^2/\text{Hz}$, PSD input level for city driving condition B

$f_n = 48.4$ Hz, minimum desired PCB natural frequency

$Q = \sqrt{48.4} = 7.0$ approximate PCB transmissibility

$$G = \sqrt{\frac{\pi}{2}(0.020)(48.4)(7.0)} = 3.26 \text{ rms PCB response} \tag{10.16}$$

The dynamic single-amplitude displacement of the PCB can be obtained from Eq. 9.19 as follows:

$$Z_1 = \frac{(9.8)(3.26)}{(48.4)^2} = 0.0136 \text{ in. rms} \tag{10.17}$$

The minimum required service life is 4032 h for city driving, as shown by Eq. 10.3; so the expected fatigue life of 114.8 h for a natural frequency of 48.4 Hz shown by Eq. 10.14 is not adequate. A higher natural frequency is required to increase the PCB fatigue life. Increase the PCB natural frequency to 125 Hz, as shown in Eq. 10.9, and repeat the previous analysis starting with Eq. 10.16:

$$G = \sqrt{\frac{\pi}{2}(0.020)(125)\sqrt{125}} = 6.62 \text{ rms, PCB response} \tag{10.18}$$

The approximate dynamic rms single-amplitude displacement of the PCB for the 125-Hz natural frequency can be obtained from Eq. 9.19 as follows:

$$Z_2 = \frac{(9.8)(6.62)}{(125)^2} = 0.0042 \text{ in. rms} \tag{10.19}$$

The approximate fatigue life or time to fail (T_F) for the 125-Hz natural frequency for condition B city driving, Eq. 10.3, can be obtained from Eq. 10.11 using the following information for the kovar wire:

$T_1 = 114.8$ h, time to fail for 48.4 Hz (Ref. Eq. 10.15)

$Z_1 = 0.0136$ in. rms, PCB displacement for 48.4 Hz (Ref. Eq. 10.17)

$Z_2 = 0.0042$ in. rms, PCB displacement for 125 Hz (Ref. Eq. 10.19)

$B = 6.4$ fatigue exponent, S–N curve with a stress concentration $k = 2$

$$T_{\text{FB}} = (114.8)\left(\frac{0.0136}{0.0042}\right)^{6.4} = 2.12 \times 10^5 \text{ hours to fail} \tag{10.20}$$

Condition B: Wire Random Vibration Fatigue Cycle Ratio, City Driving The fatigue cycle ratio showing the amount of life used up by condition B city driving can be obtained from Eqs. 10.3 and 10.20 as follows:

$$R_B = \frac{T_B}{T_{\text{FB}}} = \frac{4032}{2.12 \times 10^5} = 0.01902 \text{ wire vibration} \tag{10.21}$$

Solution C

Random Vibration for Lead Wires ESS Evaluate random vibration condition C, ESS for 1.5 h at 0.045 G^2/Hz. Environmental stress screening is an extension of the production process, where electronic equipment is stimulated with some type of dynamic environment to try to precipitate latent defects that might otherwise show up as failures in the field. When the failures occur in the factory, they can be repaired before the equipment is shipped to the consumer. The early failures, often called the infant mortality group, are eliminated so the customer receives a more reliable product with a longer operating life.

Thermal cycling and random vibration are the two most popular dynamic conditions that are being used by industry, with great success, to stimulate the electronic equipment. These environments should not be used as criteria for designing or manufacturing the electronic hardware. The sole purpose of imposing these environments on the electronics is to try to quickly force hidden defects into patent failures, which are easy to recognize and repair so there is a substantial cost savings with improved reliability.

Every stress cycle uses up some small part of the useful life of a structure. The ESS random vibration is no exception; it will also use up some of the life. The method of analysis used for parts A and B can also be used here to obtain the fatigue cycle ratio. The minimum desired PCB natural frequency to achieve a fatigue life of about 20 million stress reversals can be obtained using Eqs. 9.27 and 10.5 as follows:

$$f_d = \left[\frac{(29.4)(1.26)(0.060)(0.50)\sqrt{(\pi/2)(0.045)(1.3)}}{(0.00022)(8.0)}\right] = 66.9 \text{ Hz} \qquad (10.22)$$

The approximate fatigue life of the hybrid can be obtained from the estimated 20-million-cycle fatigue life for random vibration as shown below:

$$T_1 = \text{life} = \frac{20 \times 10^6 \text{ cycles to fail}}{(66.9 \text{ cycles/s})(3600 \text{ s/h})} = 83.0 \text{ h to fail} \qquad (10.23)$$

The rms PCB acceleration G response to random vibration can be obtained using Eq. 9.16 with the following data:

$P = 0.045\ G^2/\text{Hz}$, PSD input level for ESS condition C

$f_n = 66.9$ Hz, minimum desired PCB natural frequency

$Q = \sqrt{66.9} = 8.2$ approximate PCB transmissibility

$$G = \sqrt{\frac{\pi}{2}(0.045)(66.9)(8.2)} = 6.23 \text{ rms, PCB response} \qquad (10.24)$$

The dynamic rms single-amplitude displacement of the PCB for the 66.9-Hz natural frequency can be obtained from Eq. 6.19 as follows:

$$Z_1 = \frac{(9.8)(6.23)}{(66.9)^2} = 0.0136 \text{ in. rms} \qquad (10.25)$$

It was previously shown in Eq. 10.9 that a PCB natural frequency of about 125 Hz was required to provide an adequate random vibration fatigue life for condition A, highway driving. Therefore, the same 125-Hz PCB natural frequency must also be used for the ESS process condition C.

PCB rms response with a 125-Hz natural frequency and an ESS random vibration input PSD level of 0.045 G^2/Hz is shown:

$$G = \sqrt{\frac{\pi}{2}(0.045)(125)\sqrt{125}} = 9.94 \text{ rms, PCB response} \tag{10.26}$$

The dynamic rms single-amplitude displacement for the 125-Hz PCB natural frequency can be obtained from Eq. 9.19 as shown:

$$Z_2 = \frac{(9.8)(9.94)}{(125)^2} = 0.0062 \text{ in. rms} \tag{10.27}$$

The approximate wire fatigue life or time to fail (T_F) for the 125-Hz PCB natural frequency for condition C, ESS can be obtained from Eq. 10.11 using the following data:

$T_1 = 83.0$ h, time to fail for 66.9 Hz (Ref. Eq. 10.23)
$Z_1 = 0.0136$ in., rms PCB displacement for 66.9 Hz (Ref. Eq. 10.25)
$Z_2 = 0.0062$in., rms PCB displacement for 125 Hz (Ref. Eq. 10.27)
$b = 6.4$ fatigue exponent S–N curve with stress concentration $k = 2$

$$T_{\text{FC}} = (83.0)\left(\frac{0.0136}{0.0062}\right)^{6.4} = 1.27 \times 10^4 \text{ h to fail} \tag{10.28}$$

Condition C: Wire Random Vibration Fatigue Cycle Ratio ESS The fatigue cycle ratio showing the amount of wire life used up by condition C ESS of 1.5 h random vibration can be obtained from Eqs. 10.4 and 10.28 as follows:

$$R_C = \frac{T_C}{T_{\text{FC}}} = \frac{1.5}{1.27 \times 10^4} = 0.00012 \text{ wire vibration} \tag{10.29}$$

10.3 TOTAL VIBRATION FATIGUE DAMAGE ACCUMULATED IN LEAD WIRES AND SOLDER JOINTS

The total fatigue damage accumulated in the three random vibration events will be the sum of the three damage ratios from Eqs. 10.13, 10.21, and 10.29 as shown below. The ratio shown below will be used to apply to both the lead wires and the solder joints. Test data shows there are more lead wire failures than solder joint failures due to vibration for through-hole and for surface-mounted components on PCBs. Since this is difficult to quantify accurately, it is assumed that there are an equal number of failures in the lead wires and solder joints, so the ratio shown below will be used for both the lead wires and solder joint random vibration cycling fatigue damage accumulation:

$$R_{\mathrm{RV}} = R_A + R_B + R_C = 0.11631 + 0.01902 + 0.00012 = 0.13545$$

$$R_{\mathrm{RV}} = 0.13545 \text{ random vibration damage ratio for lead wires} \qquad (10.30)$$

$$R_{\mathrm{RV}} = 0.13545 \text{ random vibration damage ratio for solder joints} \qquad (10.31)$$

10.4 THERMAL CYCLING FATIGUE DAMAGE IN LEAD WIRES AND SOLDER JOINTS

The total number of thermal cycles expected for the life of the truck is as follows:

$$n = \left(\frac{5 \text{ deliveries}}{\text{day}}\right)\left(\frac{5 \text{ days}}{\text{week}}\right)\left(\frac{48 \text{ weeks}}{\text{year}}\right)(7 \text{ years}) = 8400 \text{ cycles} \qquad (10.32)$$

The number of thermal cycles expected in the various environments are as follows:

A. Highway driving 60% = (0.60)(8400) = 5040 thermal cycles expected (10.33)

B. City driving 40% = (0.40)(8400) = 3360 thermal cycles expected (10.34)

C. ESS production screen rapid cycles = 40 thermal cycles expected (10.35)

D. Storage for 10 years = (10)(365) = 3650 thermal cycles expected (10.36)

Solution A

Thermal Cycling, General Approach Highway Driving A temperature rise of about 60°C is expected in the engine compartment, in the area of the electronics for condition A, highway driving. With an ambient temperature of 30°C, it will result in an engine compartment temperature of 60 + 30 = 90°C. Temperatures this high will produce creep in the solder joints, which will allow the solder stresses to relax toward a lower stress level, as shown in Figs. 1.3, 1.4, 1.6, and Table 2.1. This tends to increase the average effective solder stress, based on the amount of stress relaxation expected in the solder joint. For a peak temperature of 90°C, with an expected cool down to 45°C in 48 min, it results in a temperature change of 45°C. Figure 1.6 shows a solder stress creep reduction of 33%. A correction factor of 1.33 is expected here because of the high average engine compartment temperature. This will result in the equivalent temperature rise that will be used to calculate the forces and stresses in the lead wires and solder joints due to solder creep and stress relaxation from Fig. 1.6, as shown below.

$$\text{Condition A: creep equivalent temperature rise} = 1.33 \times 45^\circ\text{C} = 60^\circ\text{C} \quad (10.37)$$

Differences in the thermal coefficient of expansion (TCE) between the hybrid and the PCB will result in expansion differences in thermal cycling conditions. This will force the lead wires on the hybrid to bend, as shown in Fig. 10.1. Bending stresses will be produced in the lead wires and shear tear-out stresses will be produced in through-hole solder joints:

Condition A: Wire Thermal Cycling Fatigue Analysis Highway Driving The wire bending stiffness is very small compared to the axial stiffness of the PCB and the body of the hybrid. Therefore, virtually all of the deflection difference, shown as X, between the PCB and the hybrid will go into the bending deflection of the lead wires. An equilibrium equation can be set up for the expansion differences, where the subscripts H and P refer to the hybrid and PCB:

$$X = (\alpha_P - \alpha_H)L\,\Delta t \quad (10.38)$$

where $\alpha_P = 13 \times 10^{-6}$ in./in./°C, TCE of polyimide glass PCB

$\alpha_H = 6 \times 10^{-6}$ in./in./°C, TCE of kovar hybrid component body

$L = \dfrac{\sqrt{(1.3)^2 + (0.30)^2}}{2} = 0.667$ in., hybrid diagonal dimension to centroid

$\Delta t = 60^\circ$C, including solder stress relaxation (Ref. Eq. 10.37)

$$X = (13 \times 10^{-6} - 6 \times 10^{-6})(0.667)(60) = 0.000280 \text{ in.} \quad (10.39)$$

The displacement shown above will also be the bending displacement of the end lead wires, similar to a fixed–fixed beam displaced laterally at one end shown in Fig. 10.2:

$$X = \frac{P_W h^3}{12EI} \quad (10.40)$$

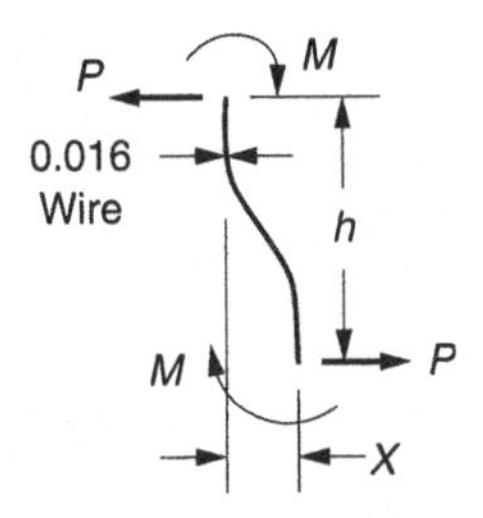

FIGURE 10.2 Component end lead wires will bend and displace the most during thermal cycling conditions.

where $X = 0.000280$ in., wire bending displacement
$h = 0.070 + 2(0.016) = 0.102$ in., wire height includes two wire diameters
$E = 20 \times 10^6$ lb/in.^{2}, kovar wire modulus of elasticity
$I = \dfrac{\pi(0.016)^4}{64} = 3.22 \times 10^{-9}$, in.^{4}, wire moment of inertia

$$P_W = \frac{(12)(20 \times 10^6)(3.22 \times 10^{-9})(0.000280)}{(0.102)^3} = 0.204 \text{ lb} \tag{10.41}$$

The bending moment in the end wire is also the overturning moment in the solder joint of a through-hole PCB and can be obtained from the sum of the moments as shown below:

$$M = \frac{P_W h}{2} = \frac{(0.204)(0.102)}{2} = 0.0104 \text{ lb} \cdot \text{in., moment} \tag{10.42}$$

The bending stress in the wire can be obtained from the standard bending stress equation. A stress concentration factor is not used here because the number of thermal cycles is not really large enough to have a significant effect on the fatigue life, so $k = 1$.

$$S_W = \frac{Mc}{I} = \frac{(0.0104)(0.016/2)}{3.22 \times 10^{-9}} = 25{,}838 \text{ lb/in.}^2 \text{ wire thermal stress} \tag{10.43}$$

Condition A: Wire Thermal Cycling Fatigue Damage Ratio Highway Driving The thermal cycling fatigue damage in the wire can be obtained from the actual number of thermal cycles (n) expected for condition A from Eq. 10.33 as 5040 cycles. Also the number of cycles (N) required to produce a fatigue failure in the kovar wire at the higher endurance stress level of 28,000 psi in Fig. 10.3 is about 100 million cycles. The wire damage due to thermal cycling is not significant as shown below:

$$R_{AW} = \frac{n}{N} = \frac{5040}{100 \times 10^6} = 0.00 \text{ wire thermal} \tag{10.44}$$

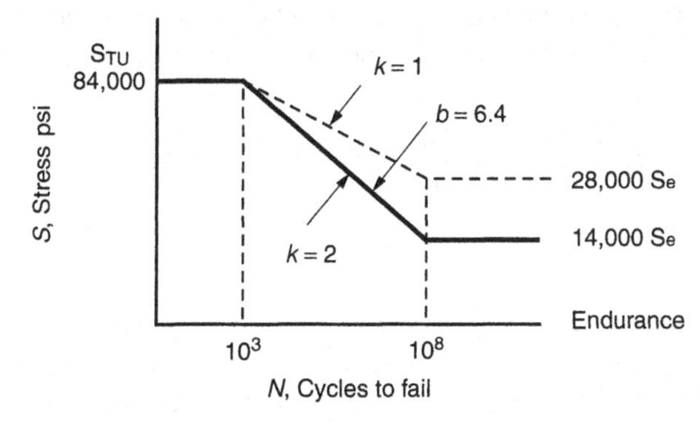

FIGURE 10.3 Typical log–log S–N fatigue life curve for kovar wire with a stress concentration k value of 2.0, which results in a b exponent slope of 6.4.

Condition A: Solder Thermal Cycling Fatigue Analysis Highway Driving The shear tear-out stress in the solder joint for through-hole mounting will appear as shown in Fig. 2.6. The shear tearout stress in the solder joint can be obtained from Eq. 2.15:

$$M = 0.0104 \text{ lb} \cdot \text{in. (Ref. Eq. 10.42)}$$

$$h = 0.060 \text{ in., PCB thickness, assumed to be the same as solder joint height}$$

$$d_{AV} = (0.016 + 0.028)/2 = 0.0220 \text{ in., average solder joint diameter}$$

$$A_S = \frac{\pi(d_{AV})^2}{4} = \frac{\pi(0.0220)^2}{4} = 3.80 \times 10^{-4} \text{ in.}^2\text{, average solder joint area}$$

$$S_{ST} = \frac{M}{hA_S} = \frac{0.0104}{(0.060)(3.80 \times 10^{-4})} = 456 \text{ lb/in.}^2 \text{ solder stress} \tag{10.45}$$

The approximate number of thermal cycles required to produce a fatigue failure in the solder joint for 63/37 tin/lead solder for condition A can be obtained from the solder S–N fatigue curve shown in Fig. 5.3, Fig. 5.6, and Eq. 5.4 as a reference:

$$N_{AS} = N_1 \left(\frac{S_1}{S_2}\right)^b \tag{10.46}$$

where $N_1 = 80{,}000$ cycles, reference point for solder failure at 200 lb/in.^{2}
$S_1 = 200$ lb/in.^{2}, reference point for solder failure
$S_2 = 456$ lb/in.^{2}, solder shear tearout stress (Ref. Eq. 10.45)
$b = 2.5$ slope of solder fatigue curve on a log– log S–N plot

$$N_{AS} = (80{,}000)\left(\frac{200}{456}\right)^{2.5} = 10{,}192 \text{ cycles to fail} \tag{10.47}$$

Condition A: Solder Thermal Cycling Fatigue Damage Ratio Highway Driving The thermal cycling fatigue damage ratio for the solder for condition A can be obtained from the actual number of thermal cycles n, expected for condition A shown by Eq. 10.3 as 5040 cycles and the number of cycles to produce a failure shown above:

$$R_{AS} = \frac{n}{N_{AS}} = \frac{5040}{10{,}192} = 0.494 \text{ solder thermal} \tag{10.48}$$

Solution B

Thermal Cycling, General Approach City Driving Condition B involves city driving, which will be more severe than condition A with an expected temperature rise of 70°C due to slower driving and more stops with the engine running. This will result in an engine compartment temperature of $70 + 30 = 100$°C. The higher temperature will produce slightly more creep than the highway driving. Figure 1.6

shows the solder creep is expected to be about 40%. This will result in a solder stress correction factor of about 1.4. The temperature drop during the 48-min delivery stop is expected to be about 50°C:

$$\text{Condition B creep equivalent temperature rise} = 1.4 \times 50 = 70^\circ\text{C} \qquad (10.49)$$

The only change in the thermal expansion (Eq. 10.38) will be the temperature rise from 60 to 70°C. Therefore, a direct ratio of the temperatures can be used for the solutions.

Condition B: Wire Thermal Cycling Fatigue Analysis City Driving There will be a wire bending stress increase of 70/60 or 1.167 times greater than the value shown in Eq. 10.43. However, the bending stress will still be under the kovar endurance stress of 28,000 psi at 100 million cycles. Therefore, the damage ratio shown in Eq. 10.44 of 0.00 will stay the same.

A quick examination of thermal cycling conditions C and D shows that the wire bending stresses will never be high enough to produce any significant damage. The wire fatigue damage ratio will then always be the same with a value of about 0.00:

$$R_{BW} = R_{CW} = R_{DW} = 0.00 \text{ wire thermal} \qquad (10.50)$$

Condition B: Solder Thermal Cycling Fatigue Analysis City Driving The solder shear tear-out stress of 456 psi shown in Eq. 10.45 will increase by the direct ratio of the temperature rise of 70/60 or 1.167. The new fatigue life of the solder can be obtained from Eq. 10.46 using this stress increase, as shown below:

$$N_{BS} = (80{,}000)\left[\frac{200}{(1.167)(456)}\right]^{2.5} = 6927 \text{ cycles to fail} \qquad (10.51)$$

Condition B: Solder Thermal Cycle Fatigue Damage Ratio City Driving The thermal cycle fatigue damage ratio for condition B can be obtained from the 3360 actual number of cycles n expected, as shown in Eq. 10.34 and the number of cycles required to produce a solder joint failure as shown above:

$$R_{BS} = \frac{n}{N_{BS}} = \left[\frac{3360}{6927}\right] = 0.485 \text{ solder thermal} \qquad (10.52)$$

Condition C: Thermal Cycling ESS General Approach Condition C, ESS, involves 40 rapid thermal cycles from −40 to +100°C. This is a change of 140°C. The solder will not have a chance to creep under these conditions. So only half of the total temperature change, or 70°C, will be used to calculate the solder joint shear tear-out stress.

Condition C: Solder Thermal Cycling Fatigue Analysis ESS The ESS condition C will increase the solder shear tear-out stress of 456 psi shown in Eq.

10.45 by a ratio of 70/60 or 1.167. The new solder fatigue life can be obtained using Eq. 10.46 as shown below:

$$N_{CS} = (80{,}000)\left[\frac{200}{[1.167](456)}\right]^{2.5} = 6927 \text{ cycles to fail} \tag{10.53}$$

Condition C: Solder Thermal Cycling Fatigue Damage Ratio ESS The thermal cycle fatigue damage ratio for the ESS condition C can be obtained from the 40 actual number of cycles n expected and the number of cycles required to produce a solder joint failure as shown above:

$$R_{CS} = \frac{n}{N_{CS}} = \frac{40}{6927} = 0.0058 \text{ solder thermal} \tag{10.54}$$

Condition D: Thermal Cycling Storage General Approach Temperature changes usually take place very slowly in storage areas, depending on the size and insulating properties of the storage facilities. Data from military storage of small electronic systems, about the size of a shoe box, in small igloos in northern climates, has shown diurnal temperature changes as high as 40°C. Storage temperatures are never high enough to produce much significant creep in the lead wire solder joints. Therefore, half of the full temperature change of 40°C will be used, resulting in an effective temperature change of 20°C for analysis purposes.

Condition D: Storage Condition Solder Thermal Cycling Fatigue Analysis The storage condition D will decrease the solder shear tear-out stress of 456 psi shown in Eq. 10.45, by a ratio of 20/60 or 0.333. The adjusted solder fatigue life can be obtained using Eq. 10.46 as shown below:

$$N_{DS} = (80{,}000)\left[\frac{200}{(0.333)(456)}\right]^{2.5} = 1.59 \times 10^5 \text{ cycles to fail} \tag{10.55}$$

Condition D: Storage Solder Thermal Cycling Fatigue Damage Ratio The thermal cycle fatigue damage ratio for the storage condition can be obtained from the 3650 thermal cycles expected, as shown in Eq. 10.36, and the number of cycles required to produce a solder joint failure as shown above:

$$R_{DS} = \frac{n}{N_{DS}} = \frac{3650}{1.59 \times 10^5} = 0.023 \text{ solder thermal} \tag{10.56}$$

10.5 TOTAL THERMAL CYCLING FATIGUE DAMAGE ACCUMULATED IN SOLDER JOINTS

The total thermal cycling fatigue damage accumulated in the solder joints can be obtained from the four thermal cycling events A through D, as shown in Eqs. 10.48,

10.52, 10.54, and 10.56 as shown below:

$$
\begin{aligned}
R_{TC} &= R_{AS} + R_{BS} + R_{CS} + R_{DS} = 0.494 + 0.485 + 0.006 + 0.023 \\
&= 1.008 \text{ solder thermal}
\end{aligned} \tag{10.57}
$$

The total thermal cycling fatigue damage ratio in the solder joints exceed a value of 1.0, slightly, which makes the design unacceptable for the solder joints. This does not include the solder fatigue damage accumulated in the random vibration event.

10.6 MINER'S WIRE VIBRATION DAMAGE PLUS WIRE THERMAL CYCLE DAMAGE

The total fatigue damage accumulated in the lead wires during the random vibration and thermal cycling events can be used to estimate the fatigue life of the lead wires, using Miner's total fatigue damage ratio. The fatigue damage ratios from the various random vibration events A through C, and the various thermal cycling events A through D are used as shown below:

$$
\begin{aligned}
R_{WV} &= \text{total wire vibration damage A through C} = 0.135 \text{ (Ref. Eq. 10.30)} \\
R_{WT} &= \text{total wire thermal cycle damage A through D} = 0.000 \text{ (Ref. Eq. 10.50)} \\
R_{WIRE} &= \text{total wire vibration plus thermal damage} = 0.135 + 0.000 \\
&= 0.135 \text{ wire vibration plus thermal}
\end{aligned} \tag{10.58}
$$

The wires are safe since the total fatigue damage ratio for vibration and thermal cycling is less than 1.0.

10.7 MINER'S SOLDER VIBRATION DAMAGE PLUS SOLDER THERMAL CYCLE DAMAGE

The total fatigue damage accumulated in the solder joints due to thermal cycling alone was shown by Eq. 10.57 to be 1.003, which is unacceptable. This did not include the fatigue damage accumulated in the solder joints due to random vibration, as shown by Eq. 10.30 to be 0.135. The total damage ratio including the random is

$$
R_{SOLDER} = 1.008 + 0.135 = 1.143 \text{ solder thermal plus vibration} \tag{10.59}
$$

The solder joints are unacceptable since the total damage ratio is more than 1.0.

10.8 RECOMMENDED DESIGN CHANGES TO IMPROVE THE SOLDER FATIGUE LIFE

1. The easiest design change to make, if there is no existing hardware, is to increase the PCB thickness from 0.060 to 0.075 in. This increases the height of the solder joint, which will reduce the shear tear-out stress in a direct ratio of the thickness, as shown in Eqs. 2.15 and 10.45. Increasing the PCB thickness to 0.075 in. will reduce the total solder thermal cycling fatigue cycle ratio:

$$R_{\mathrm{TC}} = R_{\mathrm{AS}} + R_{\mathrm{BS}} + R_{\mathrm{CS}} + R_{\mathrm{DS}} = 0.283 + 0.278 + 0.003 + 0.013 = 0.577 \tag{10.60}$$

The total solder joint cumulative damage ratio must include the random vibration damage ratio of 0.135 shown in Eq. 10.30. This results in the following damage ratio:

$$R_{\mathrm{SOLDER}} = 0.577 + 0.131 = 0.708 \text{ solder thermal plus vibration} \tag{10.61}$$

Since the above solder joint damage ratio is less than 1.0, the design is acceptable.

2. Search the technical literature to see if a smaller hybrid component is available with the same functions. A smaller component length will reduce the thermal expansions. This will then reduce the forces and stresses in the solder joints.

10.9 RECOMMENDED STRUCTURAL CHANGES TO IMPROVE SOLDER FATIGUE LIFE

1. Make the hybrid lead wires slightly longer. Longer lead wires will reduce the solder joint shear stresses. This will require raising the hybrid slightly higher above the PCB. This will not cause any thermal problems for a forced convection-cooled system. For a conduction-cooled system, it may require adding thermal pads in the air gap under the hybrid to help conduct the heat away.

2. Form a small kink in the center of each lead wire prior to assembly. This will reduce the wire bending stiffness, which will reduce the forces and stresses in the solder joints.

3. Epoxy bond a 0.025-in. thick copper–moly–copper shim on both sides of the PCB just under the hybrid. This material has a modulus of elasticity of about 40 million and a TCE of about 5 in./in./°C. It is stiff enough to reduce the thermal expansion differences between the hybrid and the PCB. This will reduce the forces and stresses in the solder joints. Very high shear forces are produced during thermal cycling, which can easily fracture good epoxy bonds, unless special precautions are taken. The metal shims must be roughened to get a grip on the epoxy bond. In addition, about eight holes with a diameter of about 0.090 in. should be drilled through the shims to allow the epoxy to fill the holes and form epoxy rivets. This

requires a significant amount of rework, which is not cheap. Tests have shown that the shims by themselves, without the holes, will not hold up very long in thermal cycling conditions. Another problem can occur when the X–Y expansions are restricted with the shims. Poisson's ratio comes into effect. This increases the Z axis expansions, which can crack plated through-holes in the PCB in the local area around the shims.

CHAPTER 11

Thermal Cycling Failures in Surface-Mounted Components

11.1 INTRODUCTION

Surface mounting techniques have many advantages over the older through-hole mounting methods that have been practiced since the printed circuit board (PCB) was first used by the electronics industry. The extensive use of pick-and-place assembly machines has substantially reduced the manufacturing costs. This technology has also resulted in a significant reduction in the size and weight of the electronic assemblies because electronic components can now be mounted on both sides of the PCBs. However, this new surface mounting technology also has a number of disadvantages. Doubling the number of electronic components mounted on a PCB can also double the amount of heat dissipated by that PCB. More heat means higher temperatures, which can result in more component failures. Therefore, better cooling methods must often be used that can increase the size, weight and manufacturing costs [25, 26].

Many of the early surface-mounted PCBs used flat pack integrated circuit components with kovar cases and thin flexible compliant electrical lead wires. No one worried about the large differences in the thermal coefficients of expansion (TCE) between the kovar metal cases and the epoxy fiberglass PCBs. This was because the thin flexible compliant electrical lead wires were able to bend to relieve any high strains produced by vibration events and by expansion differences during thermal cycling events. Low strains produced low solder joint shear stress levels with very few solder joint failures, which resulted in a very reliable PCB assembly.

When surface-mounted leadless ceramic chip carrier (LCCC) components were first used many years ago, a very large electronic component manufacturer made the prediction that these leadless devices would not be reliable for mounting on epoxy fiberglass PCBs. The thermal expansion differences of the materials were large enough to produce high strains and stresses in the solder joints, which would result in many solder joint failures. The solution to this problem was to avoid the general use of leadless electronic components larger than about 0.10 in. Larger electronic components were required to have compliant electrical lead wires if they were to be surface mounted on epoxy fiberglass PCBs, to improve the solder joint reliability.

Another large equipment manufacturer offered another solution for solving the LCCC surface mounting TCE problem. Its solution was to add thin sheets of a low TCE composite metal made from laminations of copper and invar to the multilayer PCB. This reduced the effective TCE of the PCB, bringing it closer to the LCCC. The PCB copper-plated through-holes had to be stronger, or more ductile, to prevent the increased cracking due to an increase in the Z-axis expansion perpendicular to the plane of the PCB. Poisson's strain ratios on the PCB were affected by the addition of the copper–invar laminations. When the in-plane X–Y expansions were reduced, Poisson's ratio produced an increase in the PCB Z-axis expansion, resulting in more plated through-hole copper cracks.

Military electronic document MIL-STD-454 does not permit solder to be used as a load-carrying structural element. Surface-mounted electronic devices such as LCCC components and plastic leaded chip carriers (PLCC) use solder to carry loads generated by thermal cycling, vibration, and shock, which is a violation of this document. The LCCC components have a history of high solder joint failures when they are mounted on epoxy fiberglass PCBs with no provisions for relieving high strains produced by large TCE differences in the component and PCB. The LCCC components usually have compliant J or gull wing electrical lead wires added to reduce the effects of TCE variations for increased reliability.

As this is being written, the total number of 100% surface-mounted PCBs being fabricated worldwide is still less than 30%. A very large number of PCBs are still using some through-hole-mounted components. The main reason for this is the poor reliability of the solder joints due to the large mismatch in the TCEs of some types of surface-mounted components. In addition, surface mounting techniques typically expose the electronic components to higher temperatures than through-hole-mounted components. Surface-mounted components that use vapor phase or convection oven methods for attachment to PCBs experience much higher component temperatures than through-hole components that use wave soldering attachment methods. In the vapor phase and convection oven assembly methods a local ambient temperature of about 220°C is normally used to reflow the solder. The surface-mounted components are therefore directly exposed to this hot environment. In the wave solder assembly method the solder temperature is usually about 260°C. However, the hot solder passes under the PCB so the components on the top surface of the PCB are shielded from the hot solder under the PCB. The wave-soldered components, therefore, experience temperatures of only about 140°C. This is especially important for surface mounting plastic components that have been exposed to high-humidity conditions for a couple of days before they are to be assembled to the PCBs. A substantial amount of water can be absorbed by plastic components under these conditions. When these plastic components are exposed to the high temperatures produced by vapor phase or convection oven processes, the absorbed water can turn to high-pressure steam. High pressures can produce high internal forces that may distort, warp, and crack the sensitive internal dies and wires in the surface-mounted components. It is good practice to process contaminated components to remove excessive water vapor and then store the components in vapor barrier bags before the components are surface mounted on the PCBs. This will protect the components from absorbing excessive amounts of water in high-humidity areas and prevent possible failures caused by high-pressure steam.

Many military types of PCBs are often cooled using conduction heat transfer methods. Aluminum has a high thermal conductivity, with a light weight, so it is often laminated to the PCB to reduce hot-spot temperatures. Care has to be used with aluminum because it has a high TCE and a high modulus of elasticity. When aluminum heat sinks are laminated to the PCB, the high TCE often increases the solder joint stresses on surface-mounted components, which can cause more rapid failures in thermal cycling conditions. Stresses in the thin etched copper circuit traces on the PCB will also be increased. This is usually not a problem unless the thin copper circuit traces have been damaged or overetched to an extra thin condition.

11.2 EQUILIBRIUM EQUATION FOR EVALUATING THERMAL EXPANSION FORCES AND STRESSES IN THE SOLDER JOINTS OF A SMALL LCCC

Extensive thermal cycling tests have shown that differences in the TCEs of large electronic components and their PCBs can produce high forces and stresses and rapid fatigue failures in the solder joints of leadless surface-mounted devices. Many designers and engineers believe that small leadless ceramic components can be safely mounted on plain epoxy fiberglass PCBs because small components will have small expansion displacements. This should result in small solder joint forces and stresses to produce a reliable system. Thermal cycling testing and experience has shown that the size of the leadless device is important. Components larger than about 0.10 in. should not be mounted on plain epoxy fiberglass PCBs. When the differences in the TCEs of the component and the PCB are greater than about 9 ppm/°C solder cracking problems often occur.

Thermal expansion displacements, forces, and stresses in the solder joints of surface-mounted devices can be estimated from the differences in the expansions between the components and their PCBs. An examination of Fig. 11.1 shows that differences in the thermal expansions between the component and the PCB will produce shear displacements in the solder joints. When the component thermal

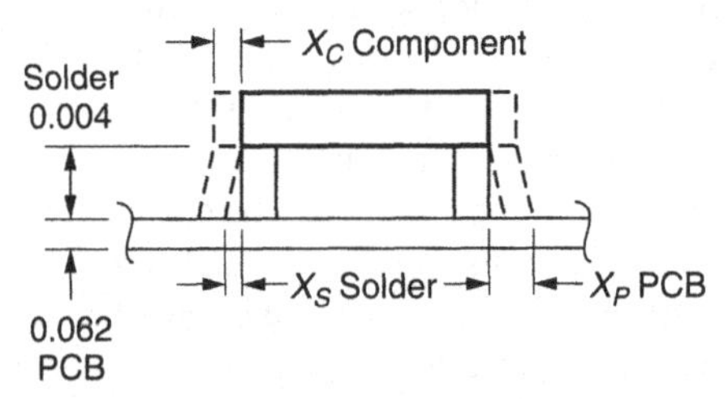

FIGURE 11.1 Differences in the TCEs between a leadless surface-mounted component and the PCB will force the solder joints to deflect in a shearing mode.

expansion is shown as X_C and the solder joint displacement is shown as X_S then the PCB expansion X_P will be the sum of these expansions:

$$X_C + X_S = X_P \tag{11.1}$$

The component will experience two expansions. The first expansion will be a free thermal expansion based on the component TCE α_C and half of the component length L_C for a symmetrical component and a temperature change Δt. The *diagonal* length is the greatest component dimension so it must eventually be used to obtain the final solder joint stress. This will be done using a correction factor based on the diagonal dimension of the component. The second expansion will be due to an axial force P_C trying to stretch the component length L_C. This is produced when the PCB thermal expansion is greater than the component thermal expansion. The effective component area A_C and the component modulus of elasticity E_C must also be considered. The total component expansion will be

$$X_C = \alpha_C L_C \, \Delta t + \frac{P_C L_C}{A_C E_C} \tag{11.2}$$

The solder joint shear displacement will be related to the solder joint shear force P_S, the solder joint height h_S, the solder joint area A_S, and the solder shear modulus G_S:

$$X_S = \frac{P_S h_S}{A_S G_S} \tag{11.3}$$

The PCB will experience two displacements—a positive one and a negative one. The positive one will be due to the free thermal expansion based on the PCB TCE α_P, the length L_P, and the temperature increase Δt. The PCB length will usually be the same as the component length. The negative displacement will be due to the stiffness of the solder and the component, which will tend to hold back and restrict the PCB expansion. These factors are related to the PCB force P_P, the length L_P, the PCB effective area A_P, and the modulus E_P:

$$X_P = \alpha_P L_P \, \Delta t - \frac{P_P L_P}{A_P E_P} \tag{11.4}$$

Substitute Eqs. 11.2–11.4 into Eq. 11.1:

$$\alpha_C L_C \, \Delta t + \frac{P_C L_C}{A_C E_C} + \frac{P_S h_S}{A_S G_S} = \alpha_P L_P \, \Delta t - \frac{P_P L_P}{A_P E_P} \tag{11.5}$$

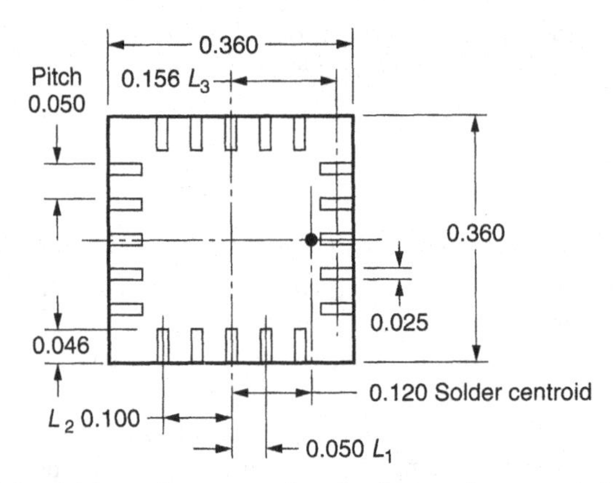

FIGURE 11.2 Solder pad geometry for a leadless surface-mounted component.

Sample Problem: Forces, Stresses and Fatigue Life in the Solder Joints on a Small Surface-Mounted LCCC for a Rapid Thermal Cycling Condition

A small LCCC with 20 input/output (I/O) solder pads is proposed for surface mounting on a 0.062-in.-thick epoxy fiberglass PCB, as shown in Fig. 11.2. A cross section through the LCCC is shown in Fig. 11.3. A solder joint height of 0.004 in. is expected. The assembly must be capable of passing a qualification proof of life test that requires 500 rapid thermal cycles over a temperature range from −50 to +90°C. Is the proposed design acceptable?

Solution An examination of Fig. 11.2 shows that various component lengths L_C must be used, based on the locations of several solder joints. These lengths represent the distances from the center of the component at its neutral axis to the centers of the solder pads. For symmetrical LCCCs that expand with respect to their centroid, only one half of the component is used. Let the component length L_1 represent the distance from the center to the first solder pad at a distance of 0.050 in. Notice there are two solder pads at this distance, one at the top and one at the bottom. Let L_2 be the component length from the center to the second solder pad at a distance of 0.10 in. Notice there are two solder pads at this distance. Let L_3 represent the length from the center of the component to the center of the five solder pads along the right

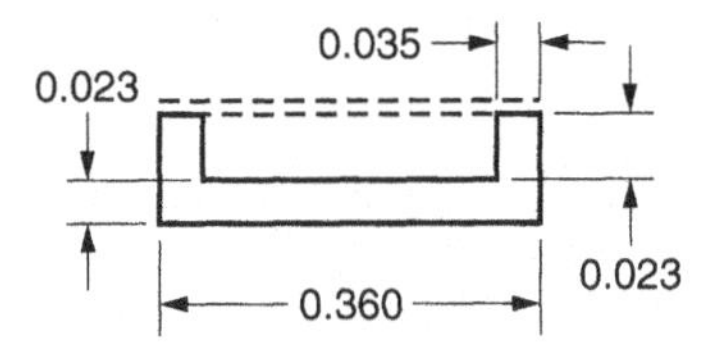

FIGURE 11.3 Cross-section view through the major load-carrying structure on the leadless surface-mounted component.

half of the component at a distance of 0.156 in. Equation (11.5) can then be modified as

$$\alpha_C L_1 \, \Delta t + \frac{P_1 L_1}{A_C E_C} + \alpha_C L_2 \, \Delta t + \frac{P_2 L_2}{A_C E_C} + \frac{P_S h_S}{A_S G_S} = \alpha_P L_3 \, \Delta t - \frac{P_3 L_3}{A_P E_P} \tag{11.6}$$

where $L_1 = 0.050$ in. distance from centroid to center of the first two solder pads
$L_2 = 0.100$ in. distance from centroid to center of second two solder pads
$L_3 = 0.156$ in. distance from centroid to center of five end solder pads
$P_1 = (0.050/0.156)P_3 = 0.3205P_3$
$P_2 = (0.100/0.156)P_3 = 0.641P_3$
$\alpha_C = 5.0 \times 10^{-6}$ in./in./°C TCE of leadless ceramic chip carrier LCCC
$\Delta t = [90 - (-50)]/2 = 70$°C rapid thermal cycle with no solder creep
$A_C = (0.023)(0.360) + (2)(0.035)(0.023) = 0.00989$ in.^{2} section area of LCCC, Fig. 11.3
$E_C = 42 \times 10^6$ lb/in.^{2} ceramic component modulus of elasticity
$h_S = 0.004$ in. typical height of solder joint
$A_S = (9)(0.046)(0.025) = 0.01035$ in.^{2} shear area of nine solder pads
$G_S = 1.10 \times 10^6$ lb/in.^{2} average 60/40 tin lead solder shear modulus
$\alpha_P = 15.0 \times 10^{-6}$ in./in./°C average TCE of PCB in X–Y plane
$A_P = (0.062)(0.360)(1.25 \text{ effective width factor}) = 0.0279$ in.^{2}
$E_P = 2.0 \times 10^6$ lb/in.^{2} average modulus epoxy fibreglass in X–Y plane

Substitute the above values into Eq. 11.6:

$$(5.0 \times 10^{-6})(0.050)(70) + \frac{(0.3205P_3)(0.050)}{(0.00989)(42 \times 10^6)} + (5.0 \times 10^{-6})(0.10)(70) + \frac{(0.641P_3)(0.10)}{(0.00989)(42 \times 10^6)} + \frac{P_S(0.004)}{(0.01035)(1.10 \times 10^6)} = (15.0 \times 10^{-6})(0.156)(70) - \frac{P_3(0.156)}{(0.0279)(2.0 \times 10^6)}$$

$$1.75 \times 10^{-5} + 3.830 \times 10^{-8}P_3 + 3.50 \times 10^{-5} + 1.536 \times 10^{-7}P_3 + 3.51 \times 10^{-7}P_S = 1.638 \times 10^{-4} - 2.795 \times 10^{-6}P_3$$

$$\sum P = 0 \quad \text{therefore} \quad P_3 = P_S$$

$$3.337 \times 10^{-6}P_3 = 1.113 \times 10^{-4} \quad \text{then} \quad P_3 = 33.36 \text{ lb} \tag{11.7}$$

Each of the nine solder joints will carry part of the shear force produced by the temperature change and the thermal coefficients of expansion. The solder joints farthest away from the centroid of the LCCC will have the greatest expansions, so they will carry the highest stress levels. The solder joints at the corners of the LCCC are the farthest away from the centroid. Assuming linear systems and uniform solder joints, the corner solder joints will then have the highest shear stresses. The P_3 force represents the *average* magnitude of the force for the nine solder joints acting at the edge of the LCCC. The *average* solder joint shear stress acting at this point can be obtained from the above force and the total solder joint area as follows:

$$\text{Average stress } S_S = P_3/A_S = 33.36/0.01035 = 3223 \text{ lb/in.}^2 \tag{11.8}$$

The *maximum* solder joint stress will be in the corner solder joints. For a square component this is at a 45° angle, so a correction factor of $\sqrt{2}$ must be applied as follows:

$$\text{Maximum stress} = \sqrt{2}(3223) = 4558 \text{ lb/in.}^2 \tag{11.9}$$

Solder can creep extensively when it is subjected to high temperatures, as shown in Fig. 1.6. In slow thermal cycling events solder creep can double the maximum stress level, as shown in Fig. 1.4. During a rapid thermal cycling event, however, the solder creep will be small so it is ignored in this sample problem.

The approximate fatigue life of the corner solder joints can be obtained using Fig. 5.3 with the use of a reference point that shows a fatigue life of about 80,000 cycles at a stress level of about 200 lb/in.^{2}. The standard fatigue equation is

$$N_1 S_1^b = N_2 S_2^b \quad \text{(Ref. Eq. 5.1)}$$

where $N_1 = 80{,}000$ fatigue cycles reference point
$S_1 = 200$ lb/in.^{2} stress reference point
$S_2 = 4558$ lb/in.^{2} solder stress (Ref. Eq. 11.9)
$b = 2.5$

$$N_2 = N_1 \left(\frac{S_1}{S_2}\right)^b = (80{,}000)\left(\frac{200}{4558}\right)^{2.5} = 32 \text{ cycles to fail} \tag{11.10}$$

The proposed design is unacceptable for the environment. Thermal cycling test data on similar surface-mounted small LCCC components have shown visible solder joint failures with less than 100 cycles. The usual fix is to add compliant wires to reduce the expansion forces and stresses. These J wires or gull wing wires can be thermal compression bonded to the sides of the component where they can make electrical contact. Wires shaped like the letter S work well, and they are less expensive to install because they can be vapor phase or convection oven soldered to the bottom of the LCCC.

Sample Problem: Adding Compliant J Wires to Reduce Forces and Stresses

Add 0.018-in. diameter copper J wires 0.040 in. long, thermal compression bonded, to the perimeter of the surface-mounted LCCC to reduce the thermal expansion forces and stresses. Will this change meet the 500 rapid thermal cycle requirement from -50 to $+90$°C?

Solution The long J wires are very compliant so almost all of the expansion forces will be reflected in the bending of these wires. Under these conditions the forces in the component body and the PCB can be ignored with very little error. This saves a substantial amount of analysis effort. The J wire can be evaluated as a beam fixed at both ends bending laterally, as shown in Fig. 10.2. This results in a wire bending spring rate of $K_W = (12E_W I_W / L_W^3)$. Equation 11.6 can now be simplified as shown below. The values will be the same as in Eq. 11.6. Only the copper wire properties have to be added:

$$\alpha_C L_1 \, \Delta t + \alpha_C L_2 \, \Delta t + \frac{P_W}{K_W} = \alpha_P L_3 \, \Delta t \tag{11.11}$$

where $E_W = 16 \times 10^6$ lb/in.^{2}, copper wire modulus of elasticity
$I_W = \pi d^4/64 = \pi(0.018)^4/64 = 5.15 \times 10^{-9}$ in.^{4}, wire moment of inertia
$L_W = 0.040$ in., length of J wire
$K_W = (12)(16 \times 10^6)(5.15 \times 10^{-9})/(0.040)^3 = 1.54 \times 10^4$ lb/in., wire spring rate

$$(5 \times 10^{-6})(0.050)(70) + (5 \times 10^{-6})(0.10)(70) + \frac{P_3}{1.54 \times 10^4}$$

$$= (15 \times 10^{-6})(0.156)(70)$$

$$1.75 \times 10^{-5} + 3.50 \times 10^{-5} + 6.49 \times 10^{-5} P_3 = 1.638 \times 10^{-4}$$

$$P_3 = \frac{1.113 \times 10^{-4}}{6.49 \times 10^{-5}} = 1.71 \text{ lb} \tag{11.12}$$

This represents the average force for the nine lead wires acting at the edge of the LCCC. The average solder joint shear stress acting at this point can be obtained from the above force and the area of 0.01035 in.^{2} for the nine LCCC solder pads shown for Eq. 11.6:

$$\text{Average stress} = P_3/A_S = 1.71/0.01035 = 165.4 \text{ lb/in.}^2 \tag{11.13}$$

The maximum force and stress are expected to occur in the corner solder joints. For a square component this is at a 45° angle, requiring a correction factor of $\sqrt{2}$:

$$\text{Maximum solder stress} = \sqrt{2}(165.4) = 234 \text{ lb/in.}^2 \tag{11.14}$$

The approximate number of thermal cycles required for a solder crack initiation can be obtained from the information shown in Eq. (11.10) as follows:

$$N_2 = (80{,}000)\left(\frac{200}{234}\right)^{2.5} = 54{,}030 \text{ cycles to fail} \tag{11.15}$$

The proposed copper J wires will easily allow the LCCC surface-mounted component to meet the 500-cycle failure free requirement.

11.3 BENDING STRESSES IN THE COPPER J WIRES

Bending stresses in the copper wires can be obtained from the bending moment generated by the deflection of the wire and the standard bending stress equation shown below. A stress concentration factor K is often used when there are sharp bends or cuts in the wire, along with several thousand fatigue stress reversals. When only several hundred stress cycles are expected and the wire has no major defects, then the stress concentration is not important. The maximum force and stress are expected to occur at the corner wires. For a square component this requires a correction factor of $\sqrt{2}$:

$$S_W = \frac{KMC}{I_W} \tag{11.16}$$

where $K = 1.0$ stress concentration is not important

$M = \sqrt{2}P_3L/2 = (1.414)(1.71)(0.040)/2 = 0.0483\ \text{lb} \cdot \text{in.}$, wire bending moment

$C = 0.018/2 = 0.009$ in. wire radius

$I_W = 5.15 \times 10^{-9}\ \text{in.}^4$, wire moment of inertia (Ref. Eq. 11.11)

$$S_W = \frac{(1.0)(0.0483)(0.009)}{5.15 \times 10^{-9}} = 84{,}408\ \text{lb/in.}^2 \tag{11.17}$$

At a first glance the high stress value may appear to be dangerous because the ultimate tensile strength of copper wire is only about $45{,}000\ \text{lb/in.}^2$. However, past experience with stress analysis of electrical lead wires in *bending* has shown that the wires very seldom fail, even with very high calculated bending stresses, as long as there are no sharp and deep cuts or scratches in the wires and many thousand stress cycles are not expected. The bending stress relation shown in Eq. 11.16 is a linear equation that is valid up to the yield strength and the elastic limit of the material. When the calculated bending stress exceeds these values, the material experiences plastic bending. Experience has shown that electrical lead wires, with no sharp and deep cuts and scratches, can withstand at least a couple of thousand bending stress cycles with calculated stresses as high as $150{,}000\ \text{lb/in.}^2$. This is not true when the lead wires are loaded in *tension*. When the lead wires are loaded in *direct tension*, the stress levels *must never exceed the ultimate tensile strength of the wire* because the wire will break.

11.4 SIMPLIFIED ANALYSIS METHOD FOR OBTAINING THERMAL CYCLING FORCES AND STRESSES IN THE LEAD WIRES AND SOLDER JOINTS OF COMPONENTS WITH MANY ELECTRICAL CONTACTS

The equilibrium relation shown in Eq. 11.6 can get very long and complex when very large devices with many lead wires and solder pads are being evaluated. A much simpler method of analysis can be used based on the location of the solder center of gravity, or centroid. The average solder force and stress can be obtained by assuming all of the solder joints are located at the solder center of gravity. Correction factors must then be used to find the maximum solder forces and stresses expected in thermal cycling events. The maximum lead wire and solder shear forces and stresses will act at the solder joints that are farthest away from the centroid of all the solder joints. The maximum forces and stresses are therefore expected to act at the corner solder joints for rectangular components, assuming symmetrical and uniform components and solder joints.

Sample Problem: Simplified Method for Finding Thermal Cycling Forces and Stresses in the Solder Joints of a Surface-Mounted Component

Use the simplified method of analysis to obtain the forces and stresses in the solder joints in the surface-mounted LCCC shown in the previous sample problem.

Solution The simplified method of analysis can be applied to the first sample problem in this chapter shown in Fig. 11.2 for the surface-mounted LCCC with 20 solder pads and no electrical lead wires. The centroid of the solder pads can be obtained from the number of pads and their location given in Table 11.1 and shown in Eq. 11.18:

$$\text{Solder centroid is at } \frac{1.080}{9} = 0.120 \text{ in.} \tag{11.18}$$

The equilibrium equation that can be used for the simplified solution is shown in Eq. 11.5. It is repeated below as a convenient reference. All the physical properties

TABLE 11.1 Centroid of Solder Pads

Number of Solder Pads	×	Pad location	=	Product
5 pads	at	0.156 in.	=	0.780
2 pads	at	0.100 in.	=	0.200
2 pads	at	0.050 in.	=	0.100
Total 9 pads				1.080

stay the same except the lengths, which must be changed to 0.120 in. for the location of the solder joint centroid. This will result in the *average* solder joint force acting at the centroid:

$$\alpha_C L_C \,\Delta t + \frac{P_C L_C}{A_C E_C} + \frac{P_S h_S}{A_S G_S} = \alpha_P L_P \,\Delta t - \frac{P_P L_P}{A_P E_P} \quad \text{(Ref. Eq. 11.5)}$$

$$(5 \times 10^{-6})(0.120)(70) + \frac{P_C(0.120)}{(0.010)(42 \times 10^6)} + \frac{P_S(0.004)}{(0.01035)(1.1 \times 10^6)}$$

$$= (15 \times 10^{-6})(0.120)(70) - \frac{P_P(0.120)}{(0.0279)(2 \times 10^6)}$$

$$4.20 \times 10^{-5} + 2.857 \times 10^{-7} P_C + 3.513 \times 10^{-7} P_S = 1.26 \times 10^{-4} - 2.15 \times 10^{-6} P_P$$

$$\sum P = 0 \quad \text{so} \quad P_C = P_S = P_P$$

$$2.789 \times 10^{-6} P = 8.40 \times 10^{-5} \quad \text{so} \quad P = 30.12 \text{ lb} \qquad (11.19)$$

The maximum solder shear force will occur in the corners of the LCCC component. The farthest solder joints will carry the greatest forces. This requires the use of two correction factors:

1. The first correction must be made for the location of the five solder joints at the side edge of the LCCC at 0.156 in. compared to the location of the solder centroid at 0.120 in.:

$$C_1 = \frac{0.156}{0.120} = 1.30 \qquad (11.20)$$

2. The second correction must be made for the location of the solder joints at the corners of the square component at an angle of 45° using $\sqrt{2}$, compared to the location of the five solder joints at the side edge of the LCCC:

$$C_2 = \sqrt{2} = 1.414 \qquad (11.21)$$

The corrected maximum solder joint shear stress can be obtained from the *average* 30.12-lb force, acting at the centroid of the solder joints as shown in Eq. 11.19. The two correction factors must be included along with the total solder shear area of 0.01035 in.^{2} for the nine solder joints, which is shown in the data for Eq. 11.6:

$$S_S = \frac{P_{\text{MAX}}}{A_S} = \frac{(1.3)(1.414)(30.12)}{0.01035} = 5349 \text{ lb/in.}^2 \qquad (11.22)$$

Comparing the results of the simplified analysis method shown above using the location of the solder centroid, shows the corner solder joints are expected to have a shear stress value of 5349 lb/in.^{2}. This is compared to a stress level of 4558 lb/in.^{2} for the corner solder joints shown in Eq. 11.9 for the longer more accurate analysis method. This shows that the simplified analysis method using the location of the solder joint centroid is slightly more conservative by a factor of $5349/4558 = 1.17$, which is about 17% higher than the correct value.

11.5 EVALUATION OF SOLDER JOINT FORCES AND STRESSES ON SURFACE-MOUNTED BALL GRID ARRAYS IN THERMAL CYCLING ENVIRONMENTS

Ball grid array (BGA) components usually consist of a silicon chip encased in ceramic. The ceramic module is often bonded to a small, thin polyimide glass daughter board slightly larger than the encapsulated ceramic module, as shown in Fig. 11.4. The bottom surface of the small daughter board contains many solder balls, usually about 0.031 in. in diameter, for making electrical connections on a larger surface-mounted plug-in type of PCB. The solder balls are typically placed on an x and y grid spaced at increments of 0.050 in. A reflow soldering process similar to vapor phase or convection oven is normally used to attach the BGA to the larger plug-in PCB. The solder balls usually have a high-temperature solder core, with a lower-temperature solder outer coat to allow the BGA to be soldered to the plug-in PCB without melting the inner solder core. The simplified method of solder joint analysis, based on the location of the solder ball centroid is very convenient because it can save a lot of time for evaluating BGA components when they contain a large number of solder balls.

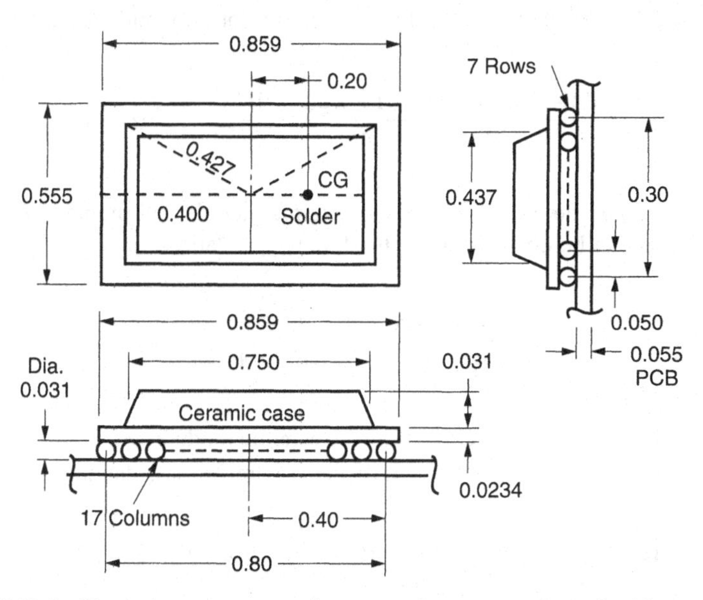

FIGURE 11.4 Physical arrangement and geometry for a rectangular ball grid array with 119 solder balls.

Sample Problem: Solder Joint Forces and Stresses in a Ball Grid Array

A small BGA with a polyimide glass daughter board measuring 0.555 × 0.859× 0.0234 in. thick has 119 solder balls, 0.031 in. in diameter mounted on the bottom surface. The solder balls are spaced 0.050 in. apart on an x and y grid of 7 rows and 17 columns, as shown in Fig. 11.4. The BGA will be surface mounted on a large plug-in type of epoxy glass PCB, 0.055 in. thick. The assembly must be capable of passing a 500 rapid thermal cycling proof of life test from −30 to +80°C. Is the proposed design acceptable?

Solution Since the ceramic module is bonded to the polyimide glass daughter board, it is convenient to evaluate this assembly as a single unit. The average composite modulus of elasticity and the average composite TCE for these two materials must then be calculated.

The average composite modulus of elasticity (E_{AV}) of the ceramic module and the polyimide glass daughter board can be obtained using the following relation:

$$E_{AV} = \frac{\sum AE}{\sum A} = \frac{A_1 E_1 + A_2 E_2}{A_1 + A_2} \tag{11.23}$$

where $A_1 = (0.437)(0.031) = 0.0135 \text{ in.}^2$, cross-section area of ceramic module
$E_1 = 42 \times 10^6 \text{ lb/in.}^2$, modulus of elasticity of ceramic module
$A_2 = (0.555)(0.0234) = 0.0130 \text{ in.}^2$, cross-section area of daughter board
$E_2 = 2.5 \times 10^6 \text{ lb/in.}^2$, modulus of elasticity for polyimide daughter board

$$E_{AV} = \frac{(0.0135)(42 \times 10^6) + (0.0130)(2.5 \times 10^6)}{0.0135 + 0.0130} = 2.262 \times 10^7 \tag{11.24}$$

The average composite TCE (α_{AV}) of the ceramic module and the polyimide glass daughter board can be obtained using the following relation:

$$\alpha_{AV} = \frac{\sum AE\alpha}{\sum AE} = \frac{A_1 E_1 \alpha_1 + A_2 E_2 \alpha_2}{A_1 E_1 + A_2 E_2} \tag{11.25}$$

where $\alpha_1 = 5 \times 10^{-6}$ in./in./°C, TCE of the ceramic module
$\alpha_2 = 13 \times 10^{-6}$ in./in./°C, TCE of the polyimide glass daughter board

$$\alpha_{AV} = \frac{(0.0135)(42 \times 10^6)(5 \times 10^{-6}) + (0.0130)(2.5 \times 10^6)(13 \times 10^{-6})}{(0.0135)(42 \times 10^6) + (0.0130)(2.5 \times 10^6)}$$
$$= 5.433 \times 10^{-6} \tag{11.26}$$

The simplified equilibrium equation that can be used to find the forces in the solder joints based on the location of the solder joint centroid is shown in Eq. 11.5 and repeated here for convenience:

$$\alpha_C L_C \, \Delta t + \frac{P_C L_C}{A_C E_C} + \frac{P_S h_S}{A_S G_S} = \alpha_P L_P \, \Delta t - \frac{P_P L_P}{A_P E_P} \quad \text{(Ref. Eq. 11.5)}$$

where $\alpha_C = \alpha_{AV} = 5.433 \times 10^{-6}$ in./in./°C, average component TCE (Ref. Eq. 11.26)

$L_C = 0.20$ in., distance from center of component to solder centroid, Fig. 11.4

$\Delta t = [(80) - (-30)]/2 = 55°\text{C}$, for a rapid temperature cycle with no solder creep

$A_C = (0.437)(0.031) + (0.0234)(0.555) = 0.0265$ in.^{2} (area of ceramic component plus area of the daughter board)

$E_C = 2.262 \times 10^7$ lb/in.^{2}, average component modulus of elasticity (Ref. Eq. 11.24)

$h_S = 0.031$ in., maximum solder ball height and diameter

$d_S = 0.0275$ in., average solder ball diameter

$A_S = (\pi/4)(0.0275)^2$ (56 balls) $= 0.0333$ in.^{2}, area 56 active solder balls on one side

$G_S = 1.10 \times 10^6$ lb/in.^{2}, solder shear modulus

$\alpha_P = 15.0 \times 10^{-6}$ in./in./°C, TCE of epoxy fibreglass PCB

$L_P = 0.200$ in., PCB length from component center to solder centroid

$A_P = (1.25 \text{ effective})(0.30)(0.55) = 0.0206$ in.^{2}, effective section area of PCB (effective width of PCB taken as 1.25 times solder ball 0.30 in. width)

$E_P = 2.0 \times 10^6$ lb/in.^{2}, modulus of epoxy fiberglass PCB

Substitute into Eq. 11.5 to find the *average* shearing force acting on the solder joints:

$$(5.433 \times 10^{-6})(0.20)(55) + \frac{P_C(0.20)}{(0.0265)(2.262 \times 10^7)} + \frac{P_S(0.031)}{(0.0333)(1.10 \times 10^6)}$$
$$= (15.0 \times 10^{-6})(0.20)(55) - \frac{P_P(0.20)}{(0.0206)(2.0 \times 10^6)}$$

$$5.976 \times 10^{-5} + 3.336 \times 10^{-7} P_C + 8.463 \times 10^{-7} P_S$$
$$= 1.650 \times 10^{-4} - 4.854 \times 10^{-6} P_P$$

$$\sum P_i = 0 \quad \text{so} \quad P_C = P_S = P_P = P$$

$$6.034 \times 10^{-6} P = 1.052 \times 10^{-4} \quad \text{so} \quad P = 17.43 \text{ lb} \tag{11.27}$$

The *average* solder joint shear stress can be obtained at the solder joint centroid:

$$S_{AV} = \frac{P}{A_S} = \frac{17.43}{0.0333} = 523 \text{ lb/in.}^2 \tag{11.28}$$

Two correction factors must be used to obtain the *maximum* expected solder joint shear stress. The maximum stress will occur at the most distant solder joints, which are at the diagonal corners of the component. The first correction factor will transfer the average solder shear stress from the centroid of the solder pattern to the solder joints at the long side edge of the rectangular component. The solder joints at the side edge are at twice the centroid distance so the first correction factor is 2.0.

$$C_1 = 2.0 \tag{11.29}$$

The second correction factor must transfer the shear stress from the side solder joints to the corner solder joints. This is based on the ratio of the diagonal distance to the corner solder joints and the distance to the side solder joints. The diagonal distance (D_S) can be obtained from the right triangle of the solder joint dimensions of 0.40 and 0.15 in. in Fig. 11.4:

$$D_S = \sqrt{(0.40)^2 + (0.15)^2} = 0.427 \text{ in.}$$

The second correction factor can be determined from the ratio of the distances:

$$C_2 = 0.427/0.400 = 1.067 \tag{11.30}$$

The maximum shear stress at the corner solder joints can be obtained as follows:

$$S_{\text{MAX}} = C_1 C_2 S_{\text{AV}} = (2.0)(1.067)(523) = 1116 \text{ lb/in.}^2 \tag{11.31}$$

The approximate number of thermal cycles required to produce a visible solder joint crack can be obtained from Eq. 5.1 and Fig. 5.3 as

$$N_2 = N_1 \left(\frac{S_1}{S_2}\right)^b = (80{,}000)\left(\frac{200}{1116}\right)^{2.5} = 1088 \text{ cycles} \tag{11.32}$$

The surface-mounted BGA component should be able to pass the 500 thermal cycle proof of life test.

CHAPTER 12

Stresses and Fatigue Life in Component Lead Wires and Solder Joints Due to Dynamic Forces and PCB Displacements

12.1 LARGE, HEAVY COMPONENTS SUPPORTED BY LEAD WIRES MOUNTED ON PCBs

Electronic components are being fabricated as leaded devices and leadless devices, with many different form factors for surface mounting and through-hole mounting on various types of printed circuit boards (PCBs). The electrical lead wires and solder joints are often the weakest structural elements that tend to fail most often due to high acceleration levels and relative displacements between the component and the PCB due to severe vibration, shock, and thermal cycling conditions.

Heavy electronic components, such as transformers, inductors, large relays, and large capacitors can literally fly off of a PCB if their lead wires or solder joints break while operating in severe vibration and shock environments. High dynamic forces can be generated in the axial lead wires of transformers mounted on plug-in types of PCBs when the natural frequency of the PCB is excited. When heavy components are exposed to high acceleration levels, high dynamic forces can be produced. Through-hole-mounted transformers that are only supported by the lead wires will transfer the high dynamic forces directly to the lead wires. The lead wires will then be loaded in tension. The wires, in turn, will transfer the dynamic loads to the PCB through the solder joints, which will be loaded in shear, as shown in Fig. 12.1. If the transformer is bolted to the PCB, as shown in Fig. 12.2, the bolt will carry a large part of the dynamic load. This will substantially reduce the dynamic forces acting on the lead wires and solder joints, increasing the fatigue life and reliability.

Large surface-mounted transformers often have their electrical lead wires extending from the sides of the transformer. The wires make a 90° bend to carry then down to the PCB for soldering. When the natural frequency of a plug-in type of PCB is excited, the PCB will be forced to bend, producing large relative dynamic displacements. This will force the lead wires on surface-mounted or through-hole-

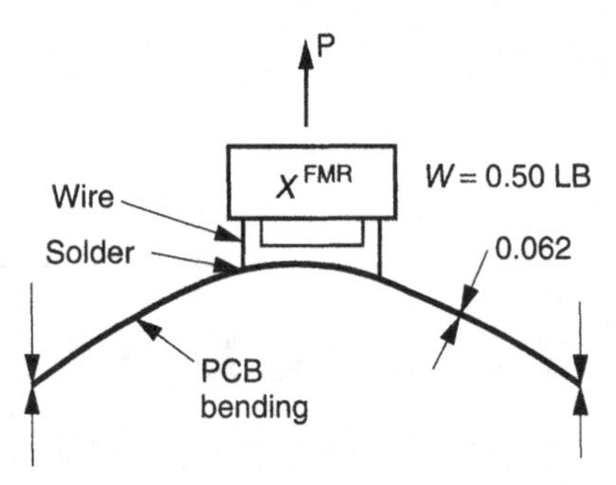

FIGURE 12.1 High acceleration forces and stresses can be produced in the lead wires and solder joints of large and heavy components operating in severe vibration and shock conditions.

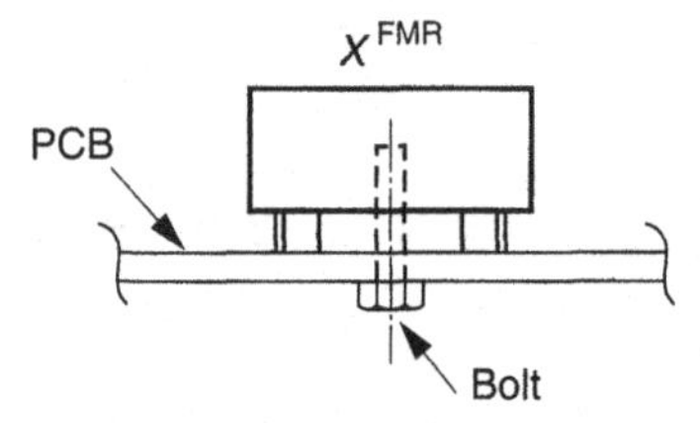

FIGURE 12.2 Large and heavy components should be bolted to the PCB to reduce high forces and stresses acting on the lead wires and solder joints in high acceleration conditions.

mounted components to bend, as shown in Fig. 12.3. High input acceleration G levels and high PCB transmissibility Q values can rapidly fatigue the lead wires and solder joints under these conditions.

High dynamic forces can also be generated in the lead wires and solder joints of a heavy transformer when the natural frequency of the transformer itself is excited

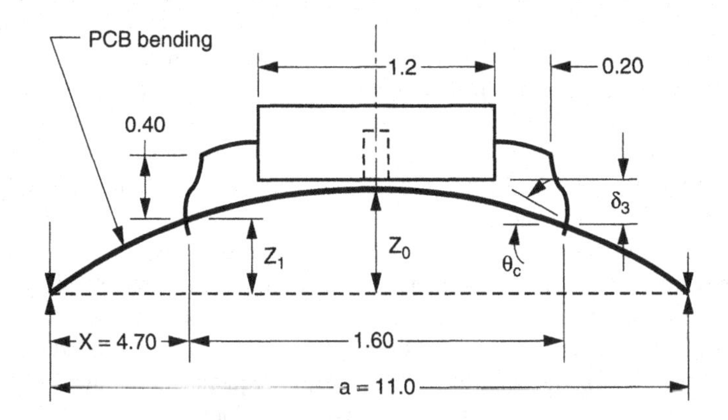

FIGURE 12.3 Large PCB bending displacements in severe vibration and shock conditions can produce high lead wire and solder joint stresses that can lead to rapid fatigue failures.

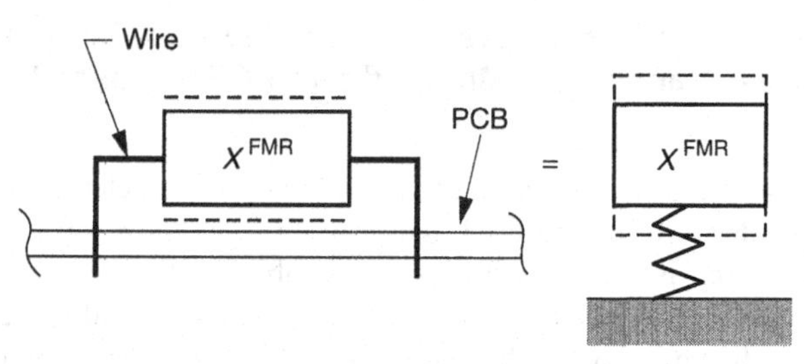

FIGURE 12.4 Large and heavy components with a few thin wires will often bounce up and down in severe vibration and shock conditions when the components are not properly secured.

while the PCB remains quiet. The transformer body can act like a large mass, and the lead wires can act as springs, as shown in Fig. 12.4. High input vibration or shock acceleration G levels with a high transformer transmissibility Q value will force the transformer to bounce up and down. This will generate bending stresses in the lead wires and shear stresses in the solder joints. Rapid fatigue failures can be produced in these members under these conditions.

Large differences in the thermal coefficients of expansion (TCE) between the component and the PCB can produce differences in the horizontal displacements between the component and the PCB. This can force the lead wires to bend and stretch, as shown in Fig. 12.5. This will produce bending stresses in the lead wires and shear tear-out stresses in the solder joints of through-hole-mounted components. Displacement differences generated by thermal cycling on surface-mounted components will produce similar bending stresses in the lead wires and direct shear stresses in the solder joints.

More details of the failure mechanisms and the approximate fatigue life expected in the lead wires and solder joints due to the vibration, shock, and thermal cycling conditions described above can be demonstrated with the sample problems in this chapter.

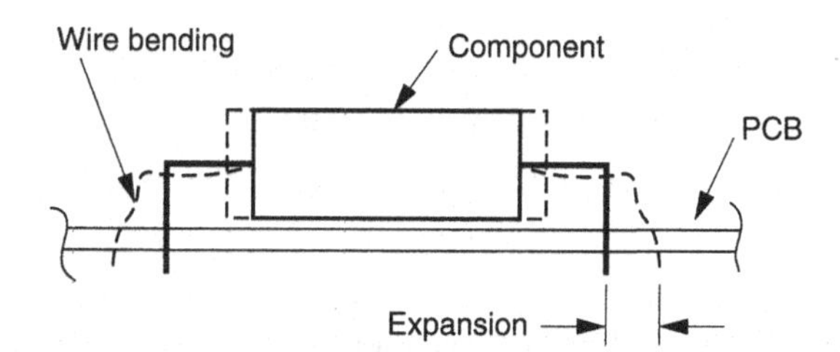

FIGURE 12.5 Differences in the TCEs between the component and the PCB can force component lead wires to bend and deflect in thermal cycling conditions.

Sample Problem: Forces, Stresses, and Fatigue Life in the Wires and Solder Joints of a Transformer Mounted on a PCB Exposed to Vibration and Shock

A 0.50-lb transformer with 4 copper wires 0.025 in. in diameter is through-hole mounted at the center of a plug-in PCB, 0.062 in. thick, as shown in Fig. 12.1. The PCB natural frequency is 185 Hz with a transmissibility of 14. Find the dynamic forces, stresses, and approximate fatigue life in the lead wires and solder joints for the environments shown below, with inputs perpendicular to the plane of the PCB [23, 27, 28]:

1. A sine resonance dwell for 30 min with an input acceleration level of 5 G peak
2. A random vibration input of 0.040 G^2/Hz for a period of 30 min
3. 500 shocks with an input of 25 G peak and a 0.001-s half sine pulse

Solution

Part 1A. Lead Wire Sine Vibration Dynamic Analysis Fatigue Life The dynamic force P_D acting on the transformer, the lead wires, and the solder joints can be obtained from the PCB natural frequency and transmissibility with Newton's force and mass acceleration relation:

$$P_D = W G_{\text{IN}} Q \tag{12.1}$$

where $W = 0.50$ lb, transformer weight
$G_{\text{IN}} = 5.0$ gravity units, input acceleration
$Q = 14$ PCB transmissibility

$$P_D = (0.50)(5.0)(14) = 35.0 \text{ lb dynamic load} \tag{12.2}$$

This axial load is acting on four transformer wires and four solder joints. The tensile stress S_t in the wires can be obtained from the 0.025-in. wire diameter and the wire area A_W as shown:

$$\begin{aligned} A_W &= (4 \text{ wires})(\pi/4)(d_W)^2 = (\pi)(0.025)^2 = 0.00196 \text{ in.}^2 \\ S_t &= P_D/A_W = 35.0/0.00196 = 17{,}857 \text{ lb/in.}^2 \end{aligned} \tag{12.3}$$

The approximate fatigue life expected for the copper lead wires will depend on the stress level in the wire as well as the condition of the wire. The copper life fatigue curve is shown in Fig. 5.4. This is a typical curve for wires with deep cuts and scratches due to poor forming dies, so it contains a fatigue stress concentration factor of 2.0. This results in a b fatigue exponent slope of 6.4. The approximate fatigue life can be obtained from Eq. 5.1 where

N_1 = number of stress cycles to fail at a stress level of 17,857 lb/in.^{2}

S_1 = calculated tensile stress of 17,857 lb/in.^{2}

N_2 = 1000 stress cycles to fail at a stress level of 45,000 lb/in.^{2}

S_2 = tensile stress of 45,000 lb/in.^{2}

b = 6.4 slope of copper fatigue curve for a stress concentration factor of 2.0

$$N_1 = N_2(S_2/S_1)^b = (1000)(45{,}000/17{,}857)^{6.4} = 3.71 \times 10^5 \text{ cycles to fail} \quad (12.4)$$

The approximate time for the wires to fail with a sine resonant dwell condition can be obtained from the natural frequency of 185 Hz:

$$\text{Time to fail} = \frac{3.71 \times 10^5 \text{ cycles to fail}}{(185 \text{ cycles/s})(60 \text{ s/min})} = 33.4 \text{ min} \quad (12.5)$$

The required sine resonant dwell period is 30 min. So many things can go wrong during a vibration qualification test that a safety factor on the life should be used for protection. A good rule of thumb is to allow enough time to complete at least five qualification tests. The minimum desired fatigue life should then be about $5 \times 33.4 = 167$ min $= 2.78$ h.

If the copper wires are clean and smooth, with no cuts and scratches, the stress concentration factor will be 1.0. This will result in a fatigue curve slope b of 10.48. This would result in an N_1 value of 1.61×10^7 stress cycles to fail. The approximate fatigue life of the copper wires would then be increased to about 24.2 h. It is obvious from this calculation that stress concentrations can have a very strong influence on the fatigue life of vibrating structural members. Therefore, it is very important to control or avoid any manufacturing processes that can introduce defects, flaws, cuts, holes, or sharp radii in critical structural elements that will be required to withstand many stress reversal cycles.

Solution

Part 1B. Solder Joint Sine Vibration Dynamic Analysis Fatigue Life The dynamic load acting on the four lead wires will be the same as the dynamic load acting on the four solder joints. The solder joints, however, will be loaded in shear.

The solder shear stress S_S is based on the wire diameter and thickness of the PCB, assuming no solder fillets where

$P_D = 35.0$ lb, total load acting on 4 wires (Ref. Eq. 12.2)

$d = 0.025$-in. wire diameter

$h = 0.062$-in. PCB thickness

$A_S = 4\pi dh = 4\pi(0.025)(0.062) = 0.0195 \text{ in.}^2$, 4 solder shear areas

$$S_S = P_D/A_S = 35.0/0.0195 = 1795 \text{ lb/in.}^2 \qquad (12.6)$$

The approximate number of fatigue cycles required to produce a solder joint failure can be obtained from Eq. 5.1 and Fig. 5.3 for eutectic solder (63/37 Sn/Pb) with the following data:

N_1 = number of stress cycles to fail at a stress level of 1795 lb/in.^2

$S_1 = 1795 \text{ lb/in.}^2$, shear stress in solder joint

$N_2 = 1000$ stress cycles reference for solder to fail at 6500 lb/in.^2

$S_2 = 6500 \text{ lb/in.}^2$, reference stress for solder to fail at 1000 cycles

$b = 4.0$ slope of solder vibration fatigue exponent on a log–log plot

$$N_1 = N_2(S_2/S_1)^b = (1000)(6500/1795)^{4.0} = 1.72 \times 10^5 \text{ cycles to fail} \qquad (12.7)$$

The approximate time for the solder joints to fail with a sine resonant dwell condition can be obtained from the natural frequency of 185 Hz:

$$\text{Time to fail} = \frac{1.72 \times 10^5 \text{ cycles to fail}}{(185 \text{ cycles/s})(60 \text{ s/min})} = 15.5 \text{ min} \qquad (12.8)$$

The solder joints are obviously more critical than the lead wires in this case. The best way to solve this problem is to bolt the transformer to the PCB as shown in Fig. 12.2, so the bolt will carry most of the dynamic force acting on the transformer.

Part 2A. Lead Wire Random Vibration Dynamic Analysis Fatigue Life The G_{rms} response of the PCB to the random vibration can be obtained from Eq. 9.16 by considering the PCB to be similar to a single-degree-of-freedom system at its natural frequency where

$P = 0.040\ G^2/\text{Hz}$, power spectral density input
$f_n = 185$ Hz, PCB natural frequency
$Q = 14$ PCB transmissibility

$$G_{\text{rms}} = \sqrt{(\pi/2)Pf_nQ} = \sqrt{(\pi/2)(0.040)(185)(14)} = 12.75 \tag{12.9}$$

The rms dynamic acceleration force acting on the transformer, the lead wires, and solder joints can be obtained from the transformer weight and Newton's force relation where $W = 0.50$ lb, transformer weight and $G_{\text{rms}} = 12.75$ acceleration level acting on the transformer:

$$P_D = WG_{\text{rms}} = (0.50)(12.75) = 6.375 \text{ lb rms} \tag{12.10}$$

The dynamic rms tensile stress acting on the four lead wires can be obtained from the wire cross-section area shown in Eq. 12.3:

$$S_t = P_D/A_W = (6.375/0.00196) = 3252 \text{ lb/in.}^2 \text{ rms} \tag{12.11}$$

In a random vibration environment the 1σ stress level is the rms stress level. This will occur about 68.3% of the time for a Gaussian distribution. A quick and easy way to obtain the approximate fatigue life of a structure is to use the 2σ stress level (this is, two times the rms stress level), which is normally expected to occur about 27.1% of the time. However, when the 2σ stress level is assumed to act 100% of the time, it provides a quick, easy, and slightly conservative method for obtaining the approximate fatigue life (see Ref. 1, p. 214). The 2σ stress value will be $2 \times 3252 = 6504$ lb/in.^{2}. Substitute this value into Eq. 12.4 to obtain the number of wire cycles to fail. A fatigue exponent b value of 6.4 was used for the copper wire. This was based on a stress concentration factor of 2.0, to account for any deep cuts or scratches due to careless manufacturing or handling practices:

$$N_1 = N_2(S_2/S_1)^b = (1000)(45{,}000/6504)^{6.4} = 2.38 \times 10^8 \text{ cycles to fail} \tag{12.12}$$

The approximate time for the wires to fail with random vibration is based on the 185-Hz PCB natural frequency:

$$\text{Time to fail} = \frac{2.38 \times 10^8 \text{ cycles to fail}}{(185 \text{ cycles/s})(3600 \text{ s/h})} = 357 \text{ h} \tag{12.13}$$

The wires will be very safe in the proposed random vibration environment.

Part 2B. Solder Joint Random Vibration Dynamic Analysis Fatigue Life The random vibration dynamic load produces shear in the four solder joints. The load will be the same as the load on the wires, as shown in Eq. 12.10. The shear area of the four solder joints was shown in Eq. 12.6. The resulting shear stress in the solder joints is

$$S_S = P_D/A_S = 6.375/0.0195 = 327 \text{ lb/in.}^2 \text{ rms} \tag{12.14}$$

The 2σ solder joint shear stress (which is two times the rms stress) can be used for the Gaussian distribution to obtain the approximate number of stress cycles to fail. The 2σ stress normally occurs about 27.1% of the time. For this application the 2σ stress level is assumed to act 100% of the time to obtain a quick and slightly conservative estimate of the number of cycles to fail (see Ref. 1, p. 214). The 2σ stress value will be $2 \times 327 = 654$ lb/in.^{2}. This value is substituted into Eq. 12.7 to obtain the number of stress cycles to fail.

$$N_1 = N_2(S_2/S_1)^b = (1000)(6500/654)^{4.0} = 9.76 \times 10^6 \text{cycles to fail} \tag{12.15}$$

The approximate time it will take the eutectic solder (63/37 Sn/Pb) to fail in random vibration can be obtained from the natural frequency of the PCB as

$$\text{Time to fail} = \frac{9.76 \times 10^6 \text{ cycles to fail}}{(185 \text{ cycles/s})(3600 \text{ s/h})} = 14.7 \text{ h} \tag{12.16}$$

The solder joints are more critical than the lead wires, but they are satisfactory in this case. If the transformer is bolted to the PCB, as shown in Fig. 12.2, the bolt will then carry most of the dynamic load acting on the transformer instead of the solder joints.

Part 3A. Lead Wire Shock Dynamic Analysis Fatigue Life The dynamic shock load acting on the four transformer lead wires due to a 25-G peak, 0.001-s half sine pulse can be obtained from Newton's force relation:

$$P_D = WG_{\text{in}}A \tag{12.17}$$

where $W = 0.50$ lb, transfer weight
$G_{\text{in}} = 25\ G$ peak input with a 0.001-s half sine shock pulse
$f_S = 185$ Hz, PCB natural frequency
$f_P = 1/(2)(0.001) = 500$ Hz
$R = (f_S/f_P) = (185/500) = 0.37$
$A = 2R = (2)(0.37) = 0.74$ (see Ref. 1, p. 301) shock amplification

$$P_D = (0.50)(25)(0.74) = 9.25 \text{ lb} \tag{12.18}$$

The dynamic tensile stress expected in the four lead wires can be obtained from the wire cross-section area of 0.00196 in.^{2}, as shown in Eq. 12.3:

$$S_t = P_D/A_W = 9.25/0.00196 = 4719\ \text{lb/in.}^2 \qquad (12.19)$$

The approximate number of stress cycles required to produce a fatigue failure in the copper wires can be obtained from Eq. 12.4:

$$N_1 = N_2(S_2/S_1)^b = (1000)(45{,}000/4719)^{6.4} = 1.85 \times 10^9\ \text{cycles} \qquad (12.20)$$

The number of stress cycles capable of doing a significant amount of damage in the wires for the 500 shock pulses will depend on the amount of damping in the structural system. The PCB will normally displace through its maximum amplitude for one or two cycles when it responds to the shock input. These first few cycles will generate the most damage in the wires. The displacement amplitude will be less and less for additional stress cycles, depending on the phase relations and the damping properties. For the purposes of fatigue life analysis, the number of stress cycles producing the maximum displacements and maximum damage were conservatively estimated as 20 for every imposed shock pulse. The total number of stress cycles expected for the 500 shock pulses will then be as follows:

$$N_S = (500)(20) = 10{,}000\ \text{shock stress cycles expected} \qquad (12.21)$$

The safety factor (SF) for the wires can be based on the number of cycles required to produce a failure and the actual number of cycles expected, as shown below:

$$\text{Wire SF} = 1.85 \times 10^9/10{,}000 = 1.85 \times 10^5 = \text{very high} \qquad (12.22)$$

Part 3B. Solder Joint Shock Dynamic Analysis Fatigue Life The dynamic shear force acting on the four solder joints was calculated to be 9.25 lb, as shown in Eq. 12.18. The shear area of the four solder joints was calculated to be 0.0195 in.^{2}, as shown in Eq. 12.6. These values can be used to find the shock-induced shear stress on the solder joints:

$$S_S = P_D/A_S = 9.25/0.0195 = 474\ \text{lb/in.}^2 \qquad (12.23)$$

The approximate number of stress cycles required to produce a fatigue failure in the solder joints can be obtained from Eq. 12.7:

$$N_1 = N_2(S_2/S_1)^b = (1000)(6500/474)^{4.0} = 3.54 \times 10^7\ \text{cycles to fail} \qquad (12.24)$$

The approximate number of critical stress cycles expected for the 500 shock environment was shown by Eq. 12.21 to be 10,000 cycles. The shock safety factor is

$$\text{Solder SF} = 3.54 \times 10^7/10{,}000 = 3540 = \text{very high} \qquad (12.25)$$

The shock environment does not require the transformer to be bolted to the PCB to provide an adequate fatigue life in the solder joints. The transformer bolt will have to be used, however, to satisfy the sine vibration fatigue life requirements.

12.2 PCB BENDING DISPLACEMENTS DURING VIBRATION PRODUCES RELATIVE MOTION AND STRESSES IN THE COMPONENT WIRES AND SOLDER JOINTS

Large dynamic displacements are often produced in plug-in types of PCBs during vibration when the natural frequency is excited, forcing the PCB to bend back and forth. The largest bending amplitudes are usually generated when the vibration direction is perpendicular to the plane of the PCB. The bending action of the PCB will force component lead wires to bend back and forth, as shown in Fig. 12.6. The ability of the components to survive severe vibration conditions will depend on many different factors. These include the component type, size, location, mounting method, natural frequency, input acceleration level, wire strain relief, and duration of the exposure. The most critical component location is at the center of a plug-in type of PCB.. The dynamic displacements are usually greatest when the fundamental or lowest natural frequency is excited. These dynamic displacements are trigonometric in nature, so the greatest curvature changes will occur at the center of the PCB.

An examination of the physical proportions of axial leaded components such as resistors, capacitors, diodes, integrated circuits, and transformers shows that the component body is much stiffer than the electrical lead wires. The points where the wires are joined to the component will tend to remain fixed and perpendicular to the component. The points where the wires are soldered to the stiffer PCB will tend to remain fixed and perpendicular to the PCB. The lead wires are much more compliant

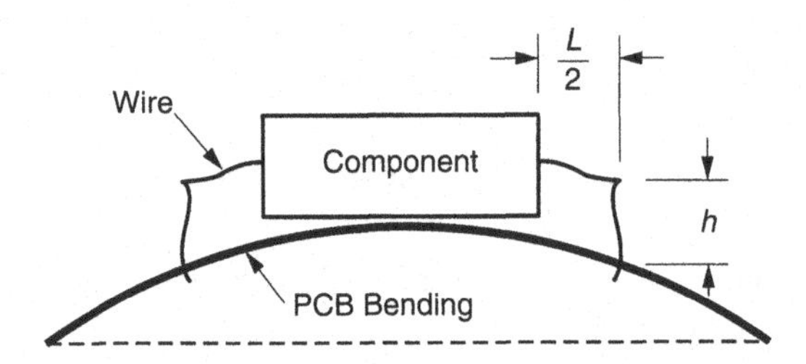

FIGURE 12.6 Large PCB bending displacements in severe dynamic environments can induce high lead wire and solder joint forces and stresses that will reduce the fatigue life.

than the component and the PCB. Therefore, virtually all of the local bending action will occur in the wires and not in the component or the PCB, as it flexes back and forth during vibration.

For the purposes of analysis, all of the bending motions generated by the vibrating PCB can be related to the bending displacements of the wire. When the combined bending action of the wire is analyzed, it will appear as shown in Fig. 12.7*a*. Using superposition methods, the wire bending can be separated into two different geometric forms. The first geometric form shown in Fig. 12.7*b* represents the wire bending displacements due to an equivalent concentrated force acting at the center of the wire where it would normally be joined at the component. The second geometric form shown in Fig. 12.7*c* represents the wire bending displacements due to the PCB bending, producing rotations of the wires where they are soldered to the PCB. (Relative dynamic PCB displacements, with lead wire and solder joint forces, moments, and stresses can be obtained from Ref. 2, pp. 145–160, along with Fig. 12.3.)

For a symmetrical system, when the component is at the center of a plug-in type PCB, the maximum relative dynamic displacement between the PCB and the component will be δ_3 at the lead wire, point 3 in Fig. 12.7*a*, due to PCB bending. The slope of the PCB at the wire θ_C can be obtained from the derivative of the displacement, as shown below:

$$Z = Z_0 \sin\frac{\pi x}{a} \quad \text{so} \quad \theta_C = \frac{dZ}{dx} = \frac{Z_0 \pi}{a}\cos\frac{\pi x}{a}$$

also

$$\delta_3 = Z_0 - Z_0 \sin\frac{\pi x}{a}$$

then

$$\delta_3 = Z_0\left(1 - \sin\frac{\pi x}{a}\right) \tag{12.26}$$

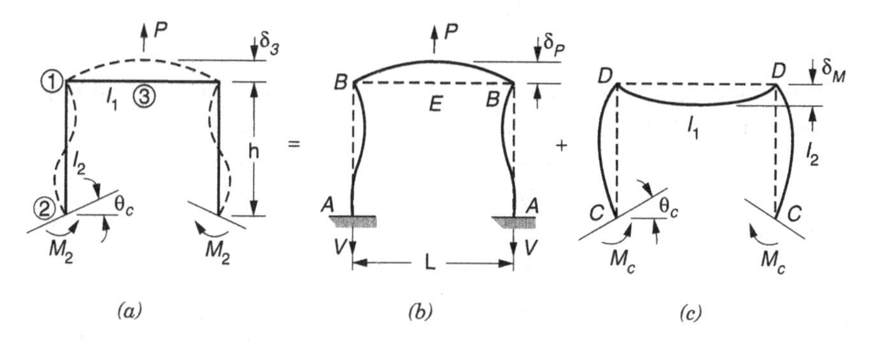

FIGURE 12.7 Mathematical model of lead wires using the superposition method to find the forces, moments, and displacements in different sections of the wire: (*a*) combined bending, (*b*) vertical load acting alone, and (*c*) bending moments acting alone.

where $x =$ in., distance from edge of PCB to lead wire at solder joint
$a =$ in., length of PCB parallel to component length
$G_{in} =$ dimensionless, input acceleration in gravity units
$Q =$ dimensionless, PCB transmissibility at the natural frequency
$f_n =$ Hz, PCB natural frequency
$Z_0 = \dfrac{9.8G_{in}Q}{(f_n)^2} =$ in., maximum single amplitude PCB displacement (12.27)

Relative dynamic wire displacement δ_3 at point 3 as a function of the force P and geometric properties of the wire are

$$\delta_3 = \frac{PL^3}{48EI_1}\left(1 - \frac{3}{2K+4}\right) - \frac{\theta_C L}{4}\left[2 - \left(\frac{3+2K}{2+K}\right)\right] \tag{12.28}$$

where $P =$ lb, dynamic force induced in the wire
$L =$ in., total wire length, horizontal section
$h =$ in., wire height, vertical section
$E =$ lb/in.^{2}, wire modulus of elasticity
$I_1 =$ in.^{4}, horizontal wire section moment of inertia
$I_2 =$ in.^{4}, vertical wire section moment of inertia
$K = hI_1/LI_2$ convenient ratio for simplifying the equation
$K = 1.0$ for a square wire frame where $h = L$ and $I_1 = I_2$
$\theta_C =$ radians, rotation angle of PCB and wire at solder joint

For a square wire frame, where the height and total length are equal, and the wire diameters are the same, the above equation simplifies into the relation shown below:

$$\delta_3 = \frac{PL^3}{96EI} - \frac{\theta_C L}{12} \tag{12.29}$$

from Eq. 12.26

$$\theta_C = \frac{Z_0 \pi}{a} \cos\frac{\pi x}{a} \tag{12.30}$$

The bending moment in the wire at the solder joint is the same as the shear tear-out moment in that solder joint, which is

$$M_2 = \frac{PL}{8K+16} + \frac{2EI_2\theta_C}{h}\left(\frac{3+2K}{2+K}\right) \tag{12.31}$$

For a square wire frame where the height and total length are equal, and the wire diameters are the same, the above equation simplifies into the following relation:

$$M_2 = \frac{PL}{24} + \frac{10EI\theta_C}{3L} \tag{12.32}$$

The maximum bending moment in the wire will occur at point 3, where the wire joins the component body:

$$M_3 = \frac{PL}{4K+8}(K+1) - \left(\frac{4EI_1\theta_C}{L} - KM_C\right) \tag{12.33}$$

and

$$M_C = \frac{2EI_2\theta_C(3+2K)}{h(2+K)} \tag{12.34}$$

For a square wire frame where the height and total length are equal, and the wire diameters are the same, the above equations simplify as

$$M_3 = \frac{PL}{6} - \left(\frac{4EI\theta_C}{L} - M_C\right) \tag{12.35}$$

and

$$M_C = \frac{10EI\theta_C}{3L} \tag{12.36}$$

Sample Problem: Dynamic Forces, Stresses, and Fatigue Life in Transformer Wires and Solder Joints Due to Relative Motion at Component as PCB Flexes in Vibration

A transformer with six axial copper lead wires, 0.028 in. diameter, is mounted at the center of a 0.075-in.-thick plug-in PCB parallel to the 11.0-in. edge, as shown in Fig. 12.8. The assembly must be capable of withstanding a 1.0-h sine vibration resonant dwell test with a 4.0-G peak input level. The PCB natural frequency is about 175 Hz with a transmissibility of about 14. Find the following:

1. The approximate fatigue life of the copper lead wires
2. The approximate fatigue life of the solder joints

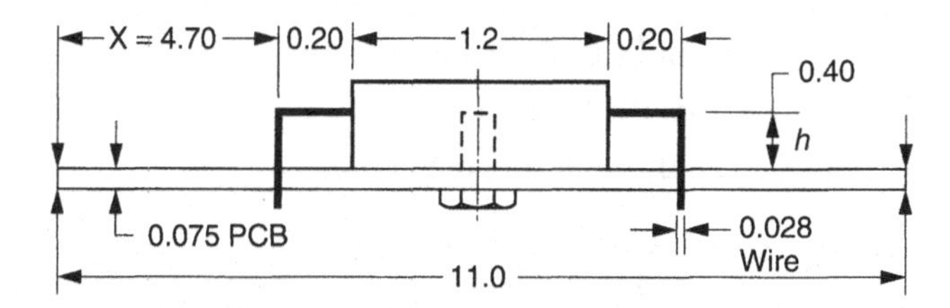

FIGURE 12.8 Dimensions of a through-hole-mounted transformer at the center of a plug-in PCB.

Solution

Part 1A. Maximum Wire Bending Stress at Point 3, Where Wire Joins Component Component wires loaded in bending are typically very compliant compared to the stiffness of the PCB and the component body. (Wires loaded in tension can be very stiff). Virtually all of the bending will, therefore, occur in the wires and not in the component or the PCB. The number of wires involved is usually not important because their relative stiffness is so low.

The dynamic displacement at the center of the PCB can be obtained from Eq. 12.27 where

$G_{in} = 4.0$ dimensionless, acceleration input level
$Q = 14$ dimensionless, PCB transmissibility at resonance
$f_n = 175$ Hz, PCB resonant frequency

$$Z_0 = \frac{9.8(4.0)(14)}{(175)^2} = 0.0179 \text{ in.} \tag{12.37}$$

Angular rotation θ_C of the PCB at the wire and solder joint can be obtained from Eq. 12.30 where

$Z_0 = 0.0179$ in., single amplitude (Ref. Eq. 12.37)
$x = 4.7$ in., distance from edge of PCB to wire
$a = 11.0$ in., length of PCB edge parallel to component

$$\theta_C = \frac{0.0179\pi}{11} \cos \frac{4.7\pi}{11} = 1.158 \times 10^{-3} \text{ rad} \tag{12.38}$$

Relative displacements δ_3 between the component body and the PCB can be obtained from Eq. 12.26 where

$Z_0 = 0.0179$ in., maximum PCB displacement
$x = 4.7$ in., distance from PCB edge to wire
$a = 11$ in., length of PCB parallel to component

$$\delta_3 = (0.0179)\left[1 - \sin \frac{\pi(4.7)}{11}\right] = 4.65 \times 10^{-4} \text{ in.} \tag{12.39}$$

The vertical force P acting on the wires can be obtained from Eq. 12.29 where

$\delta_3 = 4.65 \times 10^{-4}$ in.
$E = 16 \times 10^6$ lb/in.^{2}, copper modulus of elasticity
$d = 0.028$ in., copper wire diameter
$I = \pi d^4/64 = \pi(0.028)^4/64 = 3.017 \times 10^{-8}$ in.^{4}, wire moment of inertia
$\theta_C = 1.158 \times 10^{-3}$ rad, rotation angle (Ref. Eq. 12.38)
$h = 0.40$ in., wire height
$L = (2)(0.20) = 0.40$ in., wire length for a square wire frame

$$4.65 \times 10^{-4} = \frac{P(0.40)^3}{(96)(16 \times 10^6)(3.017 \times 10^{-8})} - \frac{(1.158 \times 10^{-3})(0.40)}{12}$$

$$4.65 \times 10^{-4} = 1.381 \times 10^{-3} P - 3.86 \times 10^{-5} \quad \text{so} \quad P = 0.365 \text{ lb} \qquad (12.40)$$

The maximum bending moment in the wire will occur at point 3, where the wire joins the component body in Fig. 12.7*a*. The moment can be obtained from Eqs. 12.35 and 12.36 where:

$P = 0.365$ lb, force acting at point 3, center of wire
$L = 0.40$ in., wire length for a square frame
$E = 16 \times 10^6$ lb/in.^{2}, wire modulus of elasticity
$I = 3.017 \times 10^{-8}$ in.^{4}, wire moment of inertia
$\theta_C = 1.158 \times 10^{-3}$ rad, slope of PCB (Ref. Eq. 12.38)

$$M_C = \frac{10(16 \times 10^6)(3.017 \times 10^{-8})(1.158 \times 10^{-3})}{3(0.40)} = 4.66 \times 10^{-3} \text{ lb} \cdot \text{in.} \qquad (12.41)$$

$$M_3 = \frac{(0.365)(0.40)}{6} - \left[\frac{4(16 \times 10^6)(3.017 \times 10^{-8})(1.158 \times 10^{-3})}{0.40} - 4.66 \times 10^{-3}\right]$$

$$M_3 = 0.0243 - 9.30 \times 10^{-4} = 0.0234 \text{ lb} \cdot \text{in.} \qquad (12.42)$$

The maximum bending stress in the wire S_W can be obtained from the standard bending stress equation shown below. A stress concentration factor must be considered in the determination of the approximate fatigue life for most structures when more than several thousand stress cycles are expected. The stress concentration factor can be used in the bending stress equation, or it can be used in the *S–N* fatigue curve to define the slope b of the fatigue line on a log–log plot. It should not be used in both places. For this sample problem, a stress concentration factor of 2 was used to characterize the effects of cuts and scratches in the wire due to poor forming dies.

A stress concentration factor of 2 results in a b value of 6.4, as shown in Fig. 5.4 for the copper wire fatigue life:

$$S_W = \frac{M_3 C}{I} = \frac{(0.0234)(0.028/2)}{3.017 \times 10^{-8}} = 10{,}858 \text{ lb/in.}^2 \tag{12.43}$$

Part 1B. Copper Wire Fatigue Life The approximate copper wire fatigue life can be obtained from the fatigue life relation shown in Eq. 5.1 where

N_1 = expected wire cycles to fail
S_1 = 10,858 lb/in.^{2}, calculated wire stress
N_2 = 1000 cycles, failure reference point at 45,000 lb/in.^{2}
S_2 = 45,000 lb/in.^{2}, stress reference point at 1000 cycles
b = 6.4 slope of fatigue curve with stress concentration of 2

$$N_1 = N_2(S_2/S_1)^b = (1000)(45{,}000/10{,}858)^{6.4}$$
$$N_1 = 8.949 \times 10^6 \text{ cycles to fail} \tag{12.44}$$

The approximate time for the copper wire to fail can be obtained from the natural frequency as

$$\text{Wire at time to fail} = \frac{8.949 \times 10^6 \text{ cycles to fail}}{(175 \text{ cycles/s})(3600 \text{ s/h})} = 14.2 \text{ h} \tag{12.45}$$

Vibration life test programs very seldom run smoothly. Many false starts often occur due to unexpected problems that tend to substantially extend the required testing time. These include poor accelerometer calibration, poor accelerometer mounting, electrical operating problems, extensive coffee breaks, vibration shaker problems, schedule problems, availability of the right personnel, equipment availability, premature failures, overtesting, bad data reduction, undertesting, and poor planning are just a few examples. A test scheduled for 1 h will often take at least 5 h to complete. It is a good practice to design the equipment to be capable of withstanding at least five full test programs. Then there will be no need to beg the vice president for a second test unit because the first test unit failed half way through the test program. This design appears to have an adequate vibration fatigue life.

Part 2A. Maximum Solder Shear Stress at Point 2 in PCB The maximum solder joint shear stress is at point 2 where the lead wire is soldered to the PCB in Fig. 12.7*a*. Two forces will produce shear in the solder joints. A vertical force V acting on the wire resulting from the vertical force P acting at point E where the wire joins the body of the component and a bending moment M_2 in the wire at the PCB producing a shear tear-out stress in the solder joint. The bending moment M_2 in the wire can be obtained from Eq. 12.32 where

$P = 0.365$ lb (Ref. Eq. 12.40)
$L = 0.40$ in., wire length for a square frame
$E = 16 \times 10^6$ lb/in.^{2}, copper wire modulus of elasticity
$I = 3.017 \times 10^{-8}$ in.^{4}, wire moment of inertia
$\theta_C = 1.158 \times 10^{-3}$ rad (Ref. Eq. 12.38)

$$M_2 = \frac{PL}{24} + \frac{10EI\theta_C}{3L} \quad \text{(Ref. Eq. 12.32)}$$

$$M_2 = \frac{(0.365)(0.40)}{24} + \frac{10(16 \times 10^6)(3.017 \times 10^{-8})(1.158 \times 10^{-3})}{3(0.40)}$$

$$M_2 = 0.00608 + 0.00466 = 0.0107 \text{ lb} \cdot \text{in.} \quad (12.46)$$

The overturning bending moment in the wire will produce a shear tear-out stress in the solder joint. Test data shows the solder joint fails in an area that is between the wire and the plated through-hole. Therefore, the solder joint shear area is based on the thickness of the PCB and the average solder joint diameter, using an average of the lead wire diameter and the plated through-hole diameter. The height of the solder joint is based on the PCB thickness without considering any solder joint fillets. The solder joint shear tear-out stress S_{ST} can be obtained from

$$S_{ST} = M_2/hA_{ST} \quad (12.47)$$

where $M_2 = 0.0107$ lb · in., overturning moment
$h = 0.075$ in., PCB thickness
$d_W = 0.028$ in., wire diameter
$d_P = 0.040$ in., plated through-hole diameter
$d_{AV} = (0.028 + 0.040)/2 = 0.034$ in., average solder diameter
$A_{ST} = \pi(d_{AV})^2/4 = \pi(0.034)^2/4 = 0.00091$ in.^{2} shear tear-out solder area
$A_S = \pi d_{AV} h = \pi(0.034)(0.075) = 0.00801$ in.^{2}, solder direct shear area

$$S_{ST} = (0.0107)/(0.075)(0.00091) = 157 \text{ lb/in.}^2 \text{ solder shear tearout stress} \quad (12.48)$$

The direct shear stress in the solder joint due to the vertical force V can be obtained from the vertical force P as follows: $V = P/2 = 0.365/2 = 0.183$ lb. The solder shear stress is then

$$S_S = V/A_S = 0.183/0.00801 = 22.8 \text{ lb/in.}^2 \text{ direct shear stress} \quad (12.48a)$$

Part 2B. Solder Joint Fatigue Life The approximate solder fatigue life can be obtained from Eq. 5.1 and Fig. 5.3 where

N_1 = number of cycles to produce a failure

$S_1 = 157 + 23 = 180\ \mathrm{lb/in.}^2$, calculated dynamic total solder shear stress

$N_2 = 1000$ reference point, cycles for solder failure at $6500\ \mathrm{lb/in.}^2$

$S_2 = 6500\ \mathrm{lb/in.}^2$, reference point, stress for solder failure at 1000 cycles

$b = 4.0$ slope of fatigue curve on log–log S–N plot

$N_1 = N_2(S_2/S_1)^b = (1000)(6500/180)^{4.0}$

$$N_1 = 1.70 \times 10^9 \quad \text{solder cycles to fail} \tag{12.49}$$

The solder joint fatigue life can be estimated from the PCB natural frequency:

$$\text{Solder time to fail} = \frac{1.70 \times 10^9 \text{ cycles to fail}}{(175 \text{ cycles/s})(3600 \text{ s/h})} = 2698 \text{ h} \tag{12.50}$$

The design appears to have a very good vibration fatigue life.

12.3 FORCES, STRESSES, AND FATIGUE LIFE OF VIBRATING COMPONENTS BOUNCING UP AND DOWN ON THEIR LEAD WIRES

Large electronic components with long thin lead wires can be excited at their natural frequency during operation in vibration environments. The component body can act like a concentrated mass and the electrical wires can act as springs. Lead wires and solder joints can experience high stresses when high vibration acceleration levels force the component to bounce up and down during resonant conditions. If the vibration exposure lasts for an extended period, fatigue failures can be expected in the wires and solder joints. Many types of large components that are mounted on PCBs with long thin wires can experience vibration fatigue failures even when the natural frequency of the PCB itself is not excited. Very rapid fatigue failures in the wires and solder joints can be expected when the natural frequencies of the component and the PCB are excited at the same time. It is often necessary to cement large or heavy components to the PCB to prevent them from bouncing up and down on their lead wires. Polyurethane adhesives that can be ultraviolet (UV) cured may be used to bond the four corners of large components to the PCB. The corners of these components do not have any lead wires so they are convenient for bonding.

Large axial leaded components with long thin wires, such as transformers, capacitors, hybrids, resistors, and fine-pitch semiconductor devices are considered here. They will have wires that extend horizontally from the component body, then make a 90° bend down to the PCB where they are soldered, as shown in Fig. 12.4. This geometry applies to through-hole and surface-mounted parts. The component will act like a 6 degree-of-freedom system when it is exposed to vibration. It will experience translation along the x, y, and z axes, and rotation about the x, y, and z axes. This means that the vibrating component can have 6 natural frequencies. However, for dynamic analysis purposes, the vibration input direction and the component motion will be restricted to move only in the vertical direction, perpendicular to the plane of the PCB. Only translation will be considered, without any rotation, to simplify the analysis. The component wires are typically very compliant in bending compared to the component body and to the PCB. Therefore, only the wire bending stiffness will be analyzed to obtain the component natural frequency, along with the wire and solder dynamic forces and stresses to estimate the fatigue life of this simple system. (See Ref. 2, pp. 145–160 for the derivation of lead wire frame displacements, forces, and moments.)

Displacements in the component wires will be similar to Figs. 12.4 and 12.7b when only the component natural frequency is excited and the PCB natural frequency is not excited at the same time. The maximum dynamic displacement δ_P will occur at the center of the horizontal segment of the wire at point E in Fig. 12.7b, where the wires join the component body, due to the concentrated force P that is produced by the vibrating component. The magnitude of this displacement is [2]:

$$\delta_P = \frac{PL^3}{48EI_1}\left(1 - \frac{3}{2K+4}\right) \tag{12.51}$$

where P = lb, dynamic force produced by vibrating component
L = in., total length of horizontal wire sections extending from component
h = in., height of vertical section
E = lb/in.^{2}, wire modulus of elasticity
I_1 = in.^{4}, horizontal wire moment of inertia
I_2 = in.^{4}, vertical wire moment of inertia
$K = hI_1/LI_2$, dimensionless, convenient ratio
$K = 1.0$ for a square frame where $h = L$ and $I_1 = I_2$
$K = 0.50$ when $L = 2h$ and $I_1 = I_2$

For a frame where I_1 equals I_2 with different length ratios for h and L, the above equation will simplify to the expressions shown below:

$$\begin{aligned} \delta_P &= \frac{PL^3}{96EI} \quad (K = 1.0) \\ \delta_P &= \frac{PL^3}{120EI} \quad (K = 0.50) \end{aligned} \tag{12.52}$$

The maximum wire bending moment will occur at point E in Fig. 12.7b as shown below [2]:

$$M_E = \frac{PL}{4K+8}(K+1) \tag{12.53}$$

$$M_E = \frac{PL}{6} \quad (K = 1.0)$$
$$M_E = \frac{3PL}{20} \quad (K = 0.50) \tag{12.54}$$

The wire bending moment at point B in Fig. 12.7b, where the wire makes a 90° bend, is shown below [2]:

$$M_B = \frac{PL}{4K+8} \tag{12.55}$$

$$M_B = \frac{PL}{12} \quad (K = 1.0)$$
$$M_B = \frac{PL}{10} \quad (K = 0.50) \tag{12.56}$$

The wire bending moment at point A where the wire is soldered to the PCB is shown below:

$$M_A = \frac{PL}{8K+16} \tag{12.57}$$

$$M_A = \frac{PL}{24} \quad (K = 1.0)$$
$$M_A = \frac{PL}{20} \quad (K = 0.50) \tag{12.58}$$

Sample Problem: Vibration Fatigue Life of Component Bouncing on Lead Wires

A 0.026-lb surface-mounted application-specific integrated circuit (ASIC) measuring 1.47 in.^{2} with 256 fine-pitch kovar lead wires 0.0080 in. wide × 0.0050 in. thick spaced on 0.020-in. centers positioned around the perimeter of the component is shown in Fig. 12.9. The component must be capable of passing a resonant dwell test for 1 h, with a 4.0-G peak sine vibration input level. The natural frequencies of the PCB and the component will be well separated, so they will not reinforce each other. Perform the following calculations:

A. Find the natural frequency of the ASIC bouncing on its lead wires.
B. Find the maximum dynamic force and stress expected in the wire.

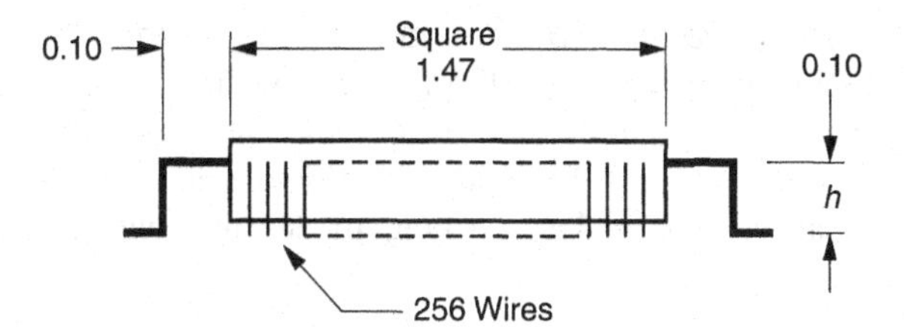

FIGURE 12.9 Dimensions of a fine-pitch-surface mounted component with 256 wires.

C. Find the approximate fatigue life expected in the wire.
D. Find the maximum dynamic stress expected in the solder joint.
E. Find the approximate fatigue life expected in the solder joint.

Solution

Part A. Natural Frequency of the ASIC Bouncing on Its Wires The natural frequency is related to the static displacement δ_{ST}, which can be obtained with the use of Eq. 12.51 by using the ASIC weight W in the place of the dynamic force P. When the wire length $L = 2h$, the resulting K value will be 0.50. The static displacement can then be obtained from Eq. 12.52 using 256 lead wires with the following information:

$W = 0.026$ lb, weight of ASIC body
$L = 2(0.10) = 0.20$ in., total effective length of horizontal wire
$h = 0.10$ in., vertical height of wire
$E = 20 \times 10^6$ lb/in.^{2}, kovar wire modulus of elasticity
$b_w = 0.008$ in., wire width
$t_w = 0.005$ in., wire thickness
$I = b_w(t_w)^3/12 = (0.008)(0.005)^3/12 = 8.33 \times 10^{-11}$ in.^{4}, wire moment of inertia

$$\delta_{ST} = \frac{WL^3}{120EI} = \frac{(0.026)(0.20)^3}{(120)(20 \times 10^6)(8.33 \times 10^{-11})(256)} = 4.10 \times 10^{-6} \text{ in.} \quad (12.59)$$

The natural frequency of the ASIC as a mass and the wires as a spring can be obtained with the use of the acceleration of gravity $g = 386$ in./s^2:

$$f_n = \frac{1}{2\pi}\sqrt{\frac{g}{\delta_{ST}}} = \frac{1}{2\pi}\sqrt{\frac{386}{4.10 \times 10^{-6}}} = 1545 \text{ Hz} \quad (12.60)$$

Part B. Maximum Dynamic Force and Stress in the Wire Newton's mass × acceleration relation can be used to find the dynamic force P acting on the component body with a 4.0-G peak input. The Q value used was $\sqrt{1545}$ or about 40. The dynamic force was assumed to be divided equally to each wire. This is not really true because resonances can produce rocking modes that will change the load distribution to the wires:

$$P = WG_{\text{in}}Q = (0.026)(4.0)(40) = 4.16 \text{ lb} \tag{12.61}$$

Bending moments in the wires will be maximum at point E in Fig. 12.7b, where the wires join the component body. This value can be obtained from Eq. 12.54 when $K = 0.50$ for $L = 2h$:

$$M_E = \frac{3PL}{20} = \frac{(3)(4.16)(0.20)}{20} = 0.125 \text{ lb} \cdot \text{in.} \tag{12.62}$$

The standard equation can be used to find the wire bending stress S_W for the 256 wires. Since fatigue is involved, some type of stress concentration must be considered. This can be applied to the wire itself, or it can be included in the b exponent slope of the fatigue curve. For this application a stress concentration factor of 2 will be used in the fatigue curve for the kovar wire, to account for any possible cuts and scratches produced by poor wire forming dies. This will result in a b exponent slope of 6.4:

$$\text{Wire } S_W = \frac{M_E C}{I} = \frac{(0.125)(0.005/2)}{(256)(8.33 \times 10^{-11})} = 14{,}654 \text{ lb/in.}^2 \tag{12.63}$$

Part C. Approximate Fatigue Life Expected in Wire Kovar wire has a fatigue curve similar to Fig. 10.3, where it has an ultimate tensile strength S_{TU} of about 84,000 lb/in.^{2} at 1000 stress cycles. Equation 5.1 can be used to find the approximate number of stress cycles required to produce a wire failure with the following data:

N_1 = number of stress cycles to fail at 14,654 lb/in.^{2}
S_1 = 14,654 lb/in.^{2}, wire bending stress
N_2 = 1000 stress cycles, reference point at 84,000 lb/in. to fail
S_2 = 84,000 lb/in.^{2}, reference point at 1000 stress cycles to fail for kovar wire

$$N_1 = N_2\left(\frac{S_2}{S_1}\right)^b = (1000)\left(\frac{84{,}000}{14{,}654}\right)^{6.4} = 7.13 \times 10^7 \text{ wire cycles to fail} \tag{12.64}$$

The resulting wire fatigue life can be obtained from the natural frequency as follows:

$$\text{Wire life} = \frac{7.13 \times 10^7 \text{ cycles to fail}}{(1545 \text{ cycles/s})(3600 \text{ s/h})} = 12.8 \text{ h} \tag{12.65}$$

The wire fatigue life appears to be satisfactory.

Part D. Maximum Dynamic Stress Expected in Solder Joint Two forces are generated in the wires that will produce solder joint stresses in the surface-mounted component. One force H acts in the horizontal direction, parallel to the plane of the PCB. This will produce a shear stress in the solder joint. The other is the vertical force P. This acts in a direction perpendicular to the plane of the PCB. It will produce a tensile peeling stress at the heel of the solder joint.

The value of the horizontal force H can be obtained from Ref. 2 (p. 149). This force was assumed to be divided equally to each of the 256 wires:

$$H = \frac{3PL}{2h(4K + 8)}$$

When $L = 2h$, then

$$K = 0.50 \quad \text{and} \quad H = \frac{3P}{10(256)}$$

Substitute the value of P from Eq. 12.61 into the above relation and solve for H:

$$H = \frac{3(4.16)}{10(256)} = 0.00487 \text{ lb in each wire} \tag{12.66}$$

Each solder pad measures 0.008 in. wide × 0.035 in. long giving an area $A = 0.000280$ in.^{2}, which results in a solder joint shear stress S_S of

$$S_S = H/A = 0.00487/0.000280 = 17.4 \text{ lb/in.}^2\text{, negligible} \tag{12.67}$$

The worst-case solder joint peel stress generated by the vertical load P was examined by considering the minimum solder joint area A_S to be equal to the wire cross-section area with wire dimensions of 0.008 × 0.0050 in., or 4.0×10^{-5} in.^{2} for 256 wires:

$$S_S = P/A_S = 4.16/(4.0 \times 10^{-5})(256) = 406 \text{ lb/in.}^2 \tag{12.68}$$

Part E. Approximate Fatigue Life Expected in Solder Joint The number of stress cycles required to produce a vibration failure in the solder joint can be obtained using the data from Eq. 12.49 as follows:

$$N_1 = N_2(S_2/S_1)^b = 1000(6500/406)^4 = 6.57 \times 10^7 \text{ cycles to fail} \qquad (12.69)$$

The solder time to fail can be obtained from the natural frequency of 1545 Hz as follows:

$$\text{Solder time to fail} = \frac{6.57 \times 10^7 \text{ cycles to fail}}{(1545 \text{ cycles/s})(3600 \text{ s/h})} = 11.8 \text{ h} \qquad (12.70)$$

The solder joint fatigue life appears to be satisfactory.

CHAPTER 13

Fatigue Life of Long Components, Tall Components, and Small Components Mounted on PCBs

13.1 VIBRATION FATIGUE LIFE OF LONG COMPONENTS MOUNTED NEAR THE FREE EDGE OF A PCB

Printed circuit boards (PCBs) operating in severe vibration environments can develop large dynamic displacements when their natural frequencies are excited and the systems are lightly damped so they produce high transmissibility Q values. When long rectangular components are mounted near the center of the PCB, the flexing action of a rectangular PCB with four sides supported will force the lead wires to strain and deform, as shown in Figs. 8.12 and 12.6. Long rectangular components, with bottom leads, surface mounted or through-hole mounted, are often placed close to and perpendicular to the free edge of a rectangular plug-in PCB that has three sides supported (hinged) and one side free. The flexing action of the PCB at its resonant condition can force the lead wires to bend and stretch, as shown in Fig. 13.1, when the component body is much stiffer than the PCB. Finite element computer studies have shown that the PCB deflection curve will approximate a trigonometric sine shape. There will be a slightly flatter section on thin PCBs, in the vicinity of the wires due to the local stiffening effects of the wires. The computer studies have also shown that about 90% of the dynamic loads in the wires will be carried in tension and only about 10% will be carried in bending.

The method of analysis was simplified by assuming 100% of the dynamic load deforming the wires will be carried in direct tension. Also, the slightly flat section on thin PCBs in the vicinity of the wires was ignored to simplify the analysis. These approximations will result in slightly higher calculated stress levels in the wires and solder joints, so the method of analysis is conservative. The relative displacement between the long rectangular component and the bending PCB will force the lead wires at the ends of the component to stretch through a small displacement δ, as shown in Fig. 13.1. The magnitude of this relative displacement can be derived using a trigonometric deflection curve as outlined below.

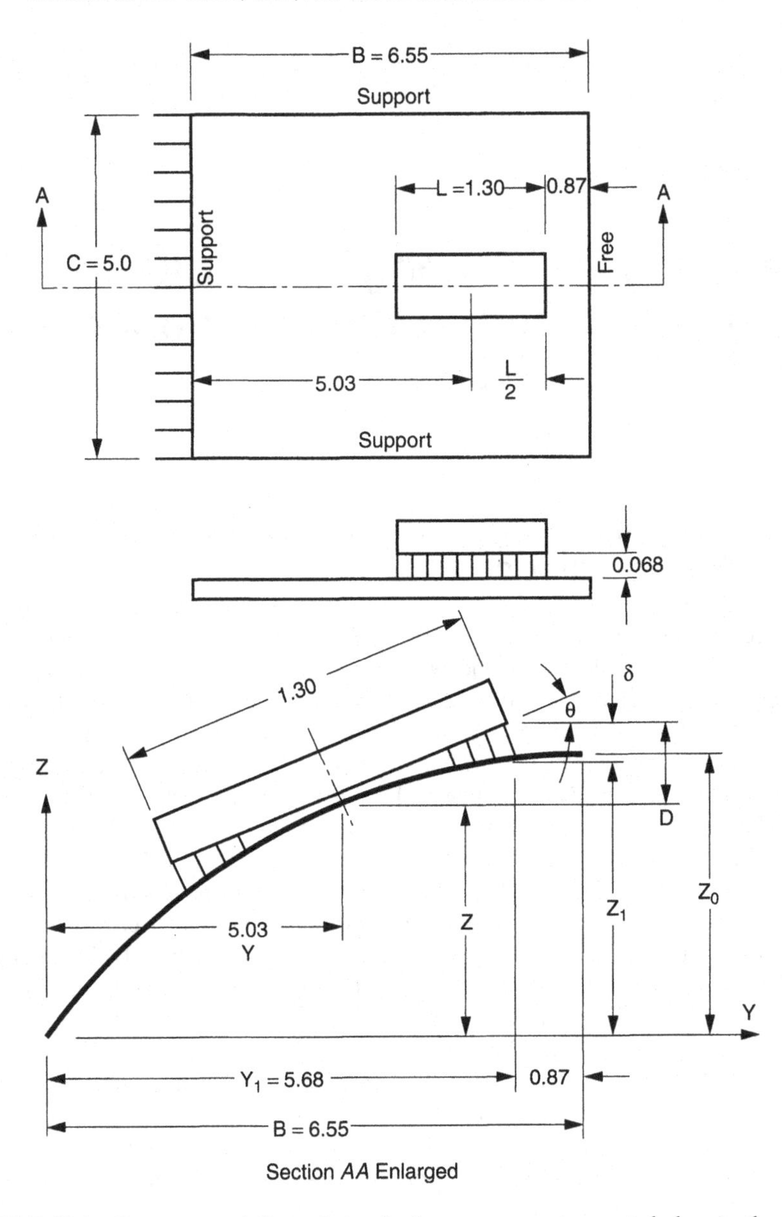

FIGURE 13.1 Geometry and dimensions of a long component mounted close to the free edge of a bending PCB when the opposite edge of the PCB is simply supported.

An examination of Fig. 13.1 shows that the following relations are valid:

$$Z + D = Z_1 + \delta \quad \text{so} \quad \delta = Z + D - Z_1 \tag{13.1}$$

The dynamic displacement Z at the center of the component can be related to the maximum expected dynamic displacement Z_0 at the center of the free edge. When B is the length of the PCB along the component, a sine wave trigonometric displacement function can be used [1]:

$$Z = Z_0 \sin \frac{\pi y}{2B} \tag{13.2}$$

This expression represents a sine wave at $\pi/2$, which is 90°. At this point the slope of the sine wave is zero. This means that the slope at the free edge of the PCB is expected to be zero. The slope at the free edge of the deflecting PCB will never be zero in any typical vibration condition. Therefore, the trigonometric displacement curve used is conservative so it will result in higher calculated displacements and stresses than will normally be expected. This is similar to using a safety factor, which is often called a factor of ignorance. Since there are many different unknown parameters that can reduce the vibration fatigue life in electronic systems, it is good practice to use safety factors to ensure a good fatigue life for systems that require a high reliability.

The dynamic displacement Z_1 at the end of the long component will be

$$Z_1 = Z_0 \sin \frac{\pi y_1}{2B} \tag{13.3}$$

The slope at the center of the long component is required. This can be obtained from the first derivative of the displacement at the center of the component:

$$\theta = \frac{dZ}{dy} = \frac{\pi}{2B} Z_0 \cos \frac{\pi y}{2B} \tag{13.4}$$

The vertical distance (in the direction of the Z axis) from the bottom center of the component to the bottom end of the component is shown as the dimension D:

$$D = \frac{L}{2} \sin \theta \quad \text{for small angles} \sin \theta = \theta \quad \text{so} \quad D = \frac{L}{2} \theta \tag{13.5}$$

Then

$$D = \frac{\pi L}{4B} Z_0 \cos \frac{\pi y}{2B} \tag{13.6}$$

Substitute Eqs. 13.2, 13.3, and 13.6 into Eq. 13.1 to solve for the relative displacement δ, which represents the relative dynamic stretch expected in the component end wires:

$$\delta = Z_0\left[\sin\frac{\pi y}{2B} + \frac{\pi L}{4B}\cos\frac{\pi y}{2B} - \sin\frac{\pi y_1}{2B}\right] \tag{13.7}$$

Sample Problem: Relative Dynamic Displacements, Stresses, and Fatigue Life in the Wires and Solder Joints of a Long Component Mounted Near the Free Edge of a PCB

A 1.3-in.-long rectangular hybrid with two rows of kovar wires, 0.012 in. diameter, is through-hole mounted close to and perpendicular to the 5.00-in. free edge of a rectangular plug-in epoxy fiberglass PCB supported (hinged) on three sides. See Fig. 13.1. The PCB measures 6.55 × 5.00 × 0.090 in. thick and weighs 0.55 lb. The PCB assembly must be capable of passing a 30-min dwell at its natural frequency using a 5.5-G peak sine vibration input level. A total of four internal copper planes, each with 2 ounces of copper (0.0028-in. thick each plane), are used for electrical and thermal purposes. The following information is required to verify the PCB reliability:

1. Fundamental natural frequency of the PCB
2. Relative dynamic stretch expected in the component end wires
3. Maximum dynamic force acting on the wires and solder joints
4. Dynamic stresses in the wires and solder joints
5. Expected fatigue life of the wires and solder joints

Solution

Part 1. Fundamental Natural Frequency of the PCB The natural frequency of a uniform rectangular plug-in PCB with three sides supported and one side free can be obtained from [1]

$$f_n = \frac{\pi}{2}\left(\frac{D}{\rho}\right)^{1/2}\left[\frac{1}{4B^2} + \frac{1}{C^2}\right] \tag{13.8}$$

where $E = 3 \times 10^6$ lb/in.^{2}, epoxy fiberglass modulus with four copper planes 2 ounces each

$h = 0.090$ in., PCB thickness

$\mu = 0.18$ dimensionless, Poisson's ratio for fiberglass PCB with four copper planes

$$D = \frac{Eh^3}{12(1-\mu^2)} = \frac{(3 \times 10^6)(0.090)^3}{12[1-(0.18)^2]} = 188 \text{ lb} \cdot \text{in., plate stiffness}$$

$W = 0.55$ lb, weight of PCB
$g = 386$ in./s^2, acceleration of gravity
$B = 6.55$ in., length of supported side
$C = 5.0$ in., length of one free side and one supported side
$\rho = \frac{W}{gab} = \frac{0.55}{(386)(6.55)(5.0)} = 4.35 \times 10^{-5}$ lb $\cdot$ s^2/in.^{3},
PCB mass per unit area

$$\text{PCB} \quad f_n = \frac{\pi}{2}\left(\frac{188}{4.35 \times 10^{-5}}\right)^{1/2}\left[\frac{1}{4(6.55)^2} + \frac{1}{(5.0)^2}\right] = 149 \text{ Hz} \qquad (13.9)$$

Part 2. Relative Dynamic Stretch Expected in Component End Wires (Eq. 13.7)

$G_{\text{in}} = 5.5$-G peak sine vibration input acceleration level
$f_n = 149$ Hz, PCB natural frequency, shown above
$Q = \sqrt{f_n} = \sqrt{149} = 12$ dimensionless, approximate PCB transmissibility
$Z_0 = \frac{9.8G_{\text{in}}Q}{(f_n)^2} = \frac{(9.8)(5.5)(12)}{(149)^2} = 0.0291$ in., displacement at center of free edge
$L = 1.3$ in., component length
$y = 5.03$ in., distance to center of component
$y_1 = 5.68$ in., distance to end of component

$$\delta = (0.0291)\left[\sin\frac{\pi(5.03)}{2(6.55)} + \frac{\pi(1.3)}{4(6.55)}\cos\frac{\pi(5.03)}{2(6.55)} - \sin\frac{\pi(5.68)}{2(6.55)}\right]$$
$$= (0.0291)[0.9343 + 0.0556 - 0.9783] = 3.376 \times 10^{-4} \text{ in.} \qquad (13.10)$$

Part 3. Maximum Dynamic Force Acting on the Wires and Solder Joints Maximum forces for the two wires at the ends of the component can be obtained from the wire spring rate and the local wire stretching displacement shown above. The wires were examined as beams loaded in tension. The effective length of the wire was used to find the spring rate. Finite element analyses of wires loaded in tension show that their effective length extends into the component body about two wire diameters and into the PCB about two diameters. The wire spring rate K_W can be obtained from structural handbooks as

$$K_W = \frac{A_W E_W}{L_W} \qquad (13.11)$$

where $d_W = 0.012$ in., wire diameter
$A_W = \pi(d_W)^2/4 = \pi(0.012)^2/4 = 1.131 \times 10^{-4}$ in.^{2}, area of one wire
$E_W = 20 \times 10^6$ lb/in.^{2}, kovar wire modulus of elasticity
$L_W = 0.068 + 4(d_W) = 0.068 + 4(0.012) = 0.116$ in., effective wire length

$$K_W = \frac{(2)(1.131 \times 10^{-4})(20 \times 10^6)}{0.116} = 3.90 \times 10^4 \text{ lb/in., two wires} \qquad (13.12)$$

The maximum dynamic force acting on the wires and solder joints can now be obtained for the two end wires that carry the maximum dynamic loads:

$$P_W = K_W\delta = (3.90 \times 10^4)(3.376 \times 10^{-4}) = 13.16 \text{ lb, two wires} \qquad (13.13)$$

Part 4A. Dynamic Axial Stresses in the Wires Dynamic stresses in the wires and solder joints can be obtained from structural handbooks. Only the two end wires and two end solder joints will develop the maximum dynamic force since they are the farthest away from the center of the component. The wire stiffness is small compared to the component bending stiffness and the PCB stiffness. A quick check can be made of the static displacement of a ceramic component body 0.22 in. thick as a cantilever beam with half the component length and a 13.16-lb end load. It shows that the stretch in two lead wires from Eq. 13.10 is more than 5 times greater than the bending displacement of the component body. When two wires carry the dynamic load, the stress in each wire will be

$$S_W = P_W/2A_W = 13.16/(2)(1.131 \times 10^{-4}) = 58{,}178 \text{ lb/in.}^2\text{, axial wire stress} \qquad (13.14)$$

Part 4B. Dynamic Shear Stresses in the Solder Joints A worst-case solder joint shear stress was examined when two wires carry the load by considering the minimum possible solder joint area A_S at the outer diameter of the lead wire, for a through-hole solder joint in a 0.090-in.-thick PCB with no solder fillets:

$$S_S = P_W/2A_S = 13.16/(2)(\pi)(0.012)(0.090) = 1939 \text{ lb/in.}^2\text{, solder shear stress} \qquad (13.15)$$

Part 5A. Fatigue Life of the Wires The approximate number of stress cycles needed to produce a fatigue failure in the kovar lead wires can be obtained from Eq.

5.1 and Fig. 10.3. A stress concentration factor of 2 was used considering wire scratches to obtain the slope b of 6.4 for the kovar wire S–N fatigue curve:

N_1 = number of cycles to produce a failure with a 58,178 lb/in.^{2} stress

S_1 = 58,178 calculated stress in kovar wire

N_2 = 1000 cycles to fail at a reference stress of 84, 000 lb/in.^{2}

S_2 = 84,000 lb/in.^{2}, reference stress to fail at 1000 cycles

b = 6.4 slope of kovar fatigue curve with a stress concentration of 2

$$N_1 = N_2\left(\frac{S_2}{S_1}\right)^b = (1000)\left(\frac{84{,}000}{58{,}178}\right)^{6.4} = 1.049 \times 10^4 \text{ cycles to fail} \tag{13.16}$$

The approximate time it will take for the lead wire to fail during a sine resonant dwell at the natural frequency of 149 Hz is:

$$\text{Wire fatigue life} = \frac{1.049 \times 10^4 \text{ cycles to fail}}{(149 \text{ cycles/s})(60 \text{ s/min})} = 1.2 \text{ min} \tag{13.17}$$

Part 5B. Fatigue Life of Solder Joints The approximate number of stress cycles required to produce a visible vibration fatigue crack failure in the solder joint can be obtained from Eq. 5.1 and Fig. 5.6. A fatigue exponent b value of 4.0 was used for the solder joint without a stress concentration since solder is soft:

N_1 = number of cycles required to produce a failure at 1939 lb/in.^{2} (Ref. Eq. 13.15)

S_1 = 1939 lb/in.^{2}, calculated solder shear stress

N_2 = 1000 cycles to fail, refernce point at 6500 lb/in.^{2} solder shear stress

S_2 = 6500 lb/in.^{2}, reference stress for solder to fail at 1000 cycles

b = 4.0 slope of solder fatigue curve for vibration

$$N_1 = N_2\left(\frac{S_2}{S_1}\right)^b = (1000)\left(\frac{6500}{1939}\right)^{4} = 1.26 \times 10^5 \text{ cycles to fail} \tag{13.18}$$

The approximate time it will take for the solder joint to fail during a sine resonant dwell at a frequency of 149 Hz is

$$\text{Solder fatigue life} = \frac{1.26 \times 10^5 \text{ cycles to fail}}{(149 \text{ cycles/s})(60 \text{ s/min})} = 14.1 \text{ min} \tag{13.19}$$

A vibration fatigue life of 30 min is required. Extensive vibration testing experience has shown that many things can go wrong during these tests, so these tests often experience many false starts. Because of this, it is good practice to design the electronic equipment so that it can survive at least five full qualification tests. This means that this equipment should have a fatigue life of at least 5×0.30 min or 150 min or 2.5 h. The wire fatigue life of only 1.2 min, shown in Eq. 13.17, is not adequate. The solder joint fatigue life of only 14.1 min is not adequate. Some design changes must be made to improve the fatigue life.

13.2 ADDING A STRAIN RELIEF IN THE WIRE TO REDUCE DYNAMIC FORCES AND STRESSES

Electrical lead wires can experience high dynamic forces and stresses in vibration conditions when the PCB is excited at its natural frequency. The bending action of the PCB can induce high axial loads and rapid fatigue failures in straight wires that extend from the bottom surface of long components, as shown in Figs. 8.12 and 13.1. These failures can usually be avoided by adding some type of offset strain relief in the wire so the axial force is changed to a bending force. The stiffness of a wire in bending is usually much less than the stiffness of the same wire in tension. When the strain relief is properly designed, the spring rate of the wire can be substantially reduced. This will reduce the dynamic force in the wire, which will increase the fatigue life. This can be demonstrated by considering the force and deflection curve in a linear system. The force P in the system can be obtained from the spring rate K and the displacement Y as

$$P = KY \tag{13.20}$$

For a given set of conditions where the natural frequency f is defined and the vibration acceleration G level is defined, the dynamic displacement Y will be constant as

$$Y = \frac{9.8G}{f^2} \tag{13.21}$$

When Y is constant, Eq. 13.20 shows that the best way to reduce the force P is to reduce the spring rate K.

Sample Problem: Add a Strain Relief to the Lead Wires in the Previous Sample Problem and Find the New Fatigue Life for the Same 5.5-G Peak Sine Vibration Input

Add an offset strain relief in the lead wires to reduce the effective spring rate of the wire with the geometry shown in Fig. 13.2.

Solution The spring rate for the new strain relief geometry was evaluated in Ref. 1 Eq. 5.82 with the use of strain energy methods. The results are

$$K = \frac{EI}{9.42R^3} \tag{13.22}$$

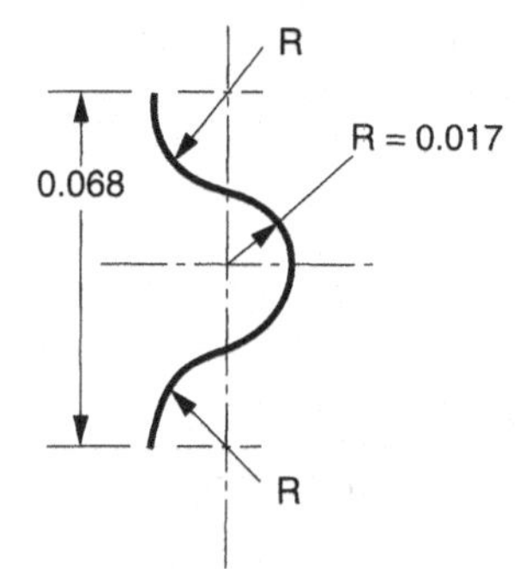

FIGURE 13.2 Component lead wire strain relief that can substantially reduce vibration and thermal cycling forces and stresses in the wires and solder joints.

where $E = 20 \times 10^6$ lb/in.^{2}, kovar wire modulus of elasticity
$d_W = 0.012$ in., wire diameter
$I_W = \pi(d_W)^4/64 = \pi(0.012)^4/64 = 1.018 \times 10^{-9}$ in.^{4}, one wire moment of inertia
$R = 0.017$ in., radius of the wire offset strain relief

$$\text{One wire spring rate} \qquad K = \frac{(20 \times 10^6)(1.018 \times 10^{-9})}{(9.42)(0.017)^3} = 440 \text{ lb/in.} \tag{13.23}$$

The dynamic force acting on the wires can be obtained from the relative displacement shown in Eq. 13.10 and the above spring rate for the same 5.5-G peak sine vibration input level. The more compliant lead wires with the strain relief will allow a slightly greater relative dynamic displacement between the PCB and the component. Since the PCB and the component are much stiffer than the wires with or without the strain relief, the small differences in the relative dynamic displacements can be ignored with little error. Again assume that the two end wires carry the maximum load:

$$P_W = 2K\delta = (2)(440)(3.376 \times 10^{-4}) = 0.297 \text{ lb, on two wires} \tag{13.24}$$

There will be an offset distance of 2 radii in the strain relief, which will generate a bending moment in the wire. The magnitude of the bending moment can be obtained from the force and the distance:

$$M_W = 2RP_W = 2(0.017)(0.297) = 0.010 \text{ lb} \cdot \text{in., on two wires} \tag{13.25}$$

The bending stress equation for each wire can be obtained from a structural handbook as shown below. When a stricture is exposed to several thousand stress reversals, stress concentrations and fatigue must be considered. The stress concentration can be included in the bending stress equation or the effects of a stress concentration can be included in the b exponent slope of the S–N fatigue curve. For this analysis a stress concentration factor of 2 will be used for the wire. This will result in a b exponent slope of 6.4 for the kovar wire on a S–N fatigue curve, as shown in Fig. 10.3:

$$S_W = \frac{M_W C}{I_W} \tag{13.26}$$

where $M_W = 0.010$ lb · in., bending moment on two wires
$C = d_W/2 = 0.012/2 = 0.006$ in., to neutral axis
$I_W = \pi(d_W)^4/64 = \pi(0.012)^4/64 = 1.018 \times 10^{-9}$ in.^{4}, one wire moment of inertia

$$S_W = \frac{(0.010)(0.006)}{(2 \text{ wires})(1.018 \times 10^{-9})} = 29{,}470 \text{ lb/in.}^2 \text{ each wire} \tag{13.27}$$

There is an axial force acting on the wire at the same time as the bending force. Both forces can combine to increase the total stress in the wire. The axial stress S_A in the wire can be obtained from the force in the wire and the wire area where $A_W = \pi(d_W)^2/4 = \pi(0.012)^2/4 = 1.131 \times 10^{-4}$ in.^{2}, area one wire:

$$S_A = P_W/(2)A_W = (0.297)/(2)(1.131 \times 10^{-4}) = 1313 \text{ lb/in.}^2 \tag{13.28}$$

The total stress acting on one wire will be the sum of the bending and the axial stress:

$$\text{Total} \qquad S_T = 29{,}470 + 1313 = 30{,}783 \text{ lb/in.}^2 \tag{13.29}$$

The approximate number of stress cycles required to produce a failure in the kovar wire can be obtained from Eq. 5.1 and Fig. 10.3 using the total wire stress shown above:

$$\text{Wire} \qquad N_1 = (1000)\left(\frac{84{,}000}{30{,}783}\right)^{6.4} = 6.17 \times 10^5 \text{ cycles to fail} \tag{13.30}$$

The approximate time required to produce a wire fatigue failure for a sine resonant dwell test can be obtained from the PCB natural frequency of 149 Hz shown in Eq. 13.9:

$$\text{Wire fatigue life} = \frac{6.17 \times 10^5 \text{ cycles to fail}}{(149 \text{ cycles/s})(60 \text{ s/min})} = 69 \text{ min} \tag{13.31}$$

Comparing the wire fatigue life of 1.2 min without a wire strain relief as shown in Eq. 13.17, with the fatigue life of 69 min with a wire strain relief shown above, means that a properly designed wire strain relief can significantly improve the PCB fatigue life in vibration environments.

Sample Problem: Find the Vibration Fatigue Life of a Long Component When the PCB Edge Opposite the Free Edge Is Changed from Supported to Fixed

The effects of changing the supported (hinged) edge of the PCB opposite the free edge to a fixed (clamped) edge with the two side edges supported in the first sample problem in this chapter can be examined *without the strain relief in the lead wire.*

Solution Use a trigonometric deflection curve for the PCB that satisfies the geometry. The cosine trigonometric function shown below can be used to satisfy the geometric deflections and slopes for the PCB with one edge fixed and the opposite edge free, as shown in Fig. 13.3. The side edges of the PCB are still simply supported. The long component is still mounted close to and perpendicular to the free edge of the PCB:

$$Z = Z_0\left(1 - \cos\frac{\pi y}{2B}\right) \tag{13.32}$$

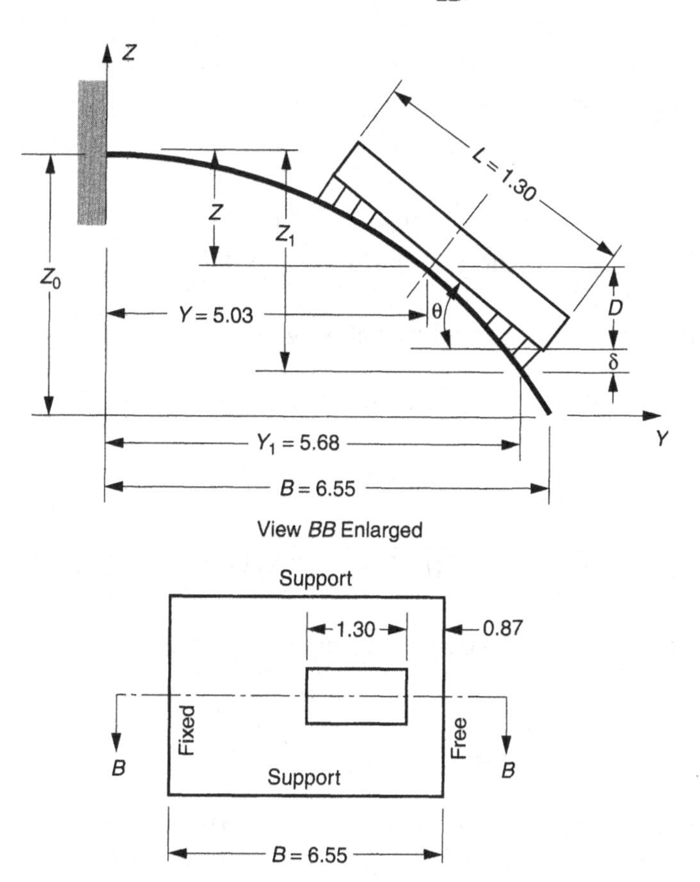

FIGURE 13.3 Geometry and dimensions of a long component mounted close to the free edge of a bending PCB when the opposite edge of the PCB is clamped.

The natural frequency of the PCB can be obtained from [1]

$$f_n = \frac{\pi}{2}\left[\frac{D}{\rho}\left(\frac{1}{C^4} + \frac{0.608}{C^2B^2} + \frac{0.126}{B^4}\right)\right]^{1/2} \tag{13.33}$$

The physical properties of the PCB were obtained from Eq. 13.9 and are repeated here for reference:

$E = 3.0 \times 10^6$ lb/in.^{2}, epoxy fiberglass PCB with four copper planes 2 ounces each
$h = 0.090$ in., PCB thickness
$\mu = 0.18$ dimensionless, Poisson's ratio with four copper planes
$D = 188$ lb in., plate stiffness factor (Ref. Eq. 13.8)
$W = 0.55$ lb, PCB weight
$g = 386$ in./s^2, acceleration of gravity
$B = 6.55$ in., PCB dimension supported edge parallel to long component
$C = 5.0$ in., PCB dimension fixed edge and free edge
$\rho = 4.35 \times 10^{-5}$ lb $\cdot$ s^2/in.^{3}, PCB mass per unit area

$$f_n = \frac{\pi}{2}\left[\frac{188}{4.35 \times 10^{-5}}\left(\frac{1}{[5.0]^4} + \frac{0.608}{[5.0]^2[6.55]^2} + \frac{0.126}{[6.55]^4}\right)\right]^{1/2} = 154 \text{ Hz} \tag{13.34}$$

An examination of Fig. 13.3 shows the following relationships:

$$Z + D = Z_1 - \delta \quad \text{so} \quad \delta = Z_1 - Z - D \tag{13.35}$$

where

$$Z = Z_0\left(1 - \cos\frac{\pi y}{2B}\right) \quad \text{and} \quad Z_1 = Z_0\left(1 - \cos\frac{\pi y_1}{2B}\right) \tag{13.36}$$

The slope of the PCB at the center of the long component can be obtained

$$\theta = \frac{dZ}{dy} = -Z_0\left(-\sin\frac{\pi y}{2B}\right)\left(\frac{\pi}{2B}\right) \tag{13.37}$$

An examination of Fig. 13.3 shows that dimension D has the value

$$D = \frac{L}{2}\sin\theta, \text{ for small angles: } \sin\theta \text{ is} \sim\theta \quad \text{so} \quad D = \frac{L}{2}\theta \tag{13.38}$$

Substitute Eqs. 13.36, 13.37, and 13.38 into Eq. 13.35 and solve for the relative dynamic displacement δ between the component and the PCB:

$$\delta = Z_0\left[\left(1 - \cos\frac{\pi y_1}{2B}\right) - \left(1 - \cos\frac{\pi y}{2B}\right) - \left(\frac{\pi L}{4B}\right)\left(\sin\frac{\pi y}{2B}\right)\right] \tag{13.39}$$

where $G_{in} = 5.5$-G peak sine vibration input

$Q = 12.2$ dimensionless, approximate transmissibility of PCB

$f_n = 154$ Hz, PCB natural frequency (Ref. Eq. 13.34)

$Z_0 = \dfrac{9.8 G_{in} Q}{(f_n)^2} = \dfrac{(9.8)(5.5)(12.2)}{(154)^2} = 0.0277$ in., single amplitude

$y = 5.03$ in., distance to center of component

$y_1 = 5.68$ in., distance to end of component

$L = 1.3$ in., length of component

$B = 6.55$ in., length of PCB parallel to component length

$$\begin{aligned}\delta &= (0.0277)\left[\left(1 - \cos\frac{\pi(5.68)}{2(6.55)}\right) - \left(1 - \cos\frac{\pi(5.03)}{2(6.55)}\right) - \left(\frac{\pi(1.3)}{4(6.55)}\right)\left(\sin\frac{\pi(5.03)}{2(6.55)}\right)\right] \\ &= (0.0277)[0.7929 - 0.6435 - 0.1456] = 1.053 \times 10^{-4} \text{ in.}\end{aligned} \tag{13.40}$$

Assume the dynamic force is still carried by two wires, without a strain relief. This can be obtained from the wire spring rate K shown in Eq. 13.12:

$$P_W = K\delta = (3.90 \times 10^4)(1.053 \times 10^{-4}) = 4.11 \text{ lb on two wires} \tag{13.41}$$

The axial stress in the wire S_W can be obtained from the wire cross-section area in Eq. 13.12 as

$$S_W = P_W/(2)A_W = 4.11/(2)(1.131 \times 10^{-4}) = 18{,}170 \text{ lb/in.}^2 \tag{13.42}$$

The approximate number of stress cycles required to produce a fatigue failure in the kovar wires can be obtained from Fig. 10.3 and Eq. 5.1. The b fatigue exponent of 6.4 has a safety factor of 2 included to account for any deep cuts or scratches in the wire:

$$N_1 = N_2\left(\frac{S_2}{S_1}\right)^b = (1000)\left(\frac{84{,}000}{18{,}170}\right)^{6.4} = 1.80 \times 10^7 \tag{13.43}$$

The approximate fatigue life expected in the wire for a resonant dwell condition can be obtained from the PCB natural frequency of 154 Hz shown in Eq. 13.34 as follows:

$$\text{Wire fatigue life} = \frac{1.80 \times 10^7 \text{ cycles to fail}}{(154 \text{ cycles/s})(60 \text{ s/min})} = 1948 \text{ min} \tag{13.44}$$

The above wire fatigue life with the fixed PCB edge is much greater than the wire fatigue life of 1.2 min, shown in Eq. 13.17 for the simply supported PCB edge for the same sine vibration input acceleration level. This is due to the higher PCB natural frequency with the fixed edge and the resulting change in the PCB curvature during the resonant condition.

13.3 STRUCTURAL MODIFICATIONS IN THE PCB THAT CAN IMPROVE THE FATIGUE LIFE

Several structural modifications were investigated to estimate their impact on reliability, effectiveness, and fatigue life, as described below.

Increasing the PCB Natural Frequency Methods for increasing the natural frequency of the PCB should be a high priority since the dynamic displacements Z of the PCB are related to the square of the frequency. If the natural frequency can be doubled, for example, by epoxy bonding a few stiffening ribs to the PCB, the dynamic displacement would be sharply reduced and the fatigue life would be increased. The PCB transmissibility Q will also increase when the natural frequency is increased, so this must be included in the evaluation. A good estimate of the Q can be made using $\sqrt{f_n}$. If the natural frequency of the PCB can be doubled, for example, the increase in the fatigue life for a given input acceleration G level can be related to the change in the displacement of the PCB as

$$\frac{Z_1}{Z_2} = \frac{Q_1/f_1^2}{Q_2/f_2^2} = \frac{\sqrt{f_1}/f_1^2}{\sqrt{f_2}/f_2^2} = \frac{(f_1)^{1/2}/f_1^2}{(f_2)^{1/2}/f_2^2} = \frac{1/(f_1)^{3/2}}{1/(f_2)^{3/2}} = \left(\frac{f_2}{f_1}\right)^{3/2} = (2)^{1.5} = 2.828 \tag{13.45}$$

The fatigue life of the kovar lead wire is related to the slope of the fatigue curve. When a stress concentration factor of 2 is used, the b exponent slope is 6.4. For linear systems, the fatigue life can be related to the displacements, or the acceleration G levels, or the stress, or the displacements, which are all linear with respect to each other. Using the displacement ratio shown above for a doubling of the PCB natural frequency, the improvement in the fatigue life can be obtained as

$$\text{Wire fatigue life improvement} = (2.828)^{6.4} = 775 \text{ times} \tag{13.46}$$

The fatigue life of the solder will also be improved, but not as much since the solder fatigue exponent b for vibration is only about 4. Using the displacement ratio shown above for a doubling of the PCB natural frequency, the improvement of the solder joint fatigue life can be obtained as

$$\text{Solder fatigue life improvement} = (2.828)^4 = 64 \text{ times} \tag{13.47}$$

Bonding Local Stiffeners to the PCB in the Immediate Area of the Long Component Thin metal shims, about 0.025 in. thick, of a stiff material with a high modulus of elasticity, such as nickel, invar, or molybdenum can effectively reduce the relative displacements between the component body and the PCB during a resonant condition. The shims work best if they can be cemented on both sides of the PCB. One shim should be cemented under the full length of the component on the top side of the PCB and one shim should be cemented on the bottom side of the PCB directly under the full length of the component if possible. The cementing process in this case is critical. Vibration dynamic forces can be very high. Vibration test data has shown that smooth metal shims that are cleaned carefully and epoxy bonded to both sides of the PCB will last only a few minutes in almost any resonant dwell test. The best epoxy cements will not adhere very well to a hard smooth surface. Think about a spider in a bathtub. It cannot climb out because every time it climbs half way up the side wall it slips and falls back to the bottom. Think about a mountain climber going up a steep slope. If he cannot grip the slope, he will slip and fall. He has to knock holes in the hard mountain rock to get a grip or he will slip. The same is true of a hard smooth metal shim. If the hard shim surface is roughened with emery, file, or wire brush, the bond will be improved, but not enough to survive a vibration resonant dwell condition for 30 min or more. Like the mountain climber, several small holes 0.060–0.10 in. diameter must be punched or drilled completely through the shim. The epoxy adhesive must be forced through the holes so they form many epoxy rivets to increase the bond strength. The epoxy rivets must be sheared before the shim will fall off. Cutting many small holes through a hard thin metal can be time consuming and expensive. Many manufacturing organizations will reject this type of rework to improve the fatigue life of a long component. They usually want a rework process that is quick, easy, and cheap to implement. The result is often a new design that ends up costing more than the proposed fix.

Adding Snubbers to Adjacent PCBs in the Areas of the Critical Component Snubbers can often be used to effectively reduce the dynamic displacements in critical areas of PCB to improve their fatigue life. Snubbers are small devices that are installed with a small gap so they strike each other when the PCBs are dynamically displaced in vibration or shock conditions. The impacting action between snubbers on adjacent PCBs substantially reduces the dynamic displacements and stresses. The best snubbers are made from epoxy fiberglass rods about 0.25 in. diameter. These rods should be epoxy bonded to critical areas on PCBs that exhibit large displacements in dynamic environments. The snubbers should be bonded to adjacent PCBs so they are in line with each other, with a very small gap between them. The best snubbers will almost touch each other so there is almost no gap. However, because of the normal manufacturing tolerances, holding very tight tolerances to ensure a small gap with a high production volume can be very expensive. The snubbers on each PCB should be located so plug-in PCBs can be inserted and removed without hitting any components on adjacent PCBs [17,29].

Adding Damping to Reduce PCB Dynamic Displacements Forget damping. It is not very effective for reducing dynamic displacements in vibration and shock environments. The best type of damping is constrained layer viscoelastic damping. This device takes up a significant amount of space on the PCB to make it effective so there is less space for electronic components. Stiffening ribs occupy much less room, they are less expensive, they are not affected by temperature variations, and they are much more effective for reducing dynamic displacements to improve the PCB fatigue life.

13.4 VIBRATION FATIGUE LIFE OF TALL COMPONENTS MOUNTED ON PCBs

Tall components often have problems in vibration when they are mounted on PCBs because they have a high center of gravity (CG). A high CG can produce high overturning moments when the natural frequencies of the components or the PCBs are excited. Tall components are often generated by stacking components such as capacitors on top of each other in an attempt to save valuable space on the PCB. Some components such as relays require a tall profile to accommodate the physical size of their internal electrical contacts and springs. Sometimes large cylindrical capacitors are mounted vertically to save space on the PCBs. Some devices such as transistors are often sensitive to the high solder temperatures used in the wave soldering operations. The solder temperatures in these processes are usually about 260°C. There have been cases where the heat from the solder wicks up the transistor lead wires into the silicon chips, which causes electrical malfunctions in temperature-sensitive devices. In low volume production applications it is a common practice to simply clip heat sinks on the transistor lead wires very close to the top surface of the PCB. These heat sinks soak up a large part of the heat being conducted up the lead wires, so less heat reaches the sensitive silicon chips. The transistors run cooler and the thermal overheating problem is solved. Adding and removing heat sinks from the transistor wires may not be a high-cost item for small production runs of less than about 100 systems. Several million systems are being assembled every year by individual automobile, telephone, television, entertainment, and computer manufacturing companies. The manual attachment and removal of thermal heat sink clips on a large scale is not a cost-effective solution. Some cheaper solution must be used.

One solution to the problem of heat conducting up the lead wires on a transistor is to simply mount the transistor higher above the top surface of the PCB. A longer heat conduction path increases the thermal resistance in that path, which reduces the hot-spot temperature. Tests with transistors mounted about 0.25 in. above the top surface of the PCB have been very successful in solving the high-temperature problems on transistors that are not exposed to vibration. The recent addition of the environmental stress screening (ESS) process for improving the electronics reliability imposes vibration and thermal cycling on the electronic equipment. The vibration has resulted in many wire failures in transistors that were mounted 0.25 in. above the PCB. The reasons for the vibration failures were analyzed so methods for solving this problem can be recommended.

Sample Problem: Random Vibration and Thermal Cycle Fatigue Life of a Transistor Mounted High on a PCB

A TO-5 transistor is mounted 0.25 in. above the top surface of a PCB using a plastic tube spacer, as shown in Fig. 13.4. The spacer is not cemented to the transistor or the PCB. The purpose of the high mount is to prevent the 260°C temperature from the wave soldering operation from conducting heat up the transistor wires and affecting the electrical operation. Raising the transistor above the PCB increases the wire thermal resistance, which reduces the transistor temperature and improves the electrical performance. However, the wires are now failing after about 15 min. of random vibration with an input power spectral density (PSD) level of 0.075 G^2/Hz. This is required for an ESS program for thermal cycling from −40 to +90°C with electrical operation, to improve the reliability of the electronic equipment before it is delivered to the customer.

1. Find the vibration forces, stresses, and approximate fatigue life in the transistor wires and solder joints.
2. Find the thermal cycling forces, stresses, and approximate fatigue life in the transistor wires and solder joints.

Solution

Part 1A. Wire Vibration Forces, Stresses, and Approximate Fatigue Life First find the natural frequency of the high mounted transistor since this will

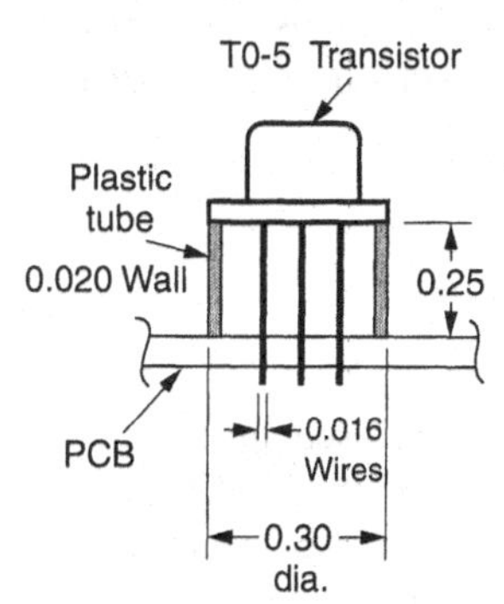

FIGURE 13.4 Heat-sensitive transistor is mounted higher above the PCB with the use of a plastic tube to reduce the high temperatures generated during the wave soldering operation.

influence the forces and stresses. This can be obtained from the static displacement of the transistor as a cantilever beam supported by three wires:

$$f_n = \frac{1}{2\pi}\sqrt{\frac{g}{y_{\mathrm{ST}}}} \tag{13.48}$$

where $g = 386$ in./s^2, acceleration of gravity
$W = 0.0028$ lb, weight of transistor
$E = 20 \times 10^6$ lb/in.^{2}, kovar wire modulus of elasticity
$d = 0.016$ in., diameter of one lead wire
$L = 0.25$ in., wire length
$I = \pi d^4/64 = (3 \text{ wires})(\pi)(0.016)^4/64 = 9.65 \times 10^{-9}$ in.^{4}, inertial for three wires

$$y_{\mathrm{ST}} = \frac{WL^3}{3EI} = \frac{(0.0028)(0.25)^3}{3(20 \times 10^6)(9.65 \times 10^{-9})} = 7.55 \times 10^{-5} \text{ in., static displacement}$$

$$f_n = \frac{1}{2\pi}\sqrt{\frac{386}{7.55 \times 10^{-5}}} = 360 \text{ Hz} \tag{13.49}$$

Most of the damage in random vibration is generated at the natural frequency of the beam acting like a single-degree-of-freedom system. The rms response to the random vibration will be as [1]

$$G_{\mathrm{rms}} = \sqrt{\frac{\pi}{2} P f_n Q} \tag{13.50}$$

where $P = 0.075$ G^2/Hz, PSD input level
$f_n = 360$ Hz
$Q = \sqrt{f_n} = \sqrt{360} = 19$ dimensionless, transmissibility observed with a sine weep

The plastic spacer under the transistor adds some damping at the resonant condition, which reduces the transmissibility about 50% based on observations made with a strobe light during sine vibration tests. The typical Q value with no spacer is usually about $2\sqrt{f_n}$ [1]:

$$G_{\mathrm{rms}} = \sqrt{\frac{\pi}{2}(0.075)(360)(19)} = 28.4 \text{ rms} \tag{13.51}$$

The Gaussian probability distribution function is most often used for electronic systems. In random vibration the rms (or 1σ) G level will occur about 68.3% of the time. The 2σ acceleration G level $= 2G_{\mathrm{rms}}$ will occur about 27.1% of the time. The 3σ acceleration G level $= 3G_{\mathrm{rms}}$ will occur about 4.33% of the time.

A good approximation of the fatigue life can be obtained by assuming the 2σ acceleration G level occurs 100% of the time [1, 3]. The dynamic force acting on the transistor can be obtained from the weight and the 2σ acceleration G level as shown:

$$P_D = W(2)G_{\text{rms}} = (0.0028)(2)(28.4) = 0.159 \text{ lb} \tag{13.52}$$

The dynamic bending stress in the wires S_W can be obtained from a structural handbook as shown below. A stress concentration factor K should be considered when several thousand stress cycles are involved. It can be used in the bending stress or in the fatigue life relation:

$$S_W = \frac{KMC}{I} \tag{13.53}$$

where $K = 1.0$ a stress concentration is not used in this equation. A value of 2 will be used in the b exponent slope of 6.4 for the kovar S–N fatigue curve instead

$a = 0.25$ in., transistor height above PCB

$P_D = 0.159$ lb, dynamic force

$M = P_D a = (0.159)(0.25) = 0.0398$ lb · in., dynamic bending moment in wires

$C = d/2 = 0.016/2 = 0.008$ in., wire radius

$I = 9.65 \times 10^{-9}$ in.^{4}, moment of inertia for three wires (Ref. Eq. 13.49)

$$\text{Wire} \qquad S_W = \frac{(1.0)(0.0398)(0.008)}{9.65 \times 10^{-9}} = 33{,}000 \text{ lb/in.}^2 \tag{13.54}$$

The approximate fatigue life can be obtained from the S–N fatigue curve for kovar wire as shown in Fig. 10.3 and Eq. 5.1. The b exponent slope value of 6.4 includes a stress concentration value of 2.0 to compensate for any deep cuts and scratches due to poor wire forming dies or to rough handling during fabrication and assembly:

$$N_1 = N_2\left(\frac{S_2}{S_1}\right)^b = (1000)\left(\frac{84{,}000}{33{,}000}\right)^{6.4} = 3.95 \times 10^5 \text{ cycles to fail} \tag{13.55}$$

The approximate time for a vibration wire failure can be obtained from the transistor natural frequency of 360 Hz as

$$\text{Time for wire failure} = \frac{3.95 \times 10^5 \text{ cycles to fail}}{(360 \text{ cycles/s})(60 \text{ s/min})} = 18.3 \text{ min} \tag{13.56}$$

This fatigue life is not adequate so some corrective action must be taken. The wires can be encapsulated or filled with room temperature vulcanizing (RTV), or the spacer can be cemented to the transistor body and to the PCB. This will reduce the transmissibility Q value, which can reduce the stress value by a factor of 2, or 50%.

This will increase the fatigue life by a factor of $(2)^{6.4}$, or 84.4 times. The new expected wire fatigue life would be as follows:

$$\text{New wire fatigue life} = 18.3 \text{ min} \times 84.4 = 1545 \text{ min} = 25.7 \text{ h} \qquad (13.57)$$

The above wire vibration fatigue life should be adequate for a reliable system.

Part 1B. Solder Joint Vibration Forces, Stresses, and Approximate Fatigue Life The solder joints vibration fatigue life will be initially evaluated *without the above proposed corrective action* to see if it is critical. The random vibration will produce overturning moments in the lead wires, which will generate shear tear-out stresses S_{ST} in the solder joints as shown in Fig. 2.6. The magnitude of these stresses will be

$$S_{ST} = \frac{M}{hA_S} \qquad \text{(Ref. Eq. 2.15)}$$

where $M = 0.0398$ lb · in., overturning moment (Ref. Eq. 13.53)
$h = 0.060$ in., solder joint height assumed to be the same as PCB thickness
$d_{AV} = (0.016 + 0.028)/2 = 0.022$ in., average diameter of wire and plated through-hole
$A_S = \pi(d_{AV})^2/4 = \pi(0.022)^2/4 = 3.80 \times 10^{-4}$ in.^{2}, average shear area of one solder joint. Three wires are used.

$$S_{ST} = \frac{0.0398}{(3)(0.060)(3.80 \times 10^{-4})} = 582 \text{ lb/in.}^2 \qquad (13.58)$$

The approximate number of stress cycles required to produce a visible solder joint vibration fatigue crack failure can be obtained from the S–N curve shown in Fig. 5.3 and Eq. 5.1, using the vibration b exponent slope of 4.0. A stress concentration is not used with solder because it is a relatively soft material that cold flows and creeps easily:

$$N_1 = N_2\left(\frac{S_2}{S_1}\right)^b = (1000)\left(\frac{6500}{582}\right)^{4.0} = 1.56 \times 10^7 \text{ cycles to fail} \qquad (13.59)$$

The approximate solder joint vibration fatigue life can be obtained from the transistor natural frequency of 360 Hz as

$$\text{Time for solder failure} = \frac{1.56 \times 10^7 \text{ cycles to fail}}{(360 \text{ cycles/s})(60 \text{ s/min})} = 722 \text{ min} \qquad (13.60)$$

The solder vibration fatigue life is adequate.

Part 2A. Wire Thermal Cycling Forces, Stresses, and Approximate Fatigue Life Temperature cycling environments will force the plastic tube to expand. Plastics often have higher thermal coefficients of expansion (TCE) than metals so this will generate forces and stresses in the transistor wires and solder joints. A displacement equilibrium equation can be set up for the plastic tube and the transistor wires as shown below. The subscripts W and P in the equation represent the wire and the plastic, respectively:

$$\alpha_W L_W \, \Delta t + \frac{P_W L_W}{A_W E_W} = \alpha_P L_P \, \Delta t - \frac{P_P L_P}{A_P E_P} \tag{13.61}$$

In cases where the length of the wire is just about the same as the length of the plastic tube, the length will drop out and the equation simplifies as

$$\alpha_W \, \Delta t + \frac{P_W}{A_W E_W} = \alpha_P \, \Delta t - \frac{P_P}{A_P E_P} \tag{13.62}$$

where $\alpha_W = 6.0 \times 10^{-6}$ in./in./°C, TCE kovar wire
$\Delta t = [90 - (-40)]/2 = 65°\text{C}$, for a rapid thermal cycle with no solder creep. See Fig. 1.6 for effects of temperature on solder creep.
$d = 0.016$ in., wire diameter
$A_W = \pi d^2/4 = (3 \text{ wires})\pi(0.016)^2/4 = 6.03 \times 10^{-4}$ in.^{2}, wire area
$E_W = 20 \times 10^6$ lb/in.^{2}, kovar wire modulus of elasticity
$\alpha_P = 50 \times 10^{-6}$ in./in./°C, TCE plastics (nylons, polyesters)
$E_P = 0.15 \times 10^6$ lb/in.^{2}, modulus plastics (nylons, polyesters)
$A_P = \pi[(0.30)^2 - (0.26)^2]/4 = 0.0176$ in.^{2}, plastic tube area

$$(6.0 \times 10^{-6})(65) + \frac{P_W}{(6.03 \times 10^{-4})(20 \times 10^6)} = (50 \times 10^{-6})(65) - \frac{P_P}{(0.0176)(0.15 \times 10^6)}$$

$$3.90 \times 10^{-4} + 8.29 \times 10^{-5} P_W = 3.25 \times 10^{-3} - 3.79 \times 10^{-4} P_P$$

$$\sum P = 0 \quad \text{so} \quad P_W = P_P = P$$

$$4.619 \times 10^{-4} P = 0.00286 \quad \text{so} \quad P = 6.19 \text{ lb on 3 wires} \tag{13.63}$$

The wire tensile stress generated by the axial force can be obtained from a handbook:

$$S_t = P/A_W = 6.19/6.03 \times 10^{-4} = 10{,}265 \text{ lb/in.}^2 \tag{13.64}$$

Find the approximate number of stress cycles required to produce a failure in the kovar wires, using Fig. 10.3 and Eq. 5.1. The fatigue exponent b value of 6.4

includes a concentration stress factor of 2 to account for any deep cuts and scratches in the wire:

$$N_1 = N_2\left(\frac{S_2}{S_1}\right)^b = (1000)\left(\frac{84{,}000}{10{,}265}\right)^{6.4} = 6.96 \times 10^8 \tag{13.65}$$

The wires will never be able to accumulate this number of stress cycles in its lifetime, so the wires are safe in thermal cycling.

Part 2B. Solder Thermal Cycling Forces, Stresses, and Approximate Fatigue Life The shear stress in the solder joints produced by the axial load can be obtained from the solder joint shear area. The minimum solder joint area is used to find the shear stress for a straight axial load. The minimum solder joint area is based on the wire diameter and PCB thickness:

$$S_S = \frac{P}{A_S} = \frac{6.19}{\pi(0.016)(3\text{ wires})(0.06)} = 684\text{ lb/in.}^2 \tag{13.66}$$

The approximate number of stress cycles required to produce a visible solder joint crack can be obtained from Fig. 5.3 and Eq. 5.1. Assuming a rapid thermal cycle at about 30°C per minute, with no dwell periods at the temperature extremes, very little time will be spent at the high temperatures so the solder creep will be small, as shown in Fig. 1.6. The expected fatigue life for this condition is

$$N_1 = N_2\left(\frac{S_2}{S_1}\right)^b = (80{,}000)\left(\frac{200}{684}\right)^{2.5} = 3700\text{ cycles life} \tag{13.67}$$

Many test groups prefer to dwell for 30 min at the temperature extremes. An examination of Fig. 1.6 shows that solder with an initial stress of about 1000 lb/in.^{2} would creep about 25% at a temperature of 90°C after a period of 30 min. Figure 1.4 shows that a creep rate of 100% can double the effective stress level. A creep rate of 25% would therefore be expected to increase the effective stress level about 25% to about (1.25)(684) = 855 lb/in.^{2}. This would reduce the expected fatigue life as

$$N_1 = (80{,}000)\left(\frac{200}{855}\right)^{2.5} = 2117\text{ cycles life} \tag{13.68}$$

Thermal cycling proof of life tests over similar temperature ranges usually require a minimum of 1000 thermal cycles with no visible solder cracks using a 35 power microscope. The solder joints are adequate.

13.5 PROBLEMS WITH THROUGH–HOLE MOUNTING SMALL AXIAL LEADED COMPONENTS ON PCBs

Small axial leaded components such as resistors, capacitors, and diodes with a small body diameter can fail in thermal cycling conditions when they are through-hole mounted on PCBs. The problem is most severe when there is a substantial mismatch in the TCE between the PCB and the component over a wide temperature range. The solder will often wick up the lead wires, directly into the wire bend radius during the wave soldering operation because the vertical leg of the wire is so small, as shown in Fig. 13.5. The vertical leg of the thin lead wire is very compliant so it will normally bend and act as a strain relief as the PCB and the component expands and contracts in thermal cycling events. When the solder wicks up into the wire bend radius, the vertical leg of the wire becomes very stiff so it can no longer bend and act like a strain relief. Very high forces can be produced with these events, which can crack the solder joint or pull the wires out of the component body.

Sample Problem: Thermal Expansion Failures in Small Axial Leaded Components

A manufacturer of small diodes has a specification that limits the maximum allowable axial lead wire force to a value of 4.0 lb to prevent the lead wire from being pulled out of the diode body. The small diode must be through-hole mounted on an epoxy fiberglass PCB where it will be attached using a wave soldering operation. The assembly must be capable of withstanding several dozen thermal cycles over a temperature range from −40 to +90°C for an ESS program. Will the proposed design work?

Solution Set up thermal expansion equilibrium equations to solve for the wire force. Small-diameter axial leaded components that are through-hole mounted on PCBs will have very short vertical wire legs unless the components are mounted high above the PCB or some type of strain relief is used to increase the effective length of the lead wire. There may be a thermal problem because of the diode heat dissipation, so the diode will be mounted flush on the PCB to improve the conduction heat transfer to reduce any hot-spot temperatures. The solder wicking

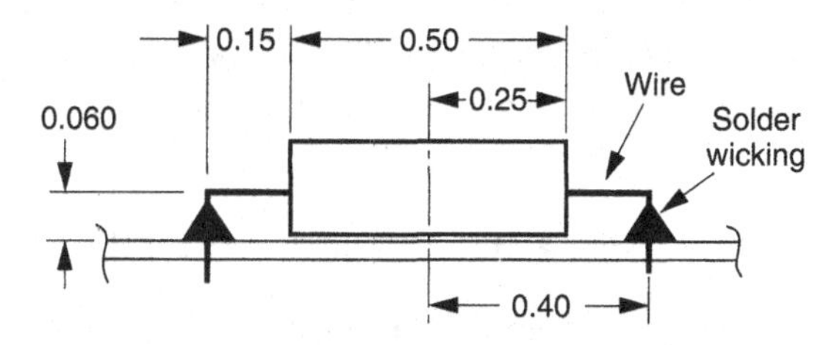

FIGURE 13.5 Solder will often wick up the wires on small axial leaded through-hole-mounted components and prevent the vertical wire legs from bending to provide a strain relief.

will result in a zero vertical wire length as shown in Fig. 13.5. Set up thermal expansion displacement equations for the system with solder wicking up into the bend radius.

Most materials will expand when there is a temperature increase. The PCB normally expands more than the diode and lead wire. As the PCB expands, it will force the component body and the wires to stretch an additional amount greater than their normal temperature expansions. At the same time the reaction force on the PCB will force the PCB to shrink slightly. This will result in the equilibrium equation shown below where α is the TCE of the various materials and the subscripts D, W, and P refer to the diode body, the wire, and the PCB, respectively:

$$\alpha_D L_D \, \Delta t + \frac{P_D L_D}{A_D E_D} + \alpha_W L_W \, \Delta t + \frac{P_W L_W}{A_W E_W} = \alpha_P L_P \, \Delta t - \frac{P_P L_P}{A_P E_P} \tag{13.69}$$

The above equation can be simplified by comparing the physical properties of the various materials involved. The axial stiffness of the diode body and the PCB will be very much greater than the axial stiffness of the thin diode lead wire. Therefore, virtually all of the stretching in the system will be in the lead wires. The stretching terms in the above equation for the diode body and the PCB can then be ignored with very little effect on the accuracy. The resulting simplified equation is

$$\frac{P_W L_{EW}}{A_W E_W} = \alpha_P L_P \, \Delta t - \alpha_D L_D \, \Delta t - \alpha_W L_W \, \Delta t \tag{13.70}$$

where $\alpha_P = 15 \times 10^{-6}$ in./in./°C, TCE of epoxy fiberglass PCB in x–y plane
$\alpha_D = 6 \times 10^{-6}$ in./in./°C, TCE of glass diode body
$\alpha_W = 6 \times 10^{-6}$ in./in./°C, TCE of kovar wire
$L_P = 0.40$ in., effective length of PCB from component center to solder joint
$L_D = 0.25$ in., half the diode body length
$L_W = 0.15$ in., horizontal length of lead wire for thermal expansion
$d_W = 0.018$ in., wire diameter
$L_{EW} = 0.15 + (2)(0.018) = 0.186$ in., effective wire length for axial force goes into component body 2 wire diameters
$\Delta t = [90 - (-40)]/2 = 65$°C, neutral to peak for a rapid temperature cycle
$A_W = \pi(d)^2/4 = \pi(0.018)^2/4 = 2.54 \times 10^{-4}$ in.^{2}, wire area
$E_W = 20 \times 10^6$ lb/in.^{2}, kovar wire modulus of elasticity

$$\frac{P_W(0.186)}{(2.54 \times 10^{-4})(20 \times 10^6)} = (15 \times 10^{-6})(0.40)(65) - (6.0 \times 10^{-6})(0.25)(65) - (6 \times 10^{-6})(0.15)(65)$$

$$P_W = \frac{3.9 \times 10^{-4} - 9.75 \times 10^{-5} - 5.85 \times 10^{-5}}{3.66 \times 10^{-5}} = 6.39 \text{ lb} \tag{13.71}$$

The axial wire force of 6.39 lb exceeds the maximum allowable axial wire force of 4.0 lb so the proposed design is not acceptable. Something has to be done to reduce the axial force in the wire to an acceptable level. A strain relief could be added to the wires in the form of a loop or a camel hump to make them more compliant. These features might interfere with the production autoloading machines. A very effective solution involves raising the component about 0.050 in. above the top surface of the PCB. This will provide some vertical length of wire free of solder for flexibility, when the solder wicks up the wire. A crimp can be formed in the wire legs to raise the component above the PCB. The high component body on thin tall wires could cause vibration problems. Thermal problems might develop if the diode heat is not conducted away by contact with the PCB. A popular solution for these problems is to use speed bumps under the component. Speed bumps are cylindrical disks of plastic that can be purchased in different sizes. A typical speed bump for this application could have a thickness of 0.050 in. with a diameter of 0.10 in. These speed bumps can be installed using a pick-and-place machine that uses a suction tube to capture one disk in a shallow container. The disk touches down on a small table where it picks up a thin polyurethane adhesive layer. The disk is then deposited at the required location on the PCB and the adhesive is quickly cured with an ultraviolet (UV) treatment.

CHAPTER 14

Wear and Interface Surface Fretting Corrosion in Electrical Contacts

14.1 INTRODUCTION

Electronic equipment is being used more and more in virtually every area of the civilized world from transportation and communication to entertainment and defense. The electronics are also becoming more complex and sophisticated so it is more difficult and expensive to maintain the equipment without special training. The safety of many people and sometimes the entire community now rely on the reliability of these complex systems, where the failure of one electrical connection may lead to extensive property damage and even death.

Investigations were made concerning the failures of electronic equipment used in military aircraft applications in 1982 by the U.S. Air Force. These investigations showed that about 40% of the failures occurred in the electrical connectors, 30% occurred in various other electrical interconnects, 20% of the failures occurred in component parts, with an additional 10% due to other causes. The connectors and interconnects from various cables and harnesses accounted for about 70% of all electrical failures. It is obvious that electrical connectors and interconnects must be selected, manufactured, installed, inspected, and maintained by skilled technicians to ensure reliable operation for long time periods. The poor performance of electrical connectors in general is one major reason why satellite communication systems avoid the use of connectors. The standard practice is to use point-to-point wiring connections that are soldered with special care, supported at many points with adequate strain relief, to ensure the integrity and reliability of these systems because it is very difficult and expensive to try to repair and maintain an electronic system in an orbiting satellite.

Electrical connectors are used extensively by virtually every manufacturer of electronic equipment because they provide a quick and simple way to assemble, maintain, and repair complex electronic assemblies. One of the most popular uses of electrical connectors is in the plug-in type of printed circuit board (PCB). The mechanical integrity and reliability of the electrical connector must be high to permit it to be inserted and removed many times without causing any damage to the electrical contacts. The connector materials, coatings, and surface finishes on the pins and sockets must be carefully selected and fabricated to provide a low electrical resistance across the contact interfaces for reliable electrical operation in many different and often severe corrosion, oxidation, vibration, and shock environments.

Investigations of many failures in connector electrical contacts used in industrial applications such as military airplanes, diesel electric train engines, automobiles, and various types of trucks have shown that combinations of oxidation, corrosion, and wear appear to cause most of the failures. The wear failures are most often associated with various dynamic forms of vibration and shock that produce relative motion between the pins and sockets of the electrical contacts. Wear-associated failures between the contacts are usually caused by high interface pressures that produce microscopic welding and shearing on the contacts. This type of failure can also be caused by thermal cycling due to relative motion resulting from differences in the thermal coefficients of expansion (TCE). This wear on the surfaces of the pins and sockets is often referred to as fretting corrosion. Fretting corrosion is generally defined as an accelerated form of atmospheric oxidation associated with high temperatures and pressures combined with alternating relative motions at the contact electrical interfaces. High vibration and shock levels can cause very rapid fretting corrosion failures on the pin and socket connector interfaces, sometimes in a matter of several minutes.

14.2 DIFFERENCES BETWEEN FRETTING CORROSION AND OXIDATION

Fretting corrosion failures should not be confused with oxidation failures on the electrical contacts, although both problems often occur at the same time. The fretting corrosion that occurs in electrical contacts due to vibration, shock, or thermal cycling is usually caused by microscopic welding and shearing at the contact interfaces. This mechanism can rapidly produce intermittent contact failures with noble metals by cutting through them to the copper base alloy where corrosion with a high electrical resistance takes place. When fretting corrosion is involved, electrical operation will return to normal about 99% of the time when the external stimulation stops. If the electrical operation does not return to normal after the stimulation is removed, the probability is very high that oxide film buildup is the problem. Oxidation-induced failures are the result of oxides that slowly build up on nonnoble metals over a longer period of time, usually from a few weeks to several years. The oxide coating electrical resistance depends on the coating material and thickness. A high voltage may be able to break through a thin oxide coating. If the oxide coating has a chance to build up its thickness over a longer time period, or if the high voltage is not available, then failures will occur. Oxide coatings on nonnoble metals will often break away when relative motion occurs between the pins and the sockets, allowing the system to work normally. When a plug-in PCB does not function properly in a static environment, it is often removed and reinserted. If the PCB functions properly after this operation, it is a pretty good indication that the original problem was due to an oxide buildup. The insertion motion and the contact interface pressure typically breaks through the thin oxide film that has formed, which reduces the interface electrical resistance, allowing the system to function properly once again [30].

Oxidation failures can only occur on nonnoble metals when oxygen is present. Oxidation will not occur on nonnoble metals that are sealed in dry nitrogen enclosures. Oxidation will not occur on nonnoble metal interfaces that are gas tight. Unless special O-ring seals are used on connectors, it is almost impossible to obtain a gas-tight joint on a typical pin-and-socket type of plug-in connector. Nonnoble metals that have been coated with a very thin coating of a noble metal such as gold can still form a thin oxide film after a period of time when oxygen is present. Oxidation-induced failures have been reported on connector pins and sockets that were coated with only 3 μin. of gold. The gold will not oxidize. Coatings this thin are very porous, so the oxidation actually occurs in the open areas of the nonnoble surfaces that are not protected by the very thin porous gold plating.

Thick layers of gold over nickel over brass electrical contacts may not prevent fretting type of corrosion failures on plug-in types of PCB connectors during operation in severe vibration environments. A production electronic enclosure with several PCB assemblies was tested with PCBs that had natural frequencies of about 250 Hz. The connector pins had only two points of contact using a flat blade pin and a tuning fork type of socket, as shown in Fig. 14.1. The contacts were plated with 50 μin. of gold over 150 μin. of nickel over brass electrical contacts. About 100 sets of contacts on each plug-in type of PCB were wired in series. A "glitch" detector that could measure an open circuit down to the nanosecond range was used to monitor the interface resistance of the connector contacts. These contacts experienced intermittent open circuit failures of 1 μs after only 15 min of vibration using a white noise random vibration from 10 to 2000 Hz with a power spectral density test level of 0.80 G^2/Hz. This generated an input acceleration level of about 40 G rms in a direction perpendicular to the planes of the PCBs. The test setup was not detailed enough to locate the exact set of contacts that failed first. Locating the particular set of contacts that failed was not considered important. What was considered important

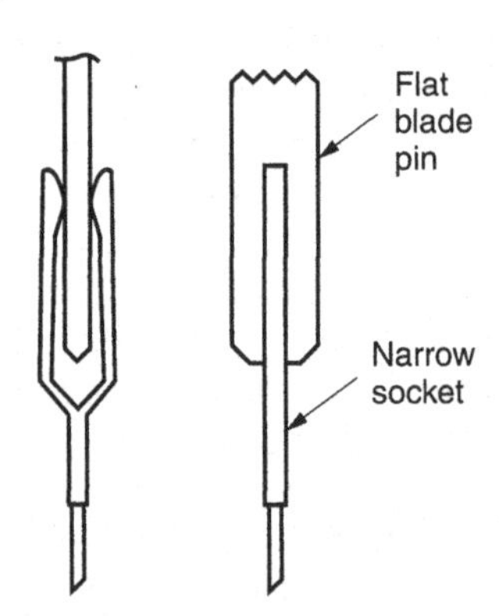

FIGURE 14.1 Electrical connector with a flat blade pin and a narrow tuning fork socket that provides two points of electrical contact.

was to find the fatigue life capability of a typical plug-in type of PCB using the blade and tuning fork type of connector contact in severe vibration environments. The second recorded intermittent connector failure occurred about 1 min after the first failure. Intermittent connector failures occurred on an average of about one a minute after the initial recorded failure during continued vibration exposure. Every time the vibration stopped, the electrical operation returned to normal. This is the classic sign of fretting corrosion types of failures. The random vibration input test levels were imposed directly on individual PCBs and not on a chassis or other type of box structure that enclosed the PCBs. The individual PCBs were instrumented with miniature accelerometers and mounted on a resonant-free vibration fixture that was bolted directly to an electrodynamic shaker head.

Microscopic examinations of the connector contact surfaces with thick gold and nickel plating are very difficult to make. Even when there is wear through the gold and nickel so the base brass alloy is making the electrical contact, the polished brass still looks like gold. Visual examinations by an inexperienced observer will usually conclude that the connectors did not cause the failures. A lot of valuable time will then be wasted trying to find cracked solder joints, broken lead wires, and defective components. An experienced observer can usually see the wear that cut through the gold and nickel layers down to the base brass metal. Another simpler method for determining the integrity of the gold and nickel plating is to use a 5% solution of *sodium sulfide* in water. This solution has a slight rotten egg odor similar to but not as strong as hydrogen sulfide. When this solution is brushed on the connector in the contact area, a small spot will turn black in a couple of seconds when wear has cut through the gold and nickel plating and there is a copper base alloy present. This is a nondestructive test. The solution can be washed off with water with no damage to the contacts. This process will not work with a 3% solution of sodium sulfide because the concentration is too low [30].

Random vibration tests were also run on several electrically operating chassis that contained 7–10 plug-in types of PCBs with the blade and tuning fork types of electrical contacts. A visual examination was made of the inside surfaces of the chassis after an hour of high-level random vibration. This revealed extremely severe fretting corrosion that looked like the surfaces had been sprayed with black paint. Wiping a finger over the black surface revealed a very fine black powder, which was believed to be copper oxide. Copper base alloy materials are normally used for connector contacts. Copper alloy interfaces will oxidize very rapidly, leaving a black residue, when they are exposed to high interface pressures with high interface velocities that are capable of producing high local interface temperatures that cause local welding (sometimes called cold welding) and shearing associated with the relative sliding interface actions. Under these conditions the fine black corrosion powder, which is an electrical insulator, can fill in between the two points of contact on the blade and tuning fork type of electrical pin-and-socket connection and produce an open electrical circuit for a period of about 1 μs. In an analog system a short-duration open circuit may never be noticed. However, in a high-speed digital multiplexing or logic circuit at least 7–10 bits of information can be lost. Lost information can cause all sorts of problems. Air bags in an automobile may not

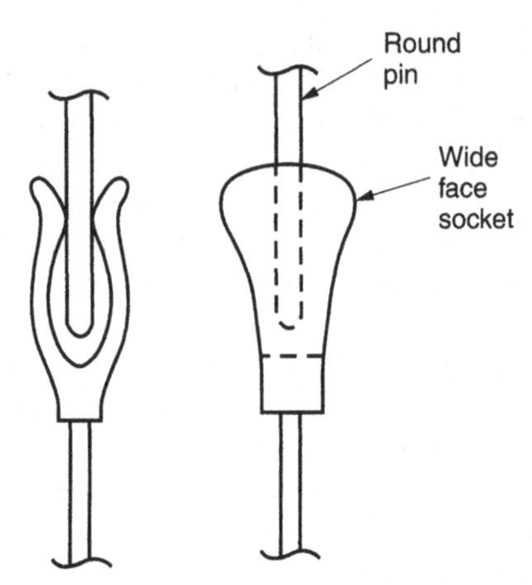

FIGURE 14.2 Electrical connector with a round pin and a wide face tuning fork socket that provides two points of electrical contact.

deploy in a crash, missiles in an airplane may not fire when the pilot presses the firing button, missiles may crash or explode when the wrong signal is received, and many other possible disasters.

Another series of failures on plug-in types of PCB connectors were recorded on several truck electronic systems that had been operating in the field for a period of about 2 years. In this case the systems were not working electrically when they were shipped back from the field. When the critical PCB was removed and reinserted, the system operated normally once again. This condition appeared to be due to the formation of an oxide at the connector contact interfaces rather than the result of connector fretting corrosion. The observed condition could also have been a combination of both conditions. The electrical contacts had only two points of contact with a round pin (instead of a flat blade pin) inserted into a wide-face tuning fork type of socket, as shown in Fig. 14.2. The electrical contacts were gold plated, but the gold had a thickness of only 3 μin. Very thin plated gold films are quite porous. Although the surface is coated with a noble metal, there is still a large copper base alloy surface area exposed to the local atmosphere, so oxidation can still occur when oxygen is present. Over a long time period a thick oxide film can be produced with a high electrical resistance that can cause electrical problems [31, 32].

14.3 PARAMETERS THAT CONTRIBUTE TO A LONG CONNECTOR CONTACT FATIGUE LIFE

Operational experiences with blade and tuning fork types of electrical connectors with only two points of electrical contact have shown that they will usually cause electrical intermittent failures of about 1 μs when they are exposed to severe vibration, shock, or thermal cycling environments. The time it takes to produce these failures depends to a great extent on the surface coatings, coating thickness, interface pressure, hardness, surface area, smoothness, displacements at the interface, interface relative velocity, lubrication, angle of contact, temperature, and, most important, *the duration of the exposure and the number of contact interfaces.* The probability of having the nonelectrical conducting corrosion oxides fill in under two connector mating interfaces at the same instant of time and causing an open circuit is very high. As the number of electrical connector contact points are increased, there is a lower probability that the oxides will fill in under every set of contacts at the same instant of time. Vibration tests with box connectors that have four points of electrical contact had no recorded intermittent failures during random vibration tests with high input acceleration levels for a couple of hours. A visual examination of the contact interfaces showed the typical severe fretting corrosion wear effects. The probability of having the corrosion oxide powder fill in between all four sets of electrical contact points at the same instant of time was very low, so no failures were observed. Several connector manufacturers fabricate electrical connectors with pins that have more than four points of electrical contact, with noble metal plating for improved reliability in harsh environments. Connectors, such as the Hypertronics and the Bendix Bristle Brush have pins with 8–16 points of electrical contact. These connectors are more expensive than the simple blade and tuning fork connectors with only two points of electrical contact, but they have shown very good results during tests in severe vibration environments [1].

14.4 METHODS FOR INCREASING CONNECTOR FATIGUE LIFE ON EXISTING PRODUCTION HARDWARE

When fretting corrosion failures are discovered in the pins and sockets of existing electrical connectors, the connectors must be refurbished with new pins and sockets if that is possible, or the connectors must be removed and replaced with new hardware. The existing defective connectors cannot be used if the intermittent failures must be eliminated. It may be cheaper to just scrap the defective hardware and concentrate on methods for improving the fatigue life of any new or existing hardware that has not been subjected to any hostile environments. In some cases expensive new hardware may have to be designed, fabricated, and tested to some severe contract requirements. Before any of these approaches are implemented, it might be easier and cheaper, if there is a customer/subcontractor relationship, to simply ask the customer for a waiver that reduces the severity of the contract requirements. The success or failure of such a request often depends on the "do not bother me" attitude of some customers and procuring agencies and any data or knowledge the subcontractor may have regarding the methods that the customer typically uses to generate his contract requirements, as described below.

Operating environments are often specified by companies and agencies that order many different types of commercial and military electronic systems for a wide variety of programs. Many companies and agencies get lazy so they draw up a set of requirements like a pair of stretch socks, where one size fits all programs. The environmental requirements for the various programs are typically plotted on charts superimposing individual profiles to show the various program characteristics. The maximum requirements of each program are then used to produce a new worst-case profile that fits every required program. The worst-case profile is then sent out to all of the subcontractors as a contract requirement. If the subcontractors have existing hardware that can pass the contract requirements there are no problems. The problems start when the subcontractor finds that their existing hardware cannot meet the new requirements, so an extensive and expensive program may have to be implemented.

A subcontractor may find that his existing hardware cannot meet the severe environmental requirements of a customer because of one critical frequency. When the subcontractor suspects that the customer has used a worst-case profile for his contract requirements, it may be possible to offer the customer a substantial price reduction for existing hardware as a trade-off for a reduction in his contract requirements at that one critical frequency. A meeting should be scheduled where the customer and the subcontractor vendor can get together to examine the various environment profiles for the different programs. In some cases the one critical area in the subcontractor vendors profile program may be lower than the worst-case profile for all of the combined superimposed programs that have been specified in the customer's contract. The subcontractor may then be able to have the energy level reduced at the one critical frequency so he can sell his existing hardware to the customer at a lower price knowing his hardware will meet the revised contract requirements, and everyone is happy.

When the vendor finds that the plug-in connectors on his existing hardware will not be able to meet the customer's vibration requirements, for example, the vendor will look for inexpensive methods that can be used to improve the vibration capability of his hardware. The vendor will have to examine the various methods he thinks he can use to improve the vibration capability of his hardware considering costs, reliability, weight, size and schedule impacts.

The easiest way to increase the vibration capability of an electronic chassis assembly is to use vibration isolators. There are many different types of wire rope and viscoelastic isolation systems available from several different companies. When these devices are properly selected and installed, the vibration and shock acceleration levels transferred to the electronic chassis are substantially reduced. This increases the vibration fatigue life of all structural elements in the assembly. Isolation systems will require extra space to accommodate the size of the isolators plus some additional sway space. There will be an increase in the cost, weight, and size for this type of system.

When there are restrictions on the size increase of the electronic assembly, then other methods for improving the vibration capability must be investigated. There is another very simple method for substantially improving the vibration capability of an enclosed electronic chassis if it has bolted covers that provides access to internal plug-in types of PCBs. The chassis can be filled with small rubber balls. The balls must be small enough to fit in between the various plug-in PCBs. The rubber balls will reduce the dynamic displacements of the PCBs when they are excited at their natural frequency. Reducing the displacements will reduce the stresses so the fatigue life will be increased. Maintenance and repair on this type of assembly is very easy. The cover is removed, the rubber balls are poured out into a container, the defective PCB is removed and replaced, the rubber balls are poured back into the chassis, and the cover is replaced. A large number of small solid rubber balls will increase the weight. This may not be acceptable. The solid rubber balls can be replaced with hollow plastic balls, which will have a smaller impact on the weight increase while still providing a good improvement in the vibration and shock capability.

Another cost-effective method for improving the vibration and shock capability of existing electronic hardware is to add snubbers to the PCBs that impact against each other to reduce their dynamic displacements in vibration and shock environments. Random vibration tests with input levels of 30 G rms and high level shock tests with 1000 G peaks were run with snubbers cemented to production PCBs. The tests proved that snubbers were very effective in reducing the dynamic displacements and stresses in the PCBs. There were significant [30] improvements in the fatigue life of the electrical connectors and in the electronic components mounted on the PCBs in these environments. The best snubbers were epoxy fiberglass dowel rods about 0.25 in. in diameter that were cemented near the centers of adjacent PCBs with very small clearances of about 0.005–0.015 in. between adjacent snubbers. The lengths of the snubbers were determined by the component heights and the spacing between the adjacent PCBs. During operation in the severe vibration and shock environments, the snubbers strike each other when the natural frequencies of the PCBs are excited, and they try to displace through large amplitudes. The impacting between the close adjacent snubbers sharply reduced the dynamic displacements in the PCBs. This decreased the amount of strain energy in the PCBs, which decreased the dynamic stresses and increased the fatigue life of the components and connectors. See Chapter 6 for more information on the use of snubbers.

Reducing the PCB dynamic displacements will reduce the dynamic stresses and increase the fatigue life of the connectors and components mounted on the PCBs. Increasing the natural frequency of the PCB will rapidly reduce the dynamic displacements. The PCB displacements are inversely related to the square of the PCB natural frequency. One popular method used for increasing the PCB natural frequency on existing hardware is to add stiffening ribs. The ribs will have to be spaced between various components across the PCB. The ribs must carry the dynamic loads into the supports to improve the PCB stiffness. Metal ribs are best suited for stiffening ribs in these applications because there is only a small chance that there will be any places that will permit a straight rib to run completely across the PCB from support to support. A continuous rib is preferred since it will add the most stiffness. Most of the ribs will probably have to be bent in several places and threaded between the various components to carry the rib across the PCB from support to support. Aluminum ribs work well and they can be bonded to the PCB using an epoxy adhesive. The ribs must have a minimum thickness of about 0.090 in. to obtain enough surface area for a good attachment to the PCB. A good rib height would be about 0.50 in. The cemented edge of the rib should be roughened to provide a good grip at the adhesive interface. A smooth metal edge will not hold up in severe vibration and shock conditions because the epoxy bond cannot grip a smooth surface. Think of a mountain climber. He has to punch holes in the steep sides of a mountain to get a good grip. If there is no grip, he will slip. Care must be used to prevent short circuits between the PCB copper traces and the aluminum rib. The proposed stiffener rib interface on the PCB may have to be coated with a hard epoxy insulator before the metal rib is attached to prevent short circuits. The proposed stiffening ribs must be attached to a hard PCB interface. The ribs must not be cemented on the top of any soft conformal coatings because the soft interface will not allow the rib to achieve the required stiffness on the PCB.

14.5 EFFECTS OF LUBRICATION ON CONNECTOR CONTACT FATIGUE LIFE

Surfaces in close contact with each other that experience relative motion, such as bearings and gears, can experience rapid wear under harsh operating conditions. The dynamic study of these effects is called tribological relations between wear, friction, and lubrication. Connector contact interface electrical resistance tests have shown that all nonnoble metals will produce some forms of oxidation or corrosion after extended exposure to harsh environments. These effects can often be delayed with the use of some type of lubrication. One group of tests on connectors with blade and tuning fork types of electrical contacts used a lubrication that consisted of a 5% white mineral oil with 95% methylene chloride that was retracted and air dried. This lubrication was used to evaluate the fretting corrosion effects on a soft connector coating material of bright tin over brass contacts, using an interface force of 50 g. Figure 14.3 shows that the interface contact resistance increased rapidly without the

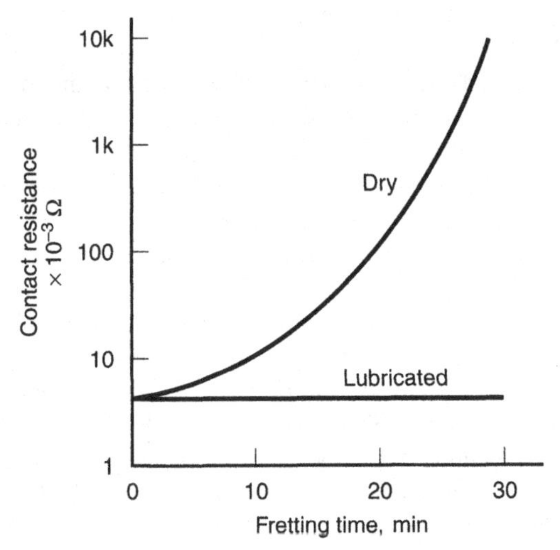

FIGURE 14.3 Curves showing the contact resistance across a bright tin-coated electrical contact with an interface force of 50 g and the fretting corrosion time for dry and lubricated contact interfaces, at 11 cycles/min.

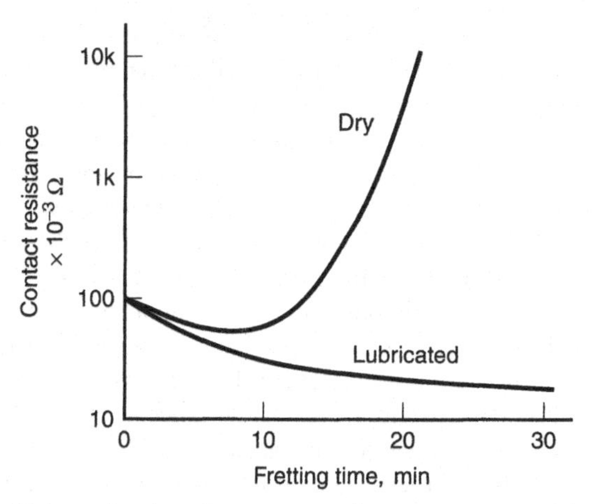

FIGURE 14.4 Curves showing the contact resistance across a nickel-coated electrical contact with an interface force of 50 g and the fretting corrosion time for dry and lubricated contact interfaces, at 11 cycles/min.

lubrication, but the interface contact resistance was held constant with the use of the lubrication. Lubrication tests were also performed on connector contacts with a hard coating material of nickel over brass with an interface force of 50 g. Figure 14.4 again shows that the interface contact resistance increased rapidly without the lubrication, and the interface contact resistance decreased with the use of the lubrication. The hardness of the surface coating material appears to have an effect on the fretting corrosion and wear. The harder materials experienced less wear, less cold welding and shearing action, and less debris generation along with less fretting corrosion damage. Extended testing of lubricated connector contacts showed that the lubrication appeared to hold the wear particles suspended in the immediate area of the electrical contacts, which tended to increase the abrasive action. Although the connector life was increased, a more rapid deterioration of the contact interfaces developed once the corrosion action got started [31].

14.6 RESULTS OF VIBRATION FATIGUE LIFE TESTING PROGRAMS ON CONNECTOR CONTACTS

Extensive vibration testing programs were implemented to evaluate the fatigue life of several different types of electrical connectors for use in systems that required a very high reliability in severe vibration environments. The first phase of the testing program investigated the fatigue life properties of a very successful, popular, inexpensive, rugged type of connector that was used extensively by naval ships and submarines. Tests on these naval vessels showed that the major vibration forcing frequencies on electronic systems were typically sinusoidal from about 5 to 50 Hz for extended periods, with low acceleration levels of about 1 *G*-peak. Shock levels could be as high as 150–200 *G*. The electrical contacts consisted of individual sets of pins and sockets with insulating sleeves that could be pressed into predrilled holes on the metal flanges of plug-in types of PCBs and master interconnecting (mother) boards. The pins and sockets were stamped out of a brass base alloy that was plated with 25–50 μin. of gold over 100–150 μin. of nickel. The connector pins had a flat shape, and the connector sockets had the shape of a tuning fork. The popular naval type of connectors were scheduled to be used with plug-in types of PCBs in high-performance fighter aircraft. The major vibration exposure for fighter airplanes was broadband random over a frequency range from about 10 to 2000 Hz. Under gunfire and aerodynamic buffeting conditions, random vibration input levels to the electronics was expected to be about 3–5 *G* rms in some wing and fuselage sections of the airplanes. The electronic chassis assembly had to pass a qualification test that consisted of broadband random vibration with an input level of 8 *G* rms. A simultaneous 5-*G* peak sine vibration input in the areas of 100, 200, 300, and 400 Hz was also required. The qualification test required these environments to be imposed for 30 min along each of three mutually perpendicular axes.

A connector contact fatigue life test program was devised to evaluate the proposed connector. A group of 14 plug-in test boards, similar to the types expected to be used in the airplanes, were fabricated with natural frequencies from 120 to 650 Hz. One group of PCBs were then instrumented and tested in individual resonant-free test fixtures bolted to electrodynamic shakers. Increasing sine vibration input levels were used starting at 5-*G* peak, increasing in 5-*G* increments to 30-*G* peak with 30-min resonant dwells. Another group of PCBs were tested using increasing random vibration input levels starting at 10 *G* rms increasing in increments to 40 *G* rms for 30 min. The maximum random vibration input level used was 0.80 G^2/Hz from 20 to 2000 Hz. Electrical contact fretting corrosion failures only occurred during the random vibration tests. There were no electrical contact failures recorded during the sine vibration tests [30].

Individual connector pins and sockets were also tested to measure the electrical resistance across the electrical contact interfaces during vibration. An electrodynamic shaker was used to apply the horizontal vibration motion as shown in Fig. 14.5. The connector contact interface force could be varied along with the frequency to show how the contact fatigue life was affected by the thickness of the gold and nickel plating and the effects of different lubricants. The method used to measure the electrical resistance across the contact interfaces for single and double contacts during vibration is shown in Fig. 14.6.

Vibration fatigue life estimates for the flat pin and tuning fork contacts are plotted in Fig. 14.7 using the vibration test results. This shows the approximate number of fatigue cycles plotted on the *X* axis, for different rms inch relative displacement amplitudes between the pin and socket contact interfaces plotted on the *Y* axis. Some *theoretical* possible methods for improving the fatigue life of the tuning fork contacts, without the use of lubrication, were investigated using the displacements expected at the electrical contacts during qualification vibration tests. The methods for improving the connector fatigue life involved very close tolerance controls on the fabrication and assembly processes, which were not believed to be practical. Some additional methods for improving the fatigue life of the tuning fork contacts were

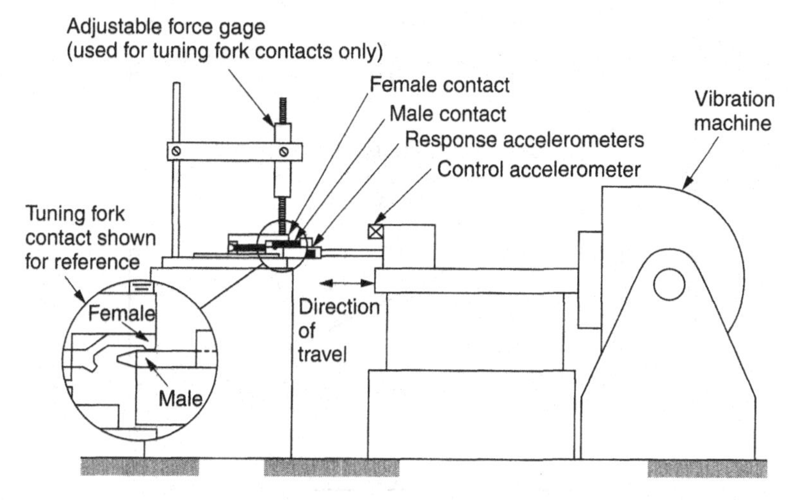

FIGURE 14.5 Connector contact test setup that can vary the frequency and the interface force across electrical contacts with the use of a vibration machine to produce fretting corrosion.

investigated. They were based on increasing the contact interface area, improved connector interface alignment, better surface finishes, and better surface smoothness. No tests were run to verify the effectiveness of the proposed improvements.

A finite element analysis program was established to promote a better understanding of the connector contact failure mechanisms, the influence of the PCBs

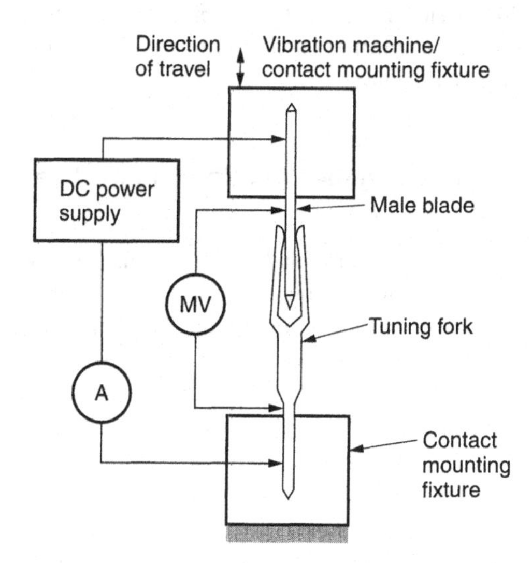

FIGURE 14.6 Test setup for measuring the contact interface resistance for an electrical connector.

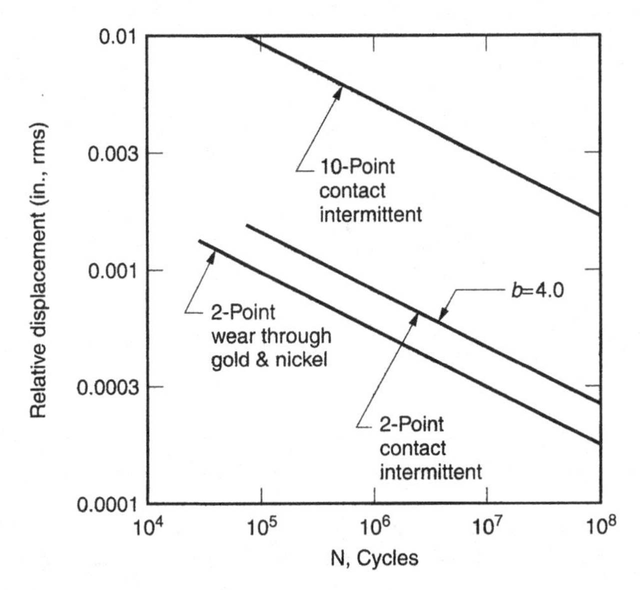

FIGURE 14.7 Log–log plot showing the fretting corrosion fatigue life of electrical connectors with 2 points of contact and connectors with 10 points of contact. The contacts on both sets of connectors have about 30 μin. of gold over 100 μin. of nickel over a brass base.

natural frequencies, mode shapes, input acceleration levels, and dynamic displacements using a high-speed computer. The goal was to establish plug-in PCB design parameters that would provide a satisfactory cost-effective fatigue life capability, with a light weight, in severe environments. The results of the computer studies were not available in time for publication.

Sample Problem: Estimating the Vibration Fatigue Life of Tuning Fork Contacts

An electrical connector that uses many blade pins and tuning fork sockets is being proposed for plug-in PCBs that will be used in a vehicle that requires a high reliability in a severe random vibration environment. The pins and sockets are plated with 30 μin. of gold over 100 μin. of nickel over a brass contact. The tuning fork sockets can only provide two points of electrical contact. A qualification test program requires a broadband random vibration input of 17 G rms for one hour in each of three mutually perpendicular axes. Will the tuning fork type of electrical contact provide the required reliability?

Solution Use the previous test data for a bench mark evaluation of the fatigue life. The previous connector vibration testing program showed that the tuning fork contacts with this gold plating over nickel over brass experienced their first fretting corrosion fatigue failures after about 15 min of random vibration using a 40-G rms input acceleration level. This information can be combined with a slightly modified Eq. 5.1 to estimate the expected fatigue life of this type of electrical connector where

T_1 = unknown time to fail with an input acceleration level of 17 G rms
T_2 = 15 min, baseline time to fail with a 40-G random vibration white noise input
G_1 = 17 G rms, required random vibration input acceleration level
G_2 = 40 G rms, baseline random vibration input level for first failure
b = 4.0 slope of connector fatigue line when plotted on a log–log paper, Fig.14.7

$$T_1(G_1)^b = T_2(G_2)^b \quad \text{so} \quad T_1 = T_2\left(\frac{G_2}{G_1}\right)^b \tag{14.1}$$

$$\text{Life} = T_1 = (15)\left(\frac{40}{17}\right)^{4.0} = 460 \text{ min} = 7.66 \text{ h} \tag{14.2}$$

The total random vibration qualification test time is only 3 h for the three axes. At a first glance it might appear that a connector life of 7.66 h would be adequate. However, testing experience has shown that qualification test programs very seldom go smoothly because it is very difficult to predict in advance where and when various problems may occur. Many tests of this type require a continuous electrical operation for each phase of the test program. If there are any problems that interrupt the continuous testing procedure, the test segment is voided and labeled as a false start. The test must be started again from time zero. Test malfunctions can be caused by many different problems. A few of these problems are outlined as follows:

1. Improper instrument calibration
2. Monitoring the wrong signal channel
3. A temporary signal distortion produced by an external transient voltage spike
4. Recording device failure
5. Accidents
6. Human errors
7. Premature component failures
8. Overtesting and undertesting

Therefore, it is a good policy to use some type of safety factor to account for problems that may require the tests to be repeated several times before they can be successfully completed. Testing experience has shown that a safety factor of about 5 should be applied to the testing time to ensure a high probability of passing the qualification test program. The actual required testing time is a total of 3 h for the three axes. Applying a safety factor of 5 on the test time would then require a vibration fatigue life of 15 h:

$$\text{Desired connector fatigue life} = 5 \times 3 = 15 \text{ h} \tag{14.3}$$

The fatigue life of 7.66 h for the proposed electrical connector using the blade pin and tuning fork socket with two points of contact is not adequate. A better choice of connector would be one that provided a box socket that had a minimum of four points of electrical contact with each pin, with 50 μin. of gold over 200 μin. of nickel plating over a brass base metal. Other connectors with more than four points of contact on the pins and sockets that have shown a good fatigue life in these environments are the Hypertronics and the Bendix Bristle Brush connectors. This does not mean that fretting corrosion will not occur. Fretting corrosion will occur in the expected random vibration environment. However, the probability of having the fretting corrosion oxides fill in between the pins and sockets at all four or more points of contact at the same instant of time for a period of 1 μs is very low. Connectors that have four or more points of electrical contact between each pin and socket will cost more than connectors with only two points of electrical contact. The increased cost will provide an increase in the connector reliability, which can be very important in high-speed multiplexing circuits. Many connector companies, such as AMP, carry a large stock of four-point connectors in different sizes and shapes.

CHAPTER 15

Case Histories of Failures and Failure Analyses

15.1 INTRODUCTION

Failures in electronic systems are often very difficult to analyze because they can be caused by many different factors. These include poor design, unusually severe environments, poor manufacturing, carelessness, accidents, improper testing, improper handling, and even sabotage. When a failure occurs, an investigation is usually made to find the cause of the failure so that corrective actions can be taken to prevent more failures. People involved with the failed hardware are usually questioned to try to pinpoint the reason or reasons for the failure. People become very defensive when they are questioned because they do not want to be accused of causing a failure. Engineering groups and manufacturing groups often end up accusing each other for causing the failures. The testing group may also get involved in the argument. Any association with a product failure can cause hard feelings and friction between working groups that can interfere with workers morale and productivity. It can result in dismissals, transfers, or quitting. Many people will deliberately lie or mislead the investigators to protect themselves and their friends. Some group leaders and managers who think they are very close to the hardware will make what they consider to be statements of fact that are not really true. The investigators usually have to sort through lies, half truths, and opinions to try to find the real cause of the failures. They do not always succeed. Some failures take years to solve. Some failures are never solved.

The failure histories, analyses, evaluations, and problems that were encountered during the various investigations described here are all true. The cases were taken from personal experiences at a number of different industries involved with the design and manufacturing of commercial, industrial, and military electronic systems.

15.2 FAILURES IN THE SMALL CANTILEVER SHAFTS OF A SPINNING GYRO

A large number of failures were occurring in the shafts of spinning gyros in the guidance systems of a major missile program during low-level vibration acceptance tests. The failures were occurring at a step in the small diameter of the shaft. A structural dynamic load, stress analysis, and fatigue life of the small shaft was made that included a high-stress concentration factor. This showed that the stress levels were not high enough to produce the testing failures. There had to be another reason for the failures in the shafts. The investigation turned to the manufacturing processes for a possible source of the failures. A theory was proposed where the cantilever shafts might be machined using a lathe with a single cutter. This cutting action could generate a bending moment in the small shaft. The rotating action of the lathe could generate many stress reversals that might initiate small cracks at the step in the shaft. The low-level vibration could then cause the small cracks to propagate and fracture the shaft.

The machining process crack initiation and propagation failure for the shaft were presented to the supervisor of the manufacturing facility. He vehemently rejected the idea that his methods and manufacturing processes could in any way affect the fatigue life of the shaft. He took great pride and great pains to explain that he knew as much, and probably a lot more than most engineers, on the causes and effects of improper machining practices on the fatigue life of manufactured parts. The shop supervisor explained in great detail how the shafts were being machined using three cutters spaced 120° apart, so there would not be any bending moments generated in the shafts during the machining process. He went on to say that the failures were probably caused by the poor engineering design, so the failures should not be blamed on his poor manufacturing processes.

The investigation turned away from the manufacturing processes to other possible sources for the failures, such as overtesting during vibration. The dynamic analysis was approached in a couple of different ways, but the results always showed that the shaft should not fail in the low-level vibration. Nothing else was uncovered that could cause the shaft failures. The investigator decided to return to the manufacturing area to see for himself how the shafts were being machined. He waited until the supervisor left the area. Then putting on his safety glasses he walked over to the area where several shafts were being machined.

What the investigator saw at one of the machining stations did not really surprise him. One of the machinists was using a single cutter to machine the shaft. The shop supervisor walked in, saw the investigator that he chastised the previous day, and came over quickly to ask him to leave the area because he was disturbing the machinist. The investigator did not move. Instead he suggested that the supervisor take a look at the machining operation in progress. One quick look and the supervisor's face turned white. He could see that a single cutter was being used for the machining operation. The supervisor demanded to know why a single cutter was being used for the machining operation, when very clear instructions were given to use three cutters for that operation. The machinist replied that using three cutters took too much time and he could not make his daily quota. However, when he used a single cutter, he could easily make more than his daily quota, and the end result with a single cutter was obviously just as good as the end result with three cutters.

The circumstances described above were presented to demonstrate that people will typically tell you what they believe to be true, even though they do not really know all of the facts in a particular case. The shop supervisor gave specific instructions to the machinists to use three cutters on that particular shaft. He did not think it was important to inform the machinists of the critical nature of that part. He assumed the machinists were all following his instructions so there was no need to check back occasionally to make sure his instructions were being followed. If the investigator really believed the supervisor at his first meeting, and if the investigator did not go back to check the machining operation himself, there is a good chance that the shaft failures would never have been solved. The investigator in this case was very lucky because the machine shop was on a double shift. If the machinist that was using a single cutter on the shafts was working on the second shift, again there is a good chance that the problem would never have been solved.

The lesson learned here is that a good investigator must always try to see as much as possible himself (or herself) before any questions are asked. Many people will give very positive answers to questions because they really think they know all the details. It is very difficult to separate good information from bad information, especially when people will lie to protect their jobs or their friends.

15.3 DEEP PENETRATION CRACK FAILURE IN AN ALUMINUM CASTING

An investigating team of four mechanical engineers was formed to try to determine the cause of a deep penetration crack in a large expensive cast aluminum housing enclosure for an electronic system that weighed about 90 lb. A visual examination of the crack showed it was located on a curved surface and it had a V shape. Each leg of the V crack was about 2 in. long with an angle of about 60° at the apex of the cracked legs. The apex of the crack penetration had pushed the casting in to a depth of about half an inch so the cast wall thickness of 0.25 in. could be seen clearly where each leg of the crack had separated from the main cast aluminum surface.

The investigation team questioned several engineers associated with the program to try to find out what caused the deep penetrating crack. The investigating team members were told that the large electronic system was going through a qualification test program that consisted of thermal cycling at high and low temperatures, shock, and vibration. The testing engineers said that a visual examination of the system after the thermal cycling and shock revealed no problems. The crack in the casting was observed after the last test, which was vibration. The testing engineers therefore concluded that the vibration exposure had somehow caused the crack in the casting. This information was passed on to the investigating team.

The investigating team decided to get another opinion from a vibration expert to see if he could shed more light on the strange vibration failure. The vibration expert took one look at the cracked casting and declared that the deep penetration crack failure could not have been caused by any type of vibration. And in fact there were only two possible ways that deep type of crack penetration could have been caused. The casting had to have been dropped from a height of several feet or the casting was deliberately and severely impacted with a sharp and heavy hammer. The idea that someone would deliberately hit the casting with a heavy hammer did not seem reasonable. The obvious conclusion was that the casting was dropped from a height of several feet on to a hard floor where it hit something hard such as a large bolt or some other hard metal object. The investigating team said that they had proof that the casting was not dropped. When the investigating team was asked what their proof was, they replied that they asked everyone associated with the project if the casting was dropped and that everyone said no, the casting was not dropped. The investigating team then concluded that since there were no problems before the vibration test, the vibration was the cause of the problem. The investigating team wanted to get the opinion of the vibration expert before they presented their investigation results to the president of the company. The vibration expert suggested that no one in their right mind would ever admit to dropping an electronic system that was worth over $150,000.

Anyone familiar with environmental testing involving mechanical shock, vibration, thermal cycling, and acoustic noise would immediately conclude that the deep penetrating crack fracture on a thick aluminum casting wall could have only been caused by a very high impact acceleration force. Any acceleration force high enough to deeply penetrate a thick aluminum cast structure could only be generated by a pyrotechnic event or by a high-velocity drop shock. Since the qualification test did not include these environments, the only logical conclusion had to be that the casting was dropped from a height of several feet. There is no way that any type of vibration environment could cause that type of deep, concentrated, local crack penetration in a thick aluminum cast structure. The sad part of this incident is that four graduate engineers did not know how to tell the difference between a vibration-imposed failure and a high-velocity drop shock-induced failure. It also points out that anyone involved in any type of failure investigation must be very careful to evaluate the questions and answers presented during the investigation. People will often lie or try to mislead an investigation if they are trying to protect their friends or if they are afraid of losing their jobs.

15.4 FAILURES OF CHIP CERAMIC RESISTORS AND CAPACITORS WITH POLYURETHANE COATINGS

Many failures were occurring on printed circuit boards (PCBs) with small ceramic chip resistors and chip capacitors during thermal cycling acceptance tests from −40 to +71°C. Some of the 0.10-in.-long chips were cracking after only one thermal cycle. A visual examination of the PCBs showed that every cracked chip had polyurethane conformal coating bridging and filling under the chip, between the chip and the PCB. The failures appeared to be caused by the high thermal coefficient of expansion (TCE) and the high modulus of elasticity of the polyurethane conformal coating. The solder joints at both ends of the small ceramic chips did not appear to be cracked or damaged in any way. Ceramic materials are generally very strong in compression, but they are very weak in tension. As the temperature is increased, expansion of the polyurethane under the chips tries to lift the chip off of the PCB, but the solder joints at each end of the chip resist this motion. This forces the chip to bend so the top of the chip is loaded in tension, where the ceramic is very weak, causing the chip to crack.

The conformal coating on the PCB had just been changed from Humiseal 1B31 acrylic to the Humiseal polyurethane coating. The conformal coating change was made to improve the moisture and humidity resistance capability of the PCBs, which were having humidity and moisture problems. Both coatings were applied by an automated spraying operation. The acrylic coatings have a solvent base so they can be sprayed on very thin. The cured acrylic coating dries to a thickness of about 0.002 in. so bridging and filling under the chips and chip cracking in thermal cycling conditions never seem to occur.

During the early curing phase of the acrylic coating, the outer exposed skin dries first. This traps the remaining solvent under the partially cured outer skin. The trapped solvent must evaporate to produce the final cure. This is accomplished by solvent blasting (a term chemists use) microscopic holes in the cured outer skin so the evaporation and the final cure can be completed. The microscopic holes in the outer skin now become entry points for water vapor to enter these holes. Water vapor is a small molecule gas (much smaller than the air molecule) that can enter these holes. The vapor can condense into a liquid under the proper conditions and cause electrical problems in sensitive electrical circuits when enough water finally penetrates the outer skin. The big advantage for the acrylic coatings is that they are very easy to apply and easy to remove when repairs or changes must be made. The main disadvantage of the acrylic materials is that they do not have very good humidity and moisture resistance

Polyurethane coatings cure by a cross-linking process called polimerizing. There are no solvents to evaporate. This results in a much tougher skin that resists water vapor entry much better than the acrylic coatings. Polyurethane coatings, however, go on much thicker, and they have a higher modulus of elasticity than the acrylic coatings. They are also more difficult to remove when rework or repairs have to be made. Thicker polyurethane coatings tend to web, bridge, and fill in under smaller components more readily. In thermal cycling environments the polyurethane coatings can generate high expansion forces. When these high forces act on brittle ceramic chip resistors and capacitors, these brittle devices will crack. The big advantage of the polyurethane coatings, when they are properly applied, is that they are tough, very abrasion resistant, and have a good resistance to humidity and moisture

Experience has shown that webbing, bridging, and filling of a high modulus conformal coating with a high TCE under ceramic chip resistors and capacitors must not be allowed. These types of ceramic components are very brittle. They will often crack when they are exposed to the high thermal expansion forces that can often be generated in thermal cycling conditions. If the polyurethane coatings do fill in under any ceramic chip components, an air hose can be used to quickly remove the excessive coating before it is cured, usually by an ultraviolet (UV) light process. This requires hand labor, which can be expensive in a production environment. However, this will probably be better than shutting down the production line until the problem can be solved.

Other conformal coatings such as parylene and silicone are often used to protect PCBs from moisture and humidity. Parylene is a vapor-deposited process, which typically has a very thin finished coating thickness of only about 0.0004-in. Webbing, bridging, and filling in under small components will very seldom occur because the coating is so thin. This material has excellent humidity and moisture resistance. It is expensive and it is difficult to remove for repairs or rework.

Silicone coatings are popular for protecting PCBs from moisture and humidity. This material is very soft and very thick when it is applied, so it always fills in under and around small components. Silicone has a high TCE, but it also has a very low modulus of elasticity so thermal expansion problems very seldom occur. It is very easy to remove when repairs or rework is required. It has a volume resistance that is about 10 times higher than acrylic or polyurethane coatings so water vapor does not penetrate the volume easily. However, the surface interface resistance is very low because the material does not have good adhesive properties. Water vapor tends to wick in more easily at the various interfaces between the coating and the PCB and at the interfaces with the various components. Extra care must be used at these interfaces to avoid moisture problems. Since the silicone coatings are very soft, they do not have a good abrasion resistance. Rough handling will often wear away the coating and allow more moisture to penetrate into sensitive electronic components and cause problems.

15.5 FAILURES IN SURFACE-MOUNTED TRANSFORMERS IN THERMAL CYCLING AND VIBRATION TESTS

A large electronic system manufacturing company was experiencing failures in its surface-mounted transformers on PCBs during thermal cycling and vibration tests. Thermal cycling failures occurred after only 12 thermal cycles from −45 to +85°C. The failures were not in the typical areas such as the lead wires or in the solder joints. These failures were in the solder pads that lifted off of the PCB surface. The solder pad lifting action usually cracked the thin copper circuit traces, which resulted in open circuits. Most of the vibration failures occurred in the lead wires with a few failures in the solder joints. Vibration did not cause any solder pad lifting in these tests. The solder pad lifting was only caused by thermal cycling.

A meeting was called with the engineering and manufacturing groups to discuss the problem and to find possible cost-effective solutions. There were 15 production engineers, 5 design engineers, and 3 mechanical analysis engineers present when the meeting started. The main topic of the meeting was centered on possible design changes required to solve the problem. The production engineers recommended scrapping 150 existing bare, unpopulated circuit boards to relocate the solder pads and fabricate new boards. This would increase the length of the transformer lead wires, which should decrease the wire forces and stresses and increase the fatigue life. The department vice president asked what would happen to the production schedule if the existing boards were scrapped. The answer was that the production would be delayed about 4 months. This upset the vice president because this company was noted for its strict rules to meet production schedules. He inquired if there were any other possible solutions that would work to eliminate the failures so the production would not be delayed. A new mechanical analysis engineer in the group spoke up and said there was no reason to scrap any parts and there was no reason to slip the schedule by 4 months. There was a very simple solution to the problem. Simply loop the lead wires to make them longer and both the vibration and thermal cycling failures will go away. At this point about 10 production engineers jumped up and started yelling all at once, so it was impossible to understand what they were saying. When they calmed down so they could be understood, they were saying that under no circumstances would they loop the wires for a production run because it would require a hand operation that was too expensive. The production people prided themselves in never using hand operations in their manufacturing facilities. The meeting broke up and everyone went back to their desks.

Another meeting took place with the production engineers and the president of the company a short time later. The president wanted to know if the production group found a solution to the transformer failure problems. The production engineers said they did not find an acceptable solution to the problem. They wanted to scrap the existing bare circuit boards, redesign them, and fabricate new boards. The entire process would set the delivery schedule back about 4 months. This upset the president and he asked once again if there was any possible way to meet the production schedule. There was a long silence, then one of the production engineers said well—! The president asked "well what"? Another long silence, then the production engineer said someone had recommended looping the wires, but that operation might require a hand operation on 150 boards. This type of manufacturing procedure was very seldom used so the manufacturing group rejected the idea. The president asked if the wire looping process would solve the failure problems. The answer was yes. The president then asked if the wire looping would permit the company to meet its present production schedule. The answer again was yes. The president stood up, very annoyed, and yelled then loop the "blankety blank" wires.

The above story is true. The engineer that recommended looping the transformer wires is the author of this book you are reading. The purpose of this story is to demonstrate that even intelligent people will often go to great lengths to avoid performing a simple task just because they had never performed that task before. Hand looping (or camel humping) the wires for one transformer on each for 150 boards is not a very expensive or difficult task, especially if it will permit the system to meet the reliability and schedule requirements. The wire looping task was eventually made very easy with the use of simple forming dies. Hand looping the wires on several million boards typically required for the automotive, communication, and entertainment industries would not be an acceptable cost-effective solution.

15.6 FAILURES IN THE LEAD WIRES OF THROUGH-HOLE FLUSH-MOUNTED TRANSFORMERS

Electrical lead wires were failing on through-hole flush-mounted transformers after about 125–150 thermal cycles from −55 to 85°C. The transformers were encapsulated (potted) in epoxy with the wires extending from the flat bottom surface of the transformer and wave soldered to an 0.062-in.-thick epoxy fiberglass PCBs. The transformers were flush mounted on the PCBs with no interface air gaps, in order to provide a low thermal resistance path for good heat conduction from the transformers to the heat sinks on the PCBs. The copper lead wires with an ultimate tensile strength of about 45,000 psi broke during the thermal cycling tests, but the solder joints did not fail.

Several investigations were made using finite element computer methods and equilibrium equations with hand calculations to determine the reasons for the failures, so some type of corrective action could be taken (see Eq. 5.28). It was determined that the wire failures were due to forces generated by thermal expansions in the transformers, the PCBs, and the lead wires in the Z axis, perpendicular to the plane of the PCB. The length of the lead wire was required to find the breaking force in the wire for a reliable design. At first glance it would appear that the wire length for a flush-mounted transformer would be zero since there was no air gap at the transformer interface to the PCB. This presented a problem. For a wire loaded in tension, the spring rate is area × modulus/length. If the wire length is zero, the wire spring rate must be infinite, which is impossible. The wire must therefore have a finite length. The wire length is required to find the thermal expansion related force and stress, so the approximate fatigue life can be obtained. The problem was solved by using a pull test machine that showed the force required to break the wires. The known force to break the wire was substituted into the equilibrium equation, and the equation was solved for the effective length of the wire. The wire effective length was then normalized in terms of equivalent wire diameters so it could be used for copper wires with different diameters in future calculations.

The equilibrium equations showed that the effective length of a transformer wire loaded in tension will extend about two diameters into the transformer body plus two diameters into the PCB. When the thickness of the PCB is less than two diameters, then the thickness of the PCB should be used instead. When the electrical lead wire is loaded in bending, the same type of analysis showed the effective length of the wire will extend about one diameter into the transformer body and one diameter into the PCB. This approximation has been applied to other components that have electrical lead wires such as resistors, capacitors, diodes, dual-in-line packages (DIPs), pin grid arrays (PGAs), leaded chip carriers, plastic leaded chip carriers, and many other similar components with good results.

The results of these investigations showed that electronic components mounted on PCBs with their lead wires must be carefully evaluated to ensure their reliability in severe thermal cycling conditions. Assemblies with a large differences in the TCE between the component and the PCB, along with a high modulus of elasticity in these parts, can experience lead wire and solder joint failures in thermal cycling conditions. These failures can be caused by very small thermal expansion magnitudes of only about 0.0003 in. This is only about one tenth the thickness of an ordinary sheet of paper. Differences in the thermal expansions can easily generate strains greater than 0.001 in./in. This sounds like a very small number so it is easy to ignore. Some engineers tend to forget that stress equals modulus of elasticity times strain. When the strain is only 0.001 in./in., and the copper modulus of elasticity is 18 million psi, the copper wire stress will be 18,000 psi. A strain of 0.001 in./in. for solder with a modulus of about 2 million will produce a solder stress of about 2000 psi. It should be remembered that solder stresses should be kept well below about 800 psi to ensure a reliable design in any environment. A solder strain of 0.001 in./in. would result in a very short fatigue life.

Failures can often be reduced or prevented by controlling the TCE of the component parts. Adding materials such as aluminum oxide, magnesium oxide, or calcium carbonate to the epoxy potting materials can reduce the TCE and increase the thermal cycling fatigue life. These materials also have a high thermal conductivity so their addition reduces the internal thermal resistance. This reduces the hot-spot temperature rise within the component so the thermal reliability is increased.

Thermal expansion failures in many different types of components that are through-hole and flush mounted on PCBs can be reduced or eliminated by adding very small feet on the bottom surface of the component at the PCB interface. These small feet only have to be about 0.005–0.010 in. high and wide so they will act like springs and deform to act like a strain relief when high thermal expansion displacements are expected. The small feet will also deform the small area on the PCB under the feet to provide a good strain relief.

Another way to reduce high thermal expansion forces and failures is to use a thin layer of a resilient material with a low modulus of elasticity at the component interface to the PCB. Soft materials such as room temperature vulcanizing (RTV) or silicone-filled gaskets made by companies such as Chomerics can be used. These soft materials will deform when thermal expansions occur and reduce the resulting stresses. When the stresses are reduced, the fatigue life will be increased. Extra care must be used in the application of soft interface materials between the component and the PCB. These materials must include an air venting path away from the PTH to prevent an air pressure buildup in the area of the PTH. The hot solder will heat the air in the PTH, which will force the air to expand. The expanding air must be vented to get a good solder joint. An air pressure increase will prevent the solder from properly wicking up the PTH. Photographs with X-rays show the solder wicks only half way up the PTH when a good air venting path from the PTH is not provided. A PTH that is only half full of solder will fail more rapidly in thermal cycling and vibration environments.

15.7 FAILURES IN SMALL AXIAL LEADED THROUGH-HOLE-MOUNTED COMPONENTS

Field failures were occurring in small axial leaded through-hole-mounted resistors, capacitors, and diodes in thermal cycling conditions. Sometimes the component body cracked and sometimes the lead wires pulled out of the component body. An examination of the failures showed they all had one thing in common. They all had high solder joints where the solder wicked up the vertical lead wire leg directly into the bend radius, as shown in Fig. 13.5. A force and stress analysis of this geometry showed that when the vertical wire leg is very compliant it bends easily and acts as a strain relief. The vertical leg of the lead wire is very stiff when it is full of solder, so it cannot bend and reduce the forces and stresses developed in the wire. The solder wicking up into the bend radius structurally short circuits the wire strain relief. This increases the forces and stresses in the electronic components, producing more rapid fatigue failures.

Meetings were held with the production engineers to discuss the results of the investigation. They requested that no design changes be made because they would alter their production process to reduce the solder wicking up the wire into the bend radius. A few months later there were many more similar component field failures. The solder was still wicking up the wire into the bend radius and causing more component failures. The production changes did not seem to work. Another solution had to be implemented to solve the component failure problem. Adding a camel hump strain relief to the wire was recommended because it has worked well on other programs. In this case, however, the production autoloading hoppers were not large enough to accommodate the camel hump in the wire. Another solution had to be used. Raising the component higher above the PCB would increase the vertical wire length so the solder could not wick up into the bend radius. This would allow the vertical wire leg to bend and reduce the forces and stresses, which would prevent more component failures.

Two production methods were proposed. The first solution was to use 0.050-in.-thick water-soluble shims that could be placed in positions under the components with the use of pick-and-place machines. This would raise the component 0.050 in. higher so the solder would not wick up into the bend radius. Distilled water is used for a cleaning process after the soldering operation, which would dissolve the shims and leave a gap under the component. This idea was rejected because the PCBs in this program were conduction cooled. The heat flow path from the component to the heat sink on the PCB would have to pass through an air gap. Air is not a good heat conductor. This would produce a high temperature rise that could shorten the component fatigue life. The second idea was to use 0.050-in.-thick epoxy fiberglass "speed bumps" under the components to raise them above the PCB. These are flat disks with a diameter of about 0.125 in. The disks could be positioned and cemented to the PCB using the pick-and-place machines. These disks have a thermal conductivity that is about 10 times greater than air, which would be thermally acceptable for this application. The installation of the speed bumps solved the component failure problem.

15.8 THERMAL CYCLING MICROPROCESSOR LEAD WIRE AND SOLDER JOINT FAILURES

Lead wires and older joints were failing on through-hole-mounted and surface-mounted 50-pin microprocessors during thermal cycling acceptance tests, so the hardware could not be shipped to the waiting customers. Thermal cycling failures in the end solder joints of large electronic component parts are quite common. These failures are most often caused by the differences in the TCE between the PCBs and the components. Temperature changes from turning the system on and off and temperature changes from day to night can also force these structures to expand and contract. These effects will produce high forces and stresses in the lead wires and solder joints when the physical properties of the various materials are not carefully monitored. The solution to this production problem was to assemble the PCBs *without the critical components in place*. The thermal cycling tests were then run with *reduced electrical operation* with no problems. After the acceptance tests were completed, the PCBs were removed from the enclosure and the *critical microprocessors were hand soldered* to the PCBs. The modified PCBs were then installed in the enclosure once again and the fully assembled electronic enclosure was immediately shipped to the waiting customer. When the production engineers were questioned about their decision to ship hardware that did not fully pass the required acceptance tests, they replied they had to ship the hardware to meet their delivery schedule. When they were asked why they did not solve the thermal cycling failure problems, they replied they did try several quick solutions, but they had not worked. And now they do not have the time, or the budget, or the necessary personnel available that can work toward a solution. (The lead wire and solder joint failure problems were solved a year later by adding a wire strain relief similar to the one shown in Fig. 13.2.)

The purpose of an acceptance test is to ensure the reliability of the electronic system when it is shipped to the customer. Installing critical components *after the completion of the acceptance test* and then sending the system to the customer can only be described as stupid. As stupid as this may sound, this really happened at a major electronic supply house. These production people were more interested in shipping hardware, and they were obviously not concerned with the quality and reliability of the hardware. They did not seem to realize or care that poor unreliable hardware could have an adverse effect on future sales of hardware and even the continued success or failure of their company. This was typical of the attitude of the American automobile and television manufacturers 25 years ago. They were turning out poor products that had very high failure rates. The American consumers stopped buying the poor-quality American products and went overseas to the better quality products from Japan and southeast Asia. The rest is history. Many American companies went out of business and many people lost their jobs because of these careless practices.

15.9 REDUCING COMPONENT FAILURES BY REDUCING THE THERMAL COEFFICIENT OF EXPANSION

Large ceramic components with a low TCE are often mounted on epoxy fiberglass PCBs with a high TCE. Large temperature changes can produce large displacement differences between the component and the PCB under these conditions. These displacement differences can result in cracked components, cracked solder joints, and broken lead wires. These types of failures can often be reduced or eliminated by reducing the effective TCE of the PCB. This will reduce the TCE mismatch between the component and the PCB, which will reduce the forces and stresses and increase the fatigue life in thermal cycling conditions.

One simple way to reduce the TCE of the PCB is to cement stiff metal shims with a low TCE in the critical area of the PCB. Materials such as molybdenum and invar have very low TCEs with a high modulus of elasticity. These materials are often combined with copper with different thickness ratios to tailor a composite material that can be cemented to the PCB in the critical areas to reduce the PCB thermal expansions. When the proper materials are selected and properly applied, the TCE of the PCB will be very close to the TCE of the component. When the TCEs are closely matched, the relative displacements between the component and the PCB will be reduced in thermal cycling conditions. This will reduce the forces and stresses and increase the fatigue life of the assembly.

The method for solving local thermal expansion problems on PCBs with the use of stiffening shims shown above sounds very simple. There is one big problem that must be solved before the above method will work. This involves cementing the local stiffening shims to the PCB in the areas immediately under the critical component on the top and bottom sides of the PCB if this is possible. Cleaning the metal shims and then epoxy cementing them to the critical areas of the PCB will not work. Metals are usually very hard and smooth. Adhesives do not adhere very well to surfaces that are hard and smooth. Tests show that epoxy adhesives will not adhere to smooth shims in thermal cycling conditions. The shims will fall off after a couple of cycles. If the smooth shim surfaces are roughened with a file or sandpaper and then epoxy bonded to the PCB, they will last a little longer, but they will still fall off. It is like a spider in the bathtub. It tries to climb up the wall but half way up it always falls back. The walls are very smooth so the spider cannot get a good grip. If there is no grip, it will slip. The shim has the same problem. The only way that the shim will hold on to the PCB is to drill or punch at least six or more holes about 0.10 in. in diameter through the shim. The number of holes will depend upon the size of the shim. The epoxy cement must be forced through the holes to form epoxy rivets when the epoxy cures. Tests show that these epoxy rivets must be sheared off before the shim will fall off the PCB. When properly applied, this local fix works very well.

One big problem with this type of solution is to get the approval from the production people to drill or punch holes through very hard materials. This depends upon the manufacturing and engineering policies of the company. In many companies the design engineering department creates the drawings that the production department must follow. It is up to the production group to come up with the required manufacturing technology and methods to produce a cost-effective part based on the engineering design. If there is a disagreement, the two groups get together to work out a solution. In some companies the production department is very strong. If the production people do not like the way the engineering department has designed a part they will simply make what they consider minor modifications to the part and build the part the way they want it *without consulting the engineering department.* When failures occur, the design engineering group often has no idea that the part was changed and that the change caused the failure. This happens quite often at several large electronic supply houses where the production groups refuse to build certain parts according to the drawings. Their usual argument is that the design is too expensive and they are reducing the costs. In the case of the stiffening shims these manufacturing groups decided the holes were too expensive and that the epoxy cement would hold the shims in place without the holes. They were wrong and the production parts failed after only a few weeks in the field. The production groups and the design engineering groups must work closely together to produce highly reliable, safe, and cost-effective products.

15.10 THE WRONG PARTY WON A MULTIMILLION DOLLAR LAWSUIT IN A SOLDER JOINT FAILURE

Two giant electronic companies were involved in a multimillion-dollar lawsuit involving solder joint failures in an automobile radio. Company A supplied a long rectangular epoxy fiberglass PCB 0.060 in. thick that was soldered to two posts on the top of a small sheet metal radio housing that was fabricated by Company B. The assembled radio was then attached to the instrument panels of several different popular automobile models. The PCB was much longer than the small radio housing so there was a long cantilevered PCB section that extended well beyond the back of the radio. The PCB was flush mounted to the top of the radio housing by means of two 16-gage copper wire posts that extended through the top of the radio housing and into the PCB where they were soldered. The copper posts performed two functions. They carried electrical signals to and from the main radio housing and they attached the PCB to the radio housing without any other fasteners.

After a few years of operation in different automobiles, a large number of failures occurred in the solder joints that held the PCB to the radio housing. Company B supplied the radios to the various automobile manufacturers so they had to spend several million dollars replacing the defective radios. Company B then filed a multimillion-dollar lawsuit against Company A for supplying defective PCBs that were experiencing extensive solder joint failures. Company B thought that the solder joint failures were caused by vibration. They thought that the long end of the cantilevered PCB was being forced to vibrate at its natural frequency when automobiles drove along bumpy roads. They obtained the services of a well-known vibration consultant. He performed calculations that appeared to show the solder joint failures were caused by a poor PCB support that produced excessive dynamic forces and stresses that eventually cracked the solder joints. He concluded that the two copper wire supports provided by Company A were not adequate and that extra supports for the PCB should have been provided by Company A. Company A counter sued with a multimillion-dollar lawsuit claiming that Company B did not solder the PCB properly to the two copper posts at the top of the radio and that extra PCB supports were not needed. Company A then obtained the services of a well-known soldering consultant who insisted that a joint that was properly soldered would never fail. (It should be obvious that this statement cannot be true. Failures will occur in any structure when the stress levels are high enough. Solder is no exception. When the solder joint stress level is high enough, it will fail.)

The case went to court with a jury. The court case lasted 3 days. All during the trial Company A had a small vibration machine located close to the jurors. A working radio was mounted on the vibration machine with the long overhanging PCB that was supported by only two solder posts. The PCB was being vibrated at its natural frequency. The displacement at the free end of the PCB was quite large and easily observed by the jurors. The vibration demonstration continued all three days with no solder joint failures. Company A proved its point that the solder connection on the copper posts could withstand extensive vibration with no failures. Company A won the multimillion-dollar lawsuit because of the way in which the vibration demonstration was presented.

Who was really right, Company A or Company B, and why? If you go to a surgeon he or she will usually cut to cure an ailment. If you go to a vibration expert he or she will look for a vibration-based solution. Only one person thought the failures were caused by thermal expansions. He was quickly eliminated because the vibration seemed so obvious. The solder joint failures on the flush-mounted epoxy fiberglass PCB is a classic example of a thermal expansion failure. Epoxy fiberglass has a TCE of about 15×10^{-6} in./in./°C in the X and Y planes of the PCB. The TCE of the epoxy fiberglass in the Z axis, perpendicular to the plane of the PCB, can go as high as 135×10^{-6} in./in./°C at temperatures near 100°C. Automobiles sitting in the sun in any of the southern states in the summertime can easily experience temperatures this high on their instrument panels where the radio is usually located.

Experience has shown that there are far more thermal-cycling-induced solder joint failures than there are vibration-induced solder joint failures. The PCB was flush-mounted to the top of the radio housing. Experience and analyses have shown that flush mounted components and epoxy fiberglass PCBs are very prone to high thermal expansions in a direction that is perpendicular to the plane of the PCB that can easily break lead wires and crack solder joints. The typical solder joint failure, often experienced in systems that have been exposed to vibration and thermal cycling, is not really a vibration failure. Experience has shown that most solder joint failures that appear to occur during vibration are really thermal cycling failures. Thermal cycling will typically initiate the solder joint crack. However, thermal cycling is usually very slow, perhaps one or two cycles per day. There is very little crack propagation with such slow cycles. Vibration can easily have over 100–200 cycles per second. Cracks can propagate very fast under these conditions. So the cracks that appear during vibration are usually caused by thermal cycling and propagated very rapidly during vibration. Of course, if the electronic system has never been exposed to any form of thermal cycling, and a failure occurs in vibration, the true failure mechanism under these conditions would be vibration.

15.11 WHY WEDGE CLAMPS OFTEN BECOME LOOSE IN THERMAL CYCLING AND VIBRATION

Wedge clamps are small devices that slide up an inclined plane to produce high interface forces, usually between plug-in PCBs and their chassis enclosures. These devices are often fabricated in three sections with a long screw that runs the length of the wedges. When the screw is tightened, it pulls the three sections of the wedge up inclined planes that expand the wedge. This produces a high force at the interface between the wedge and the edge of the PCB. Wedge clamps are used in applications where high interface pressures are required to conduct large amounts of heat with a small temperature rise from the edge of the PCB to an enclosure or heat sink. The enclosure then dumps the heat to some local external ambient. Wedge clamps can also be used to increase the natural frequency of a plug-in type of PCB. The high force generated by the wedge clamping action can restrict the rotation motion at the edges of a plug-in PCB. This increases the effective stiffness, which increases the PCB natural frequency. Wedge clamps are very popular for spacecraft, satellite, and military electronic applications. They are usually considered to be too expensive for commercial electronic applications.

Most wedge clamps are fabricated using relatively hard aluminum alloys that have been heat treated to a T-6 condition. Hard materials are required to allow the threaded bottom wedge-shaped section to rotate and slide up the inclined planes on another wedge to obtain the high wedging action force. Sometimes the wedge material is not very hard, like a T-4 condition instead of a T-6 condition. The sharp edges of the softer threaded bottom wedge sections will rotate when the long screw is tightened. This will cut into the soft sharp edge of the adjacent wedge. This action plastically deforms both sharp corners and locks the two sections of the wedge together when the long screw is tightened. This prevents the wedges on the inclined plane sections from sliding up and expanding the wedge, which locks the PCB to the chassis. Although the proper torque has been applied to the screw, the wedges have not expanded properly so the PCB is not properly clamped to the electronic enclosure. This is difficult to see because the threaded sections are at the bottom of the three-part wedge. A slight amount of vibration or a few thermal cycles will allow the locked sections of the wedges to relax. Since the wedges were not engaged properly, they become loose when the previously deformed and locked corners are relaxed. The PCB becomes loose so the good heat transfer from the PCB to the chassis is lost and the PCBs may overheat. The loose PCBs may also experience vibration or shock problems because they are not properly restrained.

When complaints are made to the wedge clamp suppliers, personal experiences have shown that they usually deny they are shipping a lower grade aluminum alloy with a T-4 condition instead of a T-6 condition. Even when they are confronted with test data from independent laboratories, they will still deny they are shipping poor hardware. The solution to this problem is to hard anodize the wedges so the surfaces are very hard. These wedges never lock up. The typical cost for this service is about one dollar per wedge assembly. This cost would be an unacceptable expense to an automobile, or communication, or entertainment supplier that delivers several million units a year. It is usually an acceptable cost for the lower volume military supplier.

15.12 PROBLEMS WITH RELAY FAILURES DUE TO CHATTERING IN VIBRATION

An important missile program was experiencing failures in some of the relays that were mounted on plug-in PCBs that were enclosed in a chassis housing. The failures occurred when the electronic system was exposed to a severe vibration environment. The failures were due to chattering in the relay contacts that shut down the electronics. Some of the relays worked and some of the relays failed. The relays that failed were not damage by the vibration chattering because they always worked perfectly well after the vibration stopped. The relays were from a special group that had critical electrical properties and an acceleration rating of 40 *G* peak. A dynamic evaluation of the PCBs showed they had a high transmissibility, which generated a high acceleration *G* level in the sensitive frequency areas on some of the relays. These relays were difficult to obtain because they were not the standard off-the-shelf product. The relays in the existing stock had to be used because there was not enough time to reorder and ship to meet the schedule. Since only a few of the relays failed during vibration, it was obvious that some of the relays were more rugged and some of the relays were more fragile. A proposal was made to set up a screening program to separate the rugged relays from the fragile relays. The question was how to implement such a program.

An examination of the typical dynamic characteristics of the relay contacts showed that they were mounted on springs that produced a natural frequency that was related to the spring rate and the mass. Manufacturing tolerances affected the physical dimensions of the springs such as the width, thickness, and weight. These tolerance variations produced relays with different natural frequencies and different sensitivities to acceleration *G* levels. The PCBs also had different natural frequencies due to manufacturing tolerances. The screening program had to consider the natural frequencies and vibration *G* level capabilities of the various relays. The screening program also had to consider the natural frequencies and transmissibility Q values of the various PCBs to avoid combinations that would produce vibration chattering in the relays. The fundamental or lowest natural frequency of the PCB usually causes the most problems. The highest transmissibility on a plug-in PCB will almost always be at the center. A high transmissibility will produce a high acceleration *G* level, which is bad for the relay because the relay may chatter. Therefore it is a bad practice to mount relays at the center of a plug-in PCB. Problems can occur in a complex electronic PCB when critical electronic components are not grouped together properly. The components that are the most critical are positioned first. Relays are not considered to be critical so they are usually mounted in any free space that is left on the PCB. This means that the relays cannot always be mounted at the sides of plug-in PCBs where the transmissibility Q levels are low.

Vibration tests were run on every relay to establish its fragility curve. The fragility curve is the inverse of the transmissibility curve. Several relays were instrumented and mounted on a rigid nonresonant vibration fixture that was bolted directly to an electrodynamic shaker head. The vibration screen started at 50 Hz with a sinusoidal vibration input of 10 G peak. The 50-Hz frequency was held constant and the input acceleration G level was increased slowly until the relay chattered or the vibration shaker reached its maximum capability of about 70 G peak. This process was repeated for the next fixed frequency of 75 Hz. This process was repeated at fixed frequency intervals of 50 Hz with increasing acceleration G levels for every relay in stock. The fragility curve for every relay was plotted with the frequency on the horizontal axis, and the acceleration G level that produced relay chattering on the vertical axis up to a frequency of 2000 Hz. The fragility curve shows the minimum acceleration G level and the frequency point where the minimum occurs for each relay. Vibration tests were then run on every PCB to establish their individual natural frequency and transmissibility. The selection process was then used where the best combination of natural frequencies and the acceleration levels for the relays were matched with the various PCB natural frequencies and transmissibility Q values. PCB assemblies could then be selected with relays that would not chatter in the required vibration environments.

Bibliography

1. Dave S. Steinberg, *Vibration Analysis for Electronic Equipment*, 2nd ed., Wiley, New York, 1988.
2. Dave S. Steinberg, *Vibration Analysis for Electronic Equipment*, Wiley, New York, 1973.
3. R. N. Wild, Some Fatigue Properties of Solders and Solder Joints, IBM Report No. 74Z00044S, July 1974.
4. MIL-HDBK-5B. *Metallic Materials and Elements for Aerospace Vehicle Structures*, Dept. of Defense, Washington DC, 1975.
5. Dave S. Steinberg, *Cooling Techniques for Electronic Equipment*, 2nd ed., Wiley, New York, 1991.
6. B. Saelman, "Calculating Tearout Strength for Cantilever Beams," *Mach. Des. Mag.*, January 1954.
7. J. W. S. Rayleigh, *The Theory of Sound*, Dover, Mineola, NY, 1945.
8. R. W. Little, Masters Thesis, University of Wisconsin, 1959.
9. Dave S. Steinberg, Preventing Thermal Cycling and Vibration Failures in Electronic Equipment, Presented at the 9th Annual IEEE Dayton Chapter Symposium, November 30, 1988.
10. Dave S. Steinberg, "Avoiding Vibration in Odd Shaped Printed Circuit Boards," *Mach. Des. Mag.*, May 20, 1976, pp. 116–119.
11. M. A. Miner, "Cumulative Damage in Fatigue," *J. Appl. Mech.*, **12**, September 1945.
12. A. M. Freudenthal, *Fatigue in Aircraft Structures*, Academic, New York, 1956.
13. McClintock and Argon, *Mechanical Behavior of Materials*, Addison Wesley, New York, 1966.
14. H. J. Grover, S. A. Gordon, and L. R. Jackson, *Fatigue of Metals and Structures*, Bureau of Aeronautics, Department of the Navy, 1954.
15. Stephen H. Crandall, *Random Vibration*, Technology Press, Wiley, New York, 1958.
16. Dave S. Steinberg, "Preventing Vibration Damage in Electronic Assemblies," *Mach. Des. Mag.*, July 8, 1976, pp. 74–77.
17. Dave S. Steinberg, "Snubbers Calm PCB Vibrations," *Mach. Des. Mag.*, March 24, 1977.
18. E. Miller, "Plastics Products Design Handbook," Chapter 11, *Vibration Dampers*, Dave S. Steinberg, Ed., Marcel Dekker, New York, 1981.
19. C. E. Crede, *Vibration and Shock Isolation*, Wiley, New York, 1957.
20. R. E. Peterson, *Stress Design Concentration Factors*, Wiley, New York, 1959.

21. MIL-STD-1276B. *Leads, Weldable for Electronic Component Parts*, May 14, 1965.
22. U.S. Air Force Study of Fighter Air Craft Type 1 Failures, U.S. Air Force, January–June 1982.
23. MIL-E-5400T. *Electronic Equipment, Airborne, General Specifications for*, October 31, 1975.
24. Arthur W. Leissa, *Vibration of Plates*, National Aeronautics and Space Administration, Washington, DC, 1969.
25. Werner Engelmaier, *Effects of Power Cycling on Leadless Chip Carrier Mounting Reliability and Technology*, Bell Laboratories, November 1982.
26. Dave S. Steinberg, "Quick Way to Predict Temperature Rise in Electronic Circuits," *Mach. Des. Mag.*, January 20, 1977.
27. MIL-STD-810B. *Environmental Test Methods*, June 15, 1967.
28. MIL-STD-202E. *Test Methods for Electronic and Electrical Component Parts*, April 16, 1973.
29. Dave S. Steinberg, *Vibration Analysis For Electronic Equipment*, 3rd ed., Wiley, New York, 2000.
30. R. A. Wilk, Wear of Connector Contacts Exposed to Relative Motion, IES Proceedings, Random Vibration Seminar, Los Angeles, March 25–26, 1982.
31. E. M. Bock and J. H. Whitley, Fretting Corrosion in Electrical Contacts, Presented at the 20th Annual Holm Seminar on Electrical Contacts, October 29–31, 1974.
32. MIL-C-45204B. *Gold Plating, Electrodeposited*, February 26, 1971.
33. E. J. Luney, and C. E. Crede, The Establishment of Vibration and Shock Tests for Airborne Electronics, Wright Air Development Center Technical Report, January 1958.
34. N. E. Lee, *Mechanical Engineering as Applied to Military Electronic Equipment*, Coles Signal Laboratory, Red Bank, NJ, June 22, 1949.
35. W. C. Stewart, "Determining Bolt Tension," *Mach. Des. Mag.*, November 1955.
36. R. W. Dicely and H. J. Long, "Torque Tension Charts for Selection and Application of Socket Head Cap Screws," *Mach. Des. Mag.*, September 5, 1957.
37. *Kent's Mechanical Engineering Handbook, Design and Production Volume*, Wiley, New York, 1950.
38. E. F. Bruhn, *Analysis and Design of Aircraft Structures*, Tristate Offset Co., Cincinnati, Ohio, 1952.
39. *Marks Handbook*, McGraw-Hill, New York, 1951.
40. R. J. Roark, *Formulas for Stress and Strain*, McGraw-Hill, New York, 1943.
41. S. L. Hoyt, *Metals and Alloys Data Book*, Reinhold, New York, 1943.
42. C. Lipson and R. Juvinal, *Handbook of Stress and Strength*, Macmillan, New York, 1963.
43. F. B. Stulen, H. N. Cummings, and W. C. Schulte, "A Design Guide Preventing Fatigue Failures," *Mach. Des. Mag.*, Part 5, June 22, 1961.
44. F. R. Shanley, *Strength of Materials*, McGraw-Hill, New York, 1957.
45. S. Timoshenko and S. W. Krieger, *Theory of Plates and Shells*, McGraw-Hill, New York, 1959.
46. P. A. Laura and B. F. Saffel, "Study of Small Amplitude Vibrations of Clamped Rectangular Plates Using Polynomial Approximations," *J. Acoust. Soc. Am.*, **41**(4), 1967.

47. M. Vet, "Vibration Analysis of Thin Rectangular Plates," *Mach. Des. Mag.*, April 13, 1967.
48. Road Shock and Vibration Environments for a Series of Wheeled and Track Laying Vehicles, Report DPS-999, March–June 1963.
49. D. S. Steinberg, "Circuit Components vs G Forces," *Mach. Des. Mag.*, October 14, 1971.
50. Steinberg & Associates, *Assessment of Vibration on Avionic Design*, prepared for Universal Energy Systems, Dayton, Ohio, August 1984.
51. S. O. Rice, "Mathematical Analysis of Random Noise," *Bell Syst. Tech. J.*, July–October, 1944.
52. NAVMAT P-9492, Navy Manufacturing Screening Program, Department of Navy, May 1979.
53. O. W. Eshbach, *Handbook of Engineering Fundamentals*, Wiley, New York, 1969.
54. *Materials Engineering Magazine*, "Materials Reference," Penton Publication, March 1978.
55. *Materials Engineering Magazine*, "Materials Selector," Penton Publication, 1991.
56. Fred C. Trumel, "Six Ways to Cope with Thermal Expansion," *Mach. Des. Mag.*, February 10, 1977.
57. S. S. Manson, "A Designers Guide to Thermal Stress," *Mach. Des. Mag.*, November 23, 1961.
58. Steve Massie, "Lead Stress Analysis of a Surface Mountable Relay," *Electron. Pack. Product. Mag.*, May 1987.
59. David A. Followell, Computer Aided Assessment of Reliability Using Finite Element Methods, McDonnell Douglas Report B2303, February 18, 1991.
60. MIL-A-87244, *Avionics Integrity Requirements*, U.S. Air Force, May 15, 1986.
61. M. A. Miner, "Cumulative Damage in Fatigue," *J. App. Mechanics*, **12**, 1945.

INDEX